THE
SCHOOL MATHEMATICS PROJECT

When the S.M.P. was founded in 1961, its main objective was to devise radically new secondary-school mathematics courses (and corresponding G.C.E. and C.S.E. syllabuses) to reflect, more adequately than did the traditional syllabuses, the up-to-date nature and usages of mathematics.

This objective has now been realized. S.M.P. *Books 1–5* form a five-year course to the O-level examination 'S.M.P. Mathematics'. *Books 3T, 4* and *5* give a three-year course to the same O-level examination (the earlier *Books T* and *T4* being now regarded as obsolete). *Advanced Mathematics Books 1–4* cover the syllabus for the A-level examination 'S.M.P. Mathematics' and five shorter texts cover the material of the various sections of the A-level examination 'S.M.P. Further Mathematics'. There are two books for 'S.M.P. Additional Mathematics' at O-level. All the S.M.P. G.C.E. examinations are available to schools through any of the Examining Boards.

Books A–H, originally designed for non-G.C.E. streams, cover broadly the same development of mathematics as do the first few books of the O-level series. Most C.S.E. Boards offer appropriate examinations. In practice, this series is being used very widely across all streams of comprehensive schools, and its first seven books, followed by *Books X, Y* and *Z* provide a course leading to the S.M.P. O-level examination.

Teachers' Guides accompany all these series of books.

The S.M.P. has produced many other texts, and teachers are encouraged to obtain each year from the Cambridge University Press, Bentley House, 200 Euston Road, London NW1 2DB, the full list of S.M.P. books currently available. In the same way, help and advice may always be sought by teachers from the Director at the S.M.P. Office, Westfield College, Hampstead, London NW3 7ST, from which may also be obtained the annual Reports, details of forthcoming in-service training courses and so on.

The completion of this first ten years of work forms a firm base on which the S.M.P. will continue to develop its research into the mathematical curriculum. The team of S.M.P. writers, numbering some forty school and university mathematicians, is continually evaluating old work and preparing for new. But at the same time, the effectiveness of the S.M.P.'s future work will depend, as it always has done, on obtaining reactions from a wide variety of teachers – and also from pupils – actively concerned in the classroom. Readers of the texts can therefore send their comments to the S.M.P., in the knowledge that they will be warmly welcomed.

This book is based on original contributions and on material from S.M.P. Books 1, 2, and 3. It has been prepared by

R. H. Baker	C. D. B. Milton	A. Thomas
A. G. Gallant	C. M. Reynolds	J. M. Truran
D. Lee	D. R. Skinner	J. V. Tyson

and edited by D. J. Holding, assisted by Mrs E. Smith and D. C. Taylor.

Many other schoolteachers have been directly involved in the further development and revision of the material and the Project gratefully acknowledges the contributions which they and their schools have made.

THE
SCHOOL
MATHEMATICS
PROJECT

BOOK 3T
[METRIC]

CAMBRIDGE
AT THE UNIVERSITY PRESS
1970

Published by the Syndics of the Cambridge University Press
Bentley House, 200 Euston Road, London NW1 2DB
American Branch: 32 East 57th Street, New York, N.Y.10022

© Cambridge University Press 1970

Library of Congress Catalogue Card Number: 66-73798

ISBN: 0 521 07818 0

First published 1970
Reprinted 1972

Printed in Great Britain
at the University Printing House, Cambridge
(Brooke Crutchley, University Printer)

PREFACE

Book 3 T is the first book in the three-year course leading to the O-level examination in 'S.M.P. Mathematics'; the remaining two books are *Books 4* and *5* both of which are already published. This book may consequently be regarded as a replacement for *Book T* and it should be understood that no replacement for *Book T4* is envisaged; with *Books 4* and *5* already in existence it has been possible to produce a text which leads directly into them.

In the writing of this text, our chief concern has been the nature of the mathematics with which pupils may be familiar before embarking on the mathematics contained here. We have been very conscious that school mathematics has been, and still is in ferment in the primary and middle years of schooling and that any secondary text must take this fully into account. It is likely that in any group of pupils using this text there will be a variation of mathematical background; accordingly we have included rather more material than is normally sufficient for one year's work. However, the great majority of pupils will find some mathematics which they have already studied, and the remainder they will find new and stimulating. In particular, all the newer aspects of the subject have been considered from their beginnings, the development therefore being more rapid in some cases than in *Books 1, 2* and *3*.

The style of presentation developed in *Books 1–5*, in which each new section is introduced by preparatory questions through which the pupil is encouraged to discover the relevant mathematics for himself, is preserved in this text. We have also suggested other ways of study, by introducing investigations and projects, notably in Chapters 1 and 11. This marks only the beginning and teachers will want to develop their own ideas; the text book should be seen as servant, not master, and departures from its collinear demands should be free and frequent.

The work on algebra has been carefully reappraised since the publication of the earlier series. It begins in Chapter 2 on relations, and includes an important section on the deductive aspects of mathematics. This is continued in Chapter 4 where a wide range of operations is considered; this chapter also contains a novel revision of negative numbers. The notation of algebra is developed comprehensively later in Chapter 14; flow diagrams for functions are introduced

v

and these are used again in Chapter 16 for rearranging formulae. Chapter 14 closes with an investigation into problems involving orderings, a subject which is reconsidered in *Book 4*.

The book opens with a general chapter on geometry which has the notion of pattern as its theme. This chapter, designed to draw together the previous experience of the pupil, prepares the way for the later development of the geometry of transformations. The structure of the successive Chapters 5, 6, 9 and 11 is such that the number of invariants is successively reduced, the stages being isometry, similarity, affine geometry (shearing being only a special case) and topology. Chapter 5 also contains an introduction to trigonometry, a subject which may already be familiar to some; here the approach is through waves and is unusual. The use of sets in geometry is considered in Chapter 8 in the context of loci and linear programming.

Computation rightly finds an early place in the book, in Chapter 3. The competent use of the slide rule for elementary operations on numbers is the aim of this chapter. The Theorem of Pythagoras is quoted in this chapter in the work on squares and square roots, and is perhaps the one significant geometrical result which it is assumed pupils will already have met. If some initial consideration of this theorem is necessary, teaching material will be found in *S.M.P. Book 2*, Chapter 14. Familiarity is also assumed with mensuration of the circle, including chords, segments and sectors. Ratio and the beginnings of calculus are studied in Chapter 10 and the closing Chapter 16 delves into the popular field of computing.

Matrices are introduced in Chapter 7 which deals chiefly with their manipulation. They are applied extensively in Chapters 13 and 15 to the geometry which has already been examined by 'pure' methods. Chapter 13 includes isometries, shearing and the general affine transformation in two dimensions (with origin invariant); in Chapter 15 matrices describing networks are combined and some of the study of relations in Chapter 2 is further developed.

Finally, statistics and probability are investigated in Chapter 12; this is another area of mathematics in which group and project work can readily be exploited.

As is usual with S.M.P. texts, answers to exercises are not printed at the end of the book but are contained in the companion Teacher's Guide, which gives a chapter-by-chapter commentary on the pupils' text.

CONTENTS

ACKNOWLEDGEMENTS

The drawings in this book are by Cecil Keeling.

The Project is grateful to Erven J. J. Tijl N. V. for permission to use the illustration by M. C. Escher which appears at the beginning of Chapter 1; to the B.B.C. for permission to use the photograph of Harry Worth in Chapter 6 and to Mullard Limited for permission to use Figure 7 of Chapter 16.

We are much indebted to the Cambridge University Press for their cooperation and help at all times in the preparation of this book.

The Project owes a great deal to its Secretaries Miss J. Sinfield and Mrs J. Whittaker for their assistance and typing in connection with this book.

A NOTE ON METRICATION

(i) All quantities of money have been expressed in pounds (£) and new pence (p).

(ii) All measures have been expressed in metric units. The fundamental units of the Système International (that is the metric system to be used in Great Britain) are the metre, the kilogram and the second. These units have been used in the book except where practical classroom considerations or an estimation of everyday practice in the years to come have suggested otherwise.

(iii) The notation used for the abbreviations of units and on some other occasions conforms to that suggested in the British Standard publications PD 5686: 1967 and BS 1991: Part 1: 1967.

1

PATTERN IN GEOMETRY

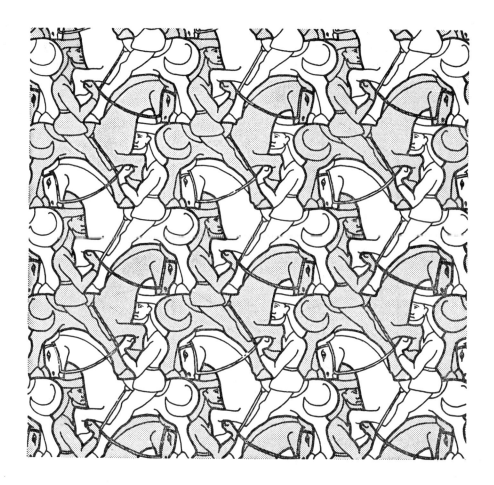

1. TESSELLATIONS

The picture at the head of this chapter, by the Dutch artist Maurits Escher, is an amazing example of how a basic shape may be used to fill a plane. This is a most fascinating subject, and we shall begin by investigating how to construct some simple plane-filling patterns (known as '*tessellations*').

(*a*) Take a sheet of plain white paper and colour one side with, say, a blue crayon. On another smaller sheet of paper or card, draw a small triangle like either of those in Figure 1; avoid making the triangle either isosceles or equilateral.

1

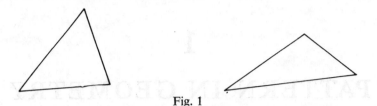

Fig. 1

By pricking through the corners of this triangle, mark out as many triangles as you can on the first sheet. Cut them out carefully and use about half of them to make a pattern like that in Figure 2, sticking them onto a page in your book. The pattern should turn out to be all white or all coloured.

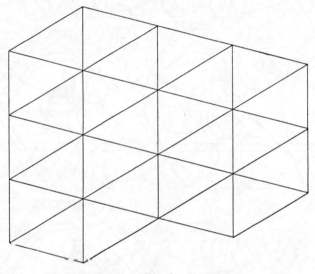

Fig. 2

(i) In your pattern outline some quadrilaterals made from two triangles.
Are they all the same shape?
How many differently shaped quadrilaterals can you find?
(ii) Examine the points where several triangles meet. How many vertices are there at such points?
On what angle property of a triangle does the pattern depend?
(b) Use the rest of the triangles to make a pattern like that in Figure 3.
(i) Is the pattern all the same colour?
(ii) Can you pick out a quadrilateral, made from two triangles, which is different from those in (a) (i)?
Are all such quadrilaterals the same way round? In how many different ways have they been placed?
(c) Can you make any other different patterns using more of the same triangles?

2

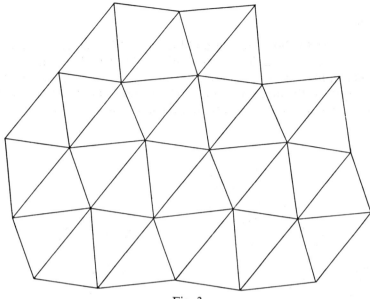

Fig. 3

Make these patterns, if you can, as you did in (*a*) and (*b*). Are you sure, in each case, that they could be continued indefinitely without leaving any spaces, assuming that you had sufficient triangles?

Investigations A

One of the following investigations should be carried out.

1. Quadrilaterals

(*a*) We often see tessellations of rectangles (bricks in walls, paving slabs, parquet flooring blocks, etc.). Sketch some of the methods of tessellating rectangles that you have seen.

(*b*) Sketch two different tessellations of parallelograms, using the same basic parallelogram for both.

(*c*) *Any* shape of quadrilateral can be the basic unit for a tessellation. Figure 4 shows how a tessellation of this kind is built up. Construct such a tessellation by cutting out quadrilaterals from card of two different colours. Arrange them so that those of the same colour are placed in the same way.

Why do the four corners of the quadrilaterals fit together exactly?

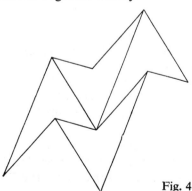

Fig. 4

3

2. *Other figures*

(*a*) Sketch a diagram to show how to build a larger triangle from a number of smaller ones of the same shape. Can you do the same thing with any other kind of figure?

(*b*) Make a sketch, like that in Figure 5, of a triangular tessellation. Mark two angles at *D* and one at *C* equal to the angle *x* at *A*. Mark one angle at *E* and two at *F* equal to the angle *y* at *B*. If $x = 60°$ and $y = 80°$, what are the angles *a*, *b*, *c*, *d* and *e*?

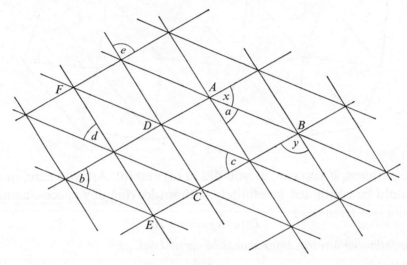

Fig. 5

(*c*) A regular pentagon is a plane figure with 5 equal sides and all its interior angles equal. What would each interior angle be? Can you have a tessellation of regular pentagons?

(*d*) What is each interior angle of a regular hexagon (6 sides)? Can you have a tessellation of regular hexagons? Can you give an example of where you have seen one?

3. *Mixed and adapted tessellations*

(*a*) Cut out some squares and some equilateral triangles. (Make the edges of the squares the same length as those of the triangles.) Find how to make a tessellation using the two shapes, and sketch it. How many different ways of doing it can you find?

(*b*) Figure 6 shows a tessellation of equilateral triangles adapted to produce a more complicated pattern. Study this adaptation carefully and try to produce another one yourself. (You may find isometric paper useful; it consists of a network of equilateral triangles. See Figure 31.)

(*c*) Examine the picture at the head of the chapter. Can you see the basic tessellation from which it is constructed? Other Escher drawings are worth study; see, for example, *The Graphic Work of M. C. Escher* (Oldbourne Press).

4

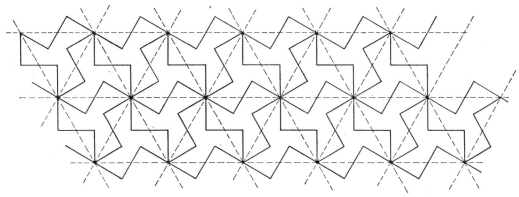

Fig. 6

Summary

Patterns which completely 'fill' the plane are called *tessellations*. A tessellation may be built up from one basic unit (for example, triangle, square, parallelogram, regular hexagon) or from a combination of two or more basic units (e.g. equilateral triangles and squares).

2. SYMMETRY

2.1 Bilateral symmetry

An ink devil is made by folding a piece of paper, opening it and then splattering ink (not too much) at random on the paper. The paper is then folded again quickly and finally opened out to reveal the finished product. Figure 7 shows an example. (Added interest can be achieved with different coloured paints.)

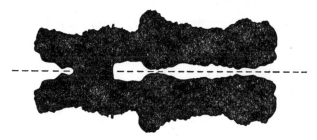

Fig. 7

The ink devil is an example of one kind of 'symmetrical' figure. The line of the fold is known as the *line of symmetry*. For every point of such a figure, there is a corresponding point on the other side of the line an equal distance from it.

This kind of symmetry is known as 'bilateral' (two-sided) symmetry, or line symmetry.

Can a figure have more than one line of symmetry? If so, give an example.

Exercise A

1. Copy the drawings in Figure 8 on to tracing paper. If you think any of them have lines of symmetry, test your opinion by folding. Sketch the drawings that do have bilateral symmetry, and dot in their line or lines of symmetry.

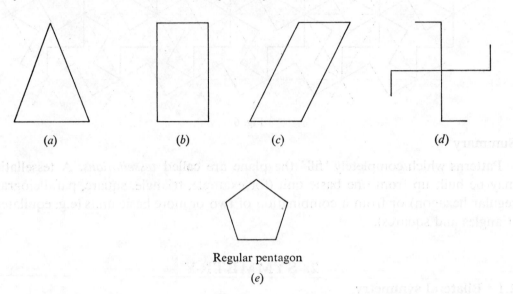

(a)   (b)   (c)   (d)

Regular pentagon

(e)

Fig. 8

2. (a) How many lines of symmetry has a circle?

   (b) Take a thin sheet of paper and trace around the outside of some circular object (not too small). Fold the paper along one of the lines of symmetry of the circle.

   (c) How can you find the centre of the circle by folding again?

3. In Figure 9, parts of certain figures are shown. Lines of symmetry are dotted. Copy and complete the figures.

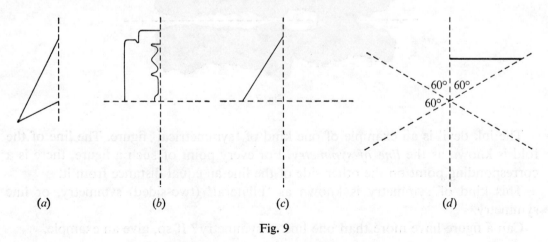

(a)   (b)   (c)   (d)

Fig. 9

(*a*) 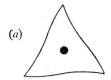   (*b*)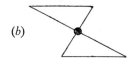

(*c*) 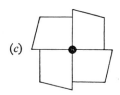   (*d*) 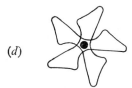   (*e*)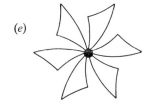

Fig. 15

(*c*) Figure 16 shows a parallelogram. If it is rotated about O through a half-turn,
    (i) What points are C, O, Y mapped onto, respectively?
    (ii) What can you say about CO and OY?
    (iii) Make a statement concerning the diagonals of a parallelogram.
    (iv) Do the diagonals of a parallelogram bisect its angles?

(*d*) Consider the letters of the alphabet as printed here:

A, B, C, D, E, F, G, H, I, J, K, L, M, N, O, P, Q, R, S, T, U, V, W, X, Y, Z.

    (i) List those letters with symmetry about a line across the page.
    (ii) List these letters with symmetry about a line up the page.
    (iii) List those letters with rotational symmetry.
    (iv) List any letters which belong to all of the first three categories.

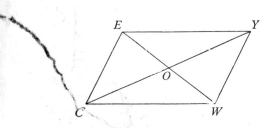

Fig. 16

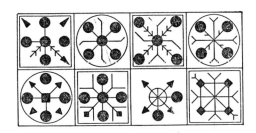

Fig. 17

2. (*a*) Classify the eight diagrams of Figure 17 as follows:

Diagram	I	
Lines of symmetry	0	
Order of rotational symmetry	I	

(b) Sketch the following figures and on them mark lines of symmetry and centres of rotational symmetry (by dotted lines and small rings respectively):

    (i) square;   (ii) rectangle;   (iii) parallelogram;   (iv) rhombus;

    (v) kite;   (vi) isosceles triangle;   (vii) equilateral triangle.

(c) The ten of spades (Figure 18) has rotational symmetry of order 2. List the other playing cards having such symmetry.

Fig. 18

3. (a) Figure 19 shows a section of an infinite tessellation of regular hexagons. Sketch the figure and mark centres of rotational symmetry. State the order in each case. Give a full description of the positions of the lines of symmetry of the infinite tessellation.

(b) Discuss the symmetries of the (infinite) tessellation, part of which is shown in Figure 20.

(c) (i) Design a figure with only rotational symmetry of order 5. Could you design a tessellation with the same symmetry?

(ii) Design a tessellation with only rotational symmetry of order 4.

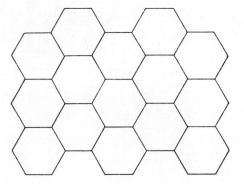

Fig. 19

10

4. Sketch, if possible, figures with the following specifications:
    (*a*) four lines of symmetry (only) at 45° to one another;
    (*b*) two parallel lines of symmetry;
    (*c*) two lines of symmetry (only) at 60° to one another.
If any of these are impossible, say so.

5. (*a*) If a rectangle has *only* two lines of symmetry, could it be a square?
    (*b*) Can a triangle have *only* two lines of symmetry? What kind of triangle must it be if it has three?
    (*c*) Make a statement about the number of lines of symmetry of a regular *n*-sided polygon.

2.2   Mediator and angle bisector

Figure 10 shows a line 'segment' (i.e. a part of an infinite line; it has two ends). The dotted line of symmetry is known as the *mediator* of the line segment. What can you say about the angle between a line segment and its mediator? If *P* is any point on the mediator what can you say about *AP* and *BP*?

Invent and illustrate a method for finding the mediator of a line segment, using compasses.

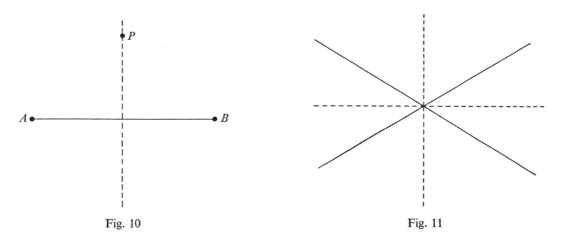

Fig. 10   Fig. 11

Figure 11 represents parts of two infinite lines. The broken lines are two lines of symmetry (in this case *angle bisectors*). Copy the figure and mark clearly any angles which are equal. Can you explain why the angle bisectors are perpendicular?

If a figure has two lines of symmetry (but no more) must they be at right-angles?

Invent a method for finding the bisectors of the angles between two lines, by drawing pairs of parallel lines only using both edges of your ruler.

2.3 Rotational symmetry

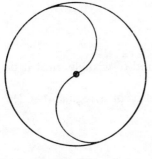
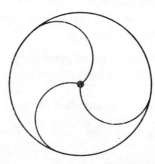

Fig. 12   Fig. 13   Fig. 14

Trace Figure 12 on two separate sheets of tracing paper. Place one exactly over the other and put a pin through the centres of the figures. Rotate the top figure until it again covers the lower figure. What fraction of a whole turn has been made? We say that the figure has been *mapped* onto itself.

Continue rotating the upper sheet until the figure is once more mapped onto itself. How does the new position of the upper sheet compare with its original position?

Figure 12 does not possess bilateral symmetry, but it does have another kind of regularity, which we have demonstrated by performing a rotation. Such figures are said to have *rotational symmetry*. The point about which these figures have to be rotated to discover the symmetry is called the *centre of rotational symmetry*.

Figure 12 made two moves before it was back in its original position; thus we say it has rotational symmetry of order 2 (sometimes also known as *half-turn symmetry*).

Figure 13 has rotational symmetry of order 3. Does it have line symmetry? Where is its centre of rotational symmetry?

Even figures with no rotational symmetry, such as Figure 14, can be mapped onto themselves by making a complete turn about *any* point. They are classified as being of order 1.

Investigations B

Carry out Investigation 1 and one of the other two.

1. (*a*) State the order of rotational symmetry of the drawings in Figure 15. Do any of these figures have lines of symmetry?

    (*b*) Sketch figures as described. If any are impossible, say so.

      (i)   No line symmetry, no rotational symmetry.

      (ii)   One line of symmetry, no rotational symmetry.

      (iii)   Two lines of symmetry, no rotational symmetry.

      (iv)   Two centres of rotational symmetry.

      (v)   Three lines of symmetry, rotational symmetry of order 3.

If a figure has two or more lines of symmetry, must it possess rotational symmetry?

Fig. 20

3. SOLIDS

3.1 Modelling with straws

When we are discussing solid figures it is a great help to have models or drawings of them to look at. Framework models are particularly easy to put together, and are a help in the drawing of solids in that all the edges are visible at once.

Frameworks can be made of wooden rods if they are to be permanent, but for temporary use it is cheaper and simpler to use drinking straws fastened together with pipe-cleaners.

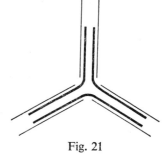

Fig. 21

For economy, the pipe-cleaners should be cut in two before they are used. Figure 21 shows how three straws are best connected at a point. Notice that it is advisable to have *two* thicknesses of pipe-cleaners in the end of each straw, otherwise the models come apart too easily.

You will see how the method is extended to more complicated joints, and with a little practice you will find them easy to make. How many lengths of pipe-cleaner will be needed at a join of six straws?

Alternatively, a bodkin or large needle can be used to thread strong cotton through the straws to fasten them together. You should try this method too, to see which you prefer.

11

Exercise B

1. Take six straws and fasten them together to form a regular framework. (Three straws will be joined at each vertex.) What is the name of the figure you have constucted? Sketch it. Could a regular framework be made with fewer straws?

2. How many straws are needed to construct the framework of a cube? Make one, and then sketch your model.

3. Construct the framework of a square-based pyramid with all its edges the same length. Sketch it.

4. Each of the eight faces of a regular octahedron is in the shape of an equilateral triangle. Construct such an octahedron and sketch it.

3.2 Drawing solid objects

Have you ever seen stereoscopic photographs or pictures? Do you know how stereoscopy works? Its purpose, of course, is to represent three-dimensional objects so that they really *look* three-dimensional; so that they look solid rather than flat. ('Stereos' is the Greek word for 'solid'.)

We live in a three-dimensional world, and so naturally much in mathematics concerns three-dimensional objects. However, when we write about such objects, we usually have to make do with two-dimensional drawings rather than models or stereoscopic representations.

You have already drawn the models you made in Exercise B. Are you satisfied with your attempts? Are they realistic? Perhaps they can be improved by using the following commonly used method of drawing.

3.3 Oblique projection

Arrange your framework of the cube so that you are looking directly at one of the faces.

(a) Do all the edges of this near face appear to be equal to one another in length? How long, in centimetres, do you think they are?

(b) Four of the edges of the cube are going directly away from you. Do they appear equal in length to the edges of the near face?

(c) Do the edges of the far face appear equal in length to each other? Do they appear equal in length to the edges of the near face?

(d) Which faces of the cube appear to be square? What shape do the other faces seem to be?

In *oblique projections* we take account of some, though not all, of the facts we have noticed when looking at the cube. The fact we neglect is that any given shape appears to become smaller when it is further away.

Imagine you are facing due north. Then in an oblique projection, any face in a vertical east-west plane is drawn with its *true shape*. The near face of any cube or cuboid is normally shown in this way. Horizontal north-south lines are drawn to a shorter scale than east-west lines and are shown as diagonals on the page. These diagonals may be drawn at any angle (usually 30° or 45°), and may be fore-shortened by any convenient amount. All sets of parallel lines are drawn parallel.

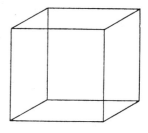

Figure 22 shows an oblique projection of the framework of a cube.

Figure 25 overleaf shows an oblique projection of a *solid* cube; notice that in this case 'hidden' edges are shown as short-dashed lines.

Fig. 22

Exercise C

1. On graph paper, draw oblique projections of:
(*a*) a framework of a square-based pyramid;
(*b*) a solid square-based pyramid.
'Hidden' lines should be short-dashed. The top vertex of the pyramid should be drawn above the centre of the base.

2. Draw oblique projections of:
(*a*) a four-legged table;
(*b*) a match-box tray;
(*c*) a cylinder with one of its circular ends facing you.

3. Draw an oblique projection of the wedge in Figure 23 when it is viewed from the east.

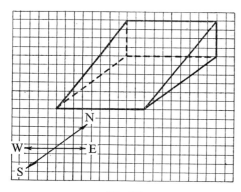

Fig. 23

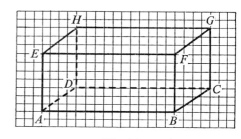

Fig. 24

4. Figure 24 shows an oblique projection of a cuboid.
(*a*) Name the faces shown in their true shape.
(*b*) What shape is the face *EFGH* in the drawing? What shape is it really?
(*c*) Name the edges parallel to *BC* in the drawing. Are they really parallel?
(*d*) Name a line which, in fact, is equal in length to *AC* but which would appear to be different in the drawing.

13

3.4 Polyhedra

Solid figures with plane faces are known as *polyhedra* (singular *polyhedron*). What is the derivation of this name? We shall investigate how to construct some polyhedra, though we shall make hollow ones rather than solid ones.

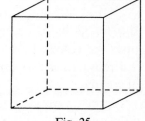

The cube (Figure 25) is a familiar example of a polyhedron. If you have not tackled the problem of making one before, your first attempt may be to cut out six equal squares and join them edge to edge with adhesive tape. But you have only to examine the construction of a chocolate box to realize that the problem can be simplified.

Fig. 25

(*a*) Six squares can be joined together edge to edge in several ways. Figure 26 shows three examples.

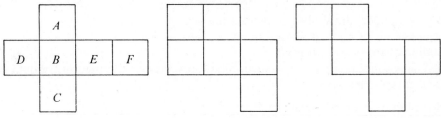

Fig. 26

In the first example, squares *A*, *D*, *E*, *C* could be folded up to form the side faces of a cube with *B* the bottom face and *F* the top face.

Can either of the other shapes in Figure 26 be folded to form a cube?

Draw as many other shapes as you can, made with six squares in this way, and decide which of them could be folded to form a cube.

Use graph paper to help you and check the doubtful cases by actually cutting the shapes out and folding them.

Each of the shapes which can be folded to form a cube is called a *net* of the cube.

(*b*) Sketch nets suitable for the construction of

(i) a wedge (see Figure 27),

(ii) a hexagonal prism (see Figure 28),

(iii) a square-based pyramid (see Figure 29).

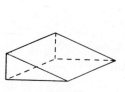

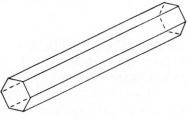

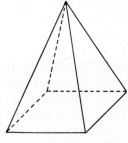

Fig. 27 Fig. 28 Fig. 29

3.5   Construction techniques

1. Accuracy in making a net is *very* important if one is to obtain a satisfactory result.

2. Although one can construct squares and rectangles using ruler, compasses and protractor (or set-square), it is best to use graph paper and prick through the corners of the net onto the material being used for the model itself.

3. Triangles are best constructed with compasses. You will probably, at some time, have constructed a flower pattern (see Figure 30). This pattern is really a network of equilateral triangles and is very useful.

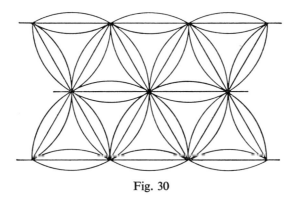

Fig. 30

An easier way of obtaining equilateral triangles is by using isometric graph paper (see Figure 31) if this is available, but you should become proficient at the other method before resorting to this.

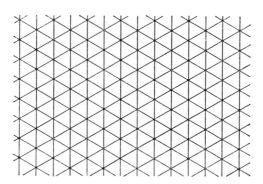

Fig. 31. Isometric graph paper.

4. Coloured card or heavyweight cartridge paper are the best materials for making models. Before folding, a compass point or other sharp instrument should be run along the lines to be folded; in this way, a clean fold is obtained. Edges may be secured with adhesive tape, but for a more professional result, use tabs on the edges

and stick with a quick drying glue (if you cannot decide where to put the tabs, put them on all free edges and cut them off when not needed).

Important. Keep one face free of tabs and secure that one last.

3.6 Some important polyhedra

A pyramid with four triangular faces is called a *tetrahedron*. If the triangles are equilateral it is called a *regular* tetrahedron. (In this case the tetrahedron looks the same no matter on what face you place it.) A net is shown in Figure 32. Can you design another net for it?

A polyhedron such as this, all of whose faces are equilateral triangles, is called a *deltahedron*. (Why?)

A convex polyhedron is one in which all the vertices (corners) point outwards.

Fig. 32

Is a cube regular? Is it convex? Is it a deltahedron?

There are five regular convex polyhedra (the so-called Platonic solids) and they are illustrated in Figure 33. Their nets are not difficult to design, but you can find them (together with many others) in *Mathematical Models* (by H. M. Cundy and A. P. Rollett, published by Oxford University Press).

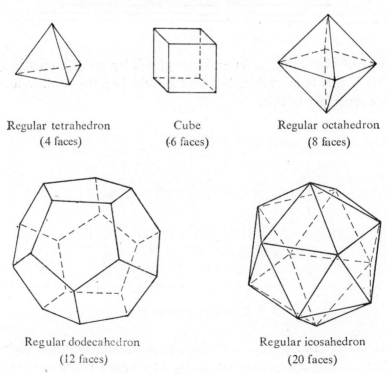

Regular tetrahedron
(4 faces)

Cube
(6 faces)

Regular octahedron
(8 faces)

Regular dodecahedron
(12 faces)

Regular icosahedron
(20 faces)

Fig. 33

Investigations C

1. Draw as many different shapes as you can, using eight equilateral triangles. Figure 34 shows two examples. How many of the shapes which you have drawn are nets for an octahedron? Only one of those in Figure 34 is such a net. Which?

(a)

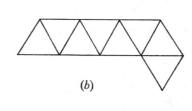

(b)

Fig. 34

2. (i) Can a deltahedron have 6 faces meeting at a vertex? Justify your answer.

(ii) Can a deltahedron have 5 faces exactly? If you think so, construct one. If you do not think so, can you give a good reason why you think it impossible?

3. Take an octahedron and stick a tetrahedron on each face. (Combine resources with some of your classmates.) You will get a 'stellated octahedron'. What solid would be formed by joining the eight outer vertices of the stellated octahedron?

4. (*A puzzle*) Make two of the figures for which the net is shown in Figure 35. Each one is half of a regular tetrahedron. The problem is to put them together to form the complete tetrahedron. Some people find it surprisingly difficult! (x may be made any convenient length, say 4 cm.)

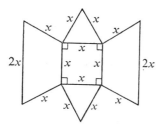

Fig. 35

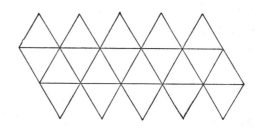

Fig. 36

5. (*Project*) Make a collection of the five Platonic polyhedra. The nets of three of these have already been discussed; the net for the icosahedron is shown in Figure 36; the remaining one is left for you to discover.

Your polyhedra should be preserved for use later in the book (Chapter 11).

6. Discuss the symmetries of some of the solids you have constructed.

3.7 Filling space with polyhedra

Figure 37 shows a close-up of part of the Giant's Causeway in Ireland, a famous geological phenomenon. Here basalt has cooled into columns of prisms of rock which are mainly hexagonal, though 3-, 4-, 5- and 7-sided columns occur. The idealized form of this space-filling arrangement is the tessellation of regular hexagons shown in Figure 19 extended to form prisms. Quite clearly any tessellation of plane figures can be extended to form space-filling prisms.

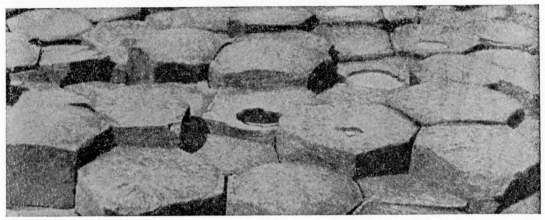

Fig. 37

On a smaller scale than the Giant's Causeway are the tiny crystals of minerals. Take some table salt and examine it closely with a magnifying glass and you should notice that the majority of the small crystals are approximately cubes.

Another substance, alum, forms crystals in the shape of a regular octahedron. It is possible with care to grow single crystals of this substance and others in the laboratory. Ask your science teacher about this.

Cubes can clearly be fitted together to fill space, but can octahedra? In actual fact, they cannot without leaving holes; consequently normal crystals of alum are distorted from their ideal shape. This ideal shape can only be obtained by careful laboratory methods.

Investigations D

Carry out the first two parts on your own. Parts 3 and 4 are best done as group projects, the class being split into two groups each to do one project.

1. (*a*) Give some examples of the shapes used in packing sweets and groceries, either for the articles themselves or the containers in which they are packed. Can you give reasons for the shapes of the articles? Are there examples in which the space within a container is completely filled?

(*b*) Give some other 'real life' examples of regular space-filling objects.

2. Examine crystals of some substances, for example, sugar, Epsom salts and copper sulphate, under a magnifying glass or microscope and describe carefully what you see.

3. A combination of regular octahedra and regular tetrahedra can be fitted together to fill space as shown in Figure 39. How many tetrahedra are there in Figure 39? How many octahedra are there?

Make a model similar to this (it is advisible to glue the tetrahedra at the bottom onto a base-board).

If this arrangement of tetrahedra and octahedra were extended indefinitely, what would be the proportion of tetrahedra to octahedra?

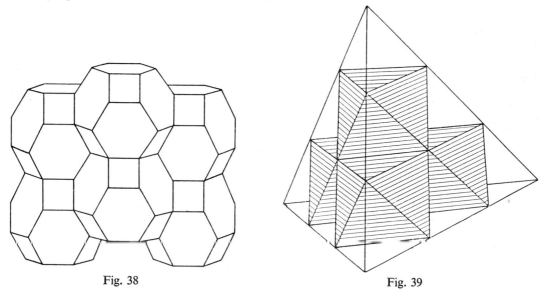

Fig. 38   Fig. 39

4. Another space-filling polyhedron is the truncated octahedron—in which the vertices of a regular octahedron are cut off to form a polyhedron with 8 hexagonal faces and 6 square faces. Six of these are shown in Figure 38. The net is not difficult to design; alternatively, it can be found in Cundy and Rollett's *Mathematical Models*.

4. GEOMETRICAL ELEMENTS

In this chapter so far we have referred frequently to the basic bits and pieces or *elements* of geometry: points, lines, planes, angles and so on. Are we really sure that we understand what they mean?

4.1. Points and lines

(*a*) Consider the two statements:

(i) a line is a set of points;

(ii) a point is the intersection of two lines.

These suggest how points and lines are related but have we said what a *point* is? Perhaps you are happy with the idea that a point 'marks a position' and if so you can leave it at that. If not, we have to say for the present that a point is an undefinable mathematical object.

(b) What is the difference between a straight line and a curved line? What happens when you join (what do we mean by 'join'?) any two points which are on:

    (i) a particular straight line;

    (ii) a particular curved line?

Unbounded

(c) How long is a line? Does it have ends? Is it necessary to distinguish between the different sorts of lines in Figure 40? If so invent names for them.

Bounded at one end

An arrow indicates that a line extends indefinitely in that direction.

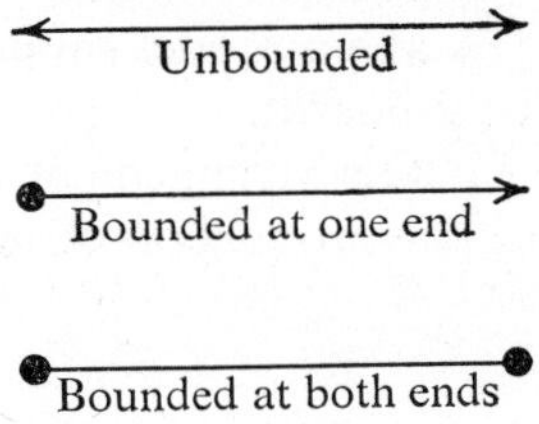

Bounded at both ends

Fig. 40

4.2   Points, lines and planes

(a) A plane, like a line, is also a set of points and you might say that it is 'flat'. What do we mean by flat?

If you join any two points in a plane to make a straight line what can you say about all the other points on that line?

Is the same true for a curved surface?

(b) Can we say that a plane is a set of lines?

What conditions must these lines obey?

We have said that two lines intersect at a point. What is the intersection of two planes?

When do two lines not intersect? Can you have two planes which do not intersect?

(c) Is a plane infinite? Does it have any boundaries? Is it possible to have half a plane?

(d) In how many ways can two points, three points, four points, etc. be joined in pairs by straight lines?

Can you continue the number sequence which you find?

Comment on any special cases which may arise.

(e) How many planes are there containing (i) a given line, (ii) two given lines?

What is the relationship between two lines which lie in one plane?

4.3   Angles between lines

We are quite clear on what we mean by the angles between two lines which intersect. How many such angles are there and what is the relation between them?

Is it possible to define the angle(s) between two lines in space which do not intersect, that is *skew* lines?

For example, what is the angle between *AB* and *CG* in Figure 41? (Imagine *AB* to be moved without turning until it coincides with *DC*.) What is the angle between *DC* and *CG*?

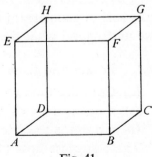

Fig. 41

To find the angle between two skew lines *l* and *m*, move one of them, *l* say, without turning it so that its image l' intersects m. Then the angle between l and m is defined to be the angle between l' and m.

What is the angle between FH and AD? Does it make any difference which of the two lines you move; that is, if, in the definition, we had moved m so that m' intersected l, would the angle between m' and l equal that between m and l'?

What is the angle between AB and FG? What happens when you move AB so that it meets DC? What can you say about the angle between these two lines? What can you say about the lines AB, DC, EF, HG?

4.4 Angles between lines and planes

Each of the three drawings in Figure 42 represents a thin pole stuck into the ground.

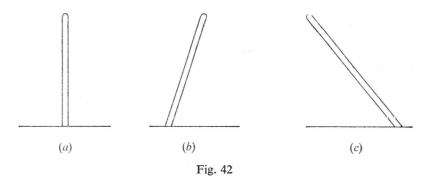

(a) (b) (c)

Fig. 42

See if you agree with these statements:

1. Figure 42(a) shows an upright pole.
2. Figure 42(b) shows a pole making an angle of about 72° with the ground.
3. Figure 42(c) shows a pole making an angle of about 50° with the ground.

If you think these statements *are* true, perhaps you have been deceived. The drawings could all be of the same pole. The apparent angle between pole and ground depends on where you look from. So if we want only *one* answer to the question 'What is the angle between pole and ground?', we shall have to decide which of the many possible angles we really mean.

Suppose that the sun is shining from directly above the pole in question. Then the shadow of the pole is called its *projection* onto the plane of the ground, and we agree that the angle between the pole and its projection is to be called *the* angle between the pole and the ground. If the pole were allowed to fall, this is the angle through which it would turn.

What would be the projection onto the ground of a vertical pole? What can you say about the angle between the ground and such a pole?

In Figure 43, l_2 is the projection of l_1 onto the plane and so the angle between l_1 and the plane is the angle marked $\theta°$. How does $\theta°$ compare in size with the angle between l_1 and l_3, where l_3 is *any* other line of the plane passing through the point of intersection of l_1 with the plane?

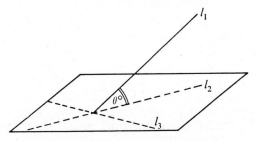

Fig. 43. Angle between line and plane.

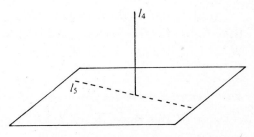

Fig. 44. Line perpendicular to plane.

In Figure 44, l_4 is perpendicular to the plane. l_5 is *any* line of the plane passing through the point of intersection l_4, with the plane. What can you say about the angle between l_4 and l_5?

4.5 Angles between planes

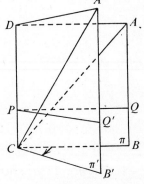

Fig. 45

Figure 45 illustrates a Christmas card which has been opened out so that it will stand up. If the card was closed to begin with, the plane π' had to be turned through a certain angle until it occupied the position shown. (Name the axis of rotation.) This angle, through which π' was turned, is defined to be *the* angle between π and π'. It is equal to $\angle BCB'$.

Is $\angle BCB'$ the same as $\angle ADA'$? Is it the same as $\angle ACA'$? What angles do BC and $B'C$ make with CD? Name an angle to which $\angle BDB'$ is equal.

P is any point between C and D. How does $\angle BPB'$ compare in size with $\angle BCB'$? If QP and $Q'P$ are both perpendicular to CD, how does $\angle QPQ'$ compare in size with $\angle BCB'$?

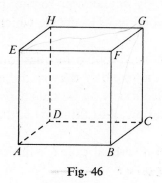

Fig. 46

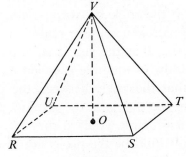

Fig. 47

Exercise D

Framework models of a cube and square-based pyramid are required. Questions 1–5 refer to the cube and pyramid lettered as in Figures 46 and 47. In Figure 47, V is vertically above the mid-point O of the square base RSTU.

1. For the *solid* cube, if you saw through *FH* and come out at *D*, do you come out at any other vertex? (Assume that the saw makes a plane cut.)

2. Describe the smaller of the two solids you obtain if you saw through *EG* and come out at *B*. What special shape is triangle *EGB*?

3. If the top of a solid square-based pyramid is sawn off, the plane of the cut being parallel to the base, what is the shape of the surface exposed?

4. Which of these pairs of lines determine a plane? (A pair of lines is said to determine a plane if there is one and only one plane containing them both.) Which pairs are skew?

(*a*) *AB* and *CD*;	(*b*) *AB* and *HG*;	(*c*) *AB* and *GC*;	(*d*) *EB* and *HC*;
(*e*) *GF* and *DC*;	(*f*) *GB* and *ED*;	(*g*) *VU* and *VS*;	(*h*) *VS* and *RS*;
(*i*) *VU* and *TS*;	(*j*) *RT* and *VO*.		

5. Which of these sets of four points determine a plane?

(*a*) *A, B, C, D*;	(*b*) *A, B, G, H*;	(*c*) *A, B, G, C*;	(*d*) *D, E, F, G*,
(*e*) *B, D, F, H*;	(*f*) *C, D, F, H*;	(*g*) *R, S, T, U*;	
(*h*) *R, S, T, V*;	(*i*) *R, S, O, V*;	(*j*) *O, S, U, V*.	

Questions 6–11 refer to the cube shown in Figure 48.

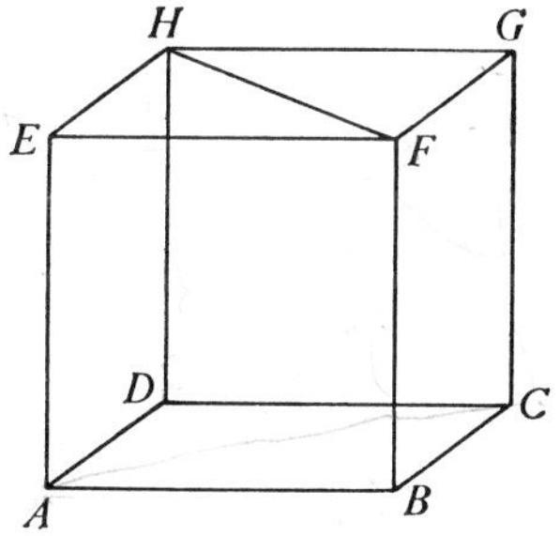

6. What are the angles in degrees between:

(*a*) *AB* and *BE*; (*b*) *DC* and *BE*; (*c*) *FC* and *EA*;
(*d*) *FG* and *AD*; (*e*) *FG* and *DH*?

7. How many lines of the cubical framework are perpendicular to *AD*?

8. Name eight lines which make an angle of 45° with *AC*.

9. Name the projection of:

(*a*) *FD* onto *ABCD*; (*b*) *FD* onto *BCGF*; (*c*) *EC* onto *CDHG*;
(*d*) *AC* onto *EFGH*; (*e*) *EA* onto *ABCD*.

Fig. 48

10. The angle between *FD* and the plane *BCGF* is ∠*DFC*. State the angles between:

(*a*) *FD* and *EAHD*; (*b*) *BH* and *EADH*; (*c*) *FC* and *ABCD*.

Give in degrees the angles between:

(*d*) *GC* and *ABCD*; (*e*) *GH* and *ABCD*.

11. By calling any new point required *X*, state the projection of

(*a*) *EF* onto *ACGE*; (*b*) *AF* onto *AEGC*.

23

12. Referring to the square-based pyramid illustrated in Figure 49, where *VO* is perpendicular to the base:

(*a*) state the projection of

(i) *RV* onto *RSTU*;   (ii) *RV* onto *VSU*;

(*b*) state the line of intersection of the planes *RTV* and *USV*;

(*c*) state in degrees the angle between the planes *RSTU* and *RTV*;

(*d*) state in letters the angle between the planes *RSTU* and *RSV*.

Questions 13–15 *refer to the cube shown in Figure* 50.

13. State the lines of intersection of the planes:

(*a*) *ABCD* and *BCGF*;   (*b*) *ABCD* and *BCHE*;   (*c*) *ABCD* and *BDHF*;

(*d*) *BDHF* and *ACGE*;   (*e*) *ADHE* and *EFGH*.

14. Give in each case the angle in degrees between the pairs of planes in Question 13.

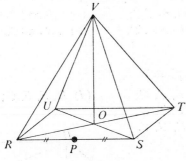

Fig. 49

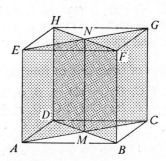

Fig. 50

15. Would it be true to say that the angle between the planes:

(*a*) *ADGF* and *EFGH* is ∠*AFE*;   (*b*) *EGB* and *EFGH* is ∠*BNF*;

(*c*) *ABGH* and *ABCD* is ∠*GBD*;   (*d*) *CDEF* and *ABGH* is 90°;

(*e*) *ABCD* and *ACGE* is 45°?

Summary

A line bounded at both ends *A* and *B* is more strictly called the *line segment AB*.

A line starting at *A* and passing through *B*, continuing indefinitely is sometimes referred to as a *half-line* or *ray*.

Points that lie on the same line are said to be *collinear*.

Points or lines that lie in the same plane are said to be *coplanar*.

Lines which are not parallel and do not intersect are called *skew* lines.

Figures 51, 52 and 53 illustrate some of the facts about points. Note carefully that two lines in space must be *either* parallel *or* intersecting *or* skew, and that, whereas a pair of parallel or intersecting lines determines a plane, a pair of skew lines does not.

24

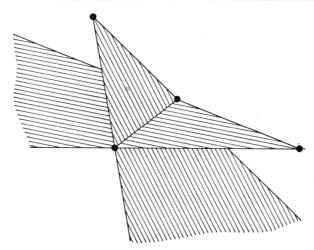

Fig. 51. Three non-collinear points determine a plane; but four points are
not necessarily coplanar.

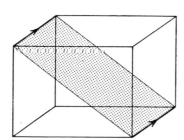

Fig. 52. Two parallel lines determine
a plane.

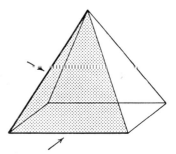

Fig. 53. Two intersecting lines
determine a plane.

The angle between a line and a plane is the angle between the line and its projection
onto the plane.

A line is perpendicular to a plane if the projection of the line onto the plane is a
single point.

The *angle between two planes* is defined to be the angle between any pair of lines,
one in each plane, which meet on and are at right-angles to the line of intersection
of the planes. (Look at the printed lines on a partly opened ruled exercise book.)

2
RELATIONS AND FUNCTIONS

Fate makes our relations, choice makes our friends.

DELILE, *La Pitié*

1. SETS AND RELATIONS

1.1 Sets and members

You are all familiar with the idea of a set in everyday life as a collection of objects, for example, a *pack* of cards, a *herd* of cows, a *fleet* of ships or a *gang* of boys.

(*a*) What special names do we give to: a set of flowers; a set of stamps; a set of geese; a set of porpoises?

The objects in a set may be anything we please. It is important that we state clearly to *which* set we are referring. For instance, 'the set of books in your satchel' or 'the set of books in your desk' describe exactly *which* set we mean.

The different objects which form a set are called the *members* or *elements* of the set in just the same way, say, as the people who form a choir are called members of the choir.

(*b*) What are the members of the set of vowels?

(*c*) Describe a set which contains the following as some of its members:

April, August, November.

Because the words 'the set whose members are' will occur frequently, we use a special shorthand: curly brackets.

26

'The set whose members are the first five even numbers' is written

{the first five even numbers}.

This could equally well be written as

(i) {2, 4, 6, 8, 10}   or   (ii) {6, 2, 10, 8, 4}

although, here, we could also read (i) as

'the set whose members are 2, 4, 6, 8 and 10'.

(d) What are the members of {outdoor games played at your school}?

(e) How could you describe {Sunday, Monday, Tuesday, Wednesday, Thursday, Friday, Saturday}?

When a set is to be referred to more than once, it is convenient to give it a label, usually a capital letter. If, for example, we let V stand for the set of vowels, then we write $V = $ {the vowels}   or $V = $ {a, e, i, o, u}.

(f) If $E = $ {2, 4, 6, 8}, which of the following phrases best describes the members of E:

    (i)   four even numbers;

    (ii)   some small even numbers;

    (iii) the even numbers between 1 and 9?

(g) There are two distinct ways of describing a set:

    (i) by giving a list of its members;

    (ii) by giving a rule, or description.

Care has to be taken when using the second method that the rule gives only the members required, as (f) above shows. State a rule which will produce {April, June, September, November}.

(h) Consider the set $A = $ {the odd numbers}. It is not possible to list all the members of A (why not?) and we may write $A = $ {1, 3, 5, 7, ...} where the dots indicate 'and so on'. Which of the following numbers is a member of A:

(i) 230;   (ii) 231;   (iii) 232?

A special shorthand, $\in$, is used for the phrase 'is a member of', so we may write

'$77 \in A$' as shorthand for '77 is a member of A'.

1.2   Relations

It is often said that mathematics is a language. What does this mean? Here are three mathematical sentences:

(i) $y = x$;   (ii) $3 < 4$;   (iii) $7 \in$ {prime numbers}.

All of these are like ordinary English sentences since they make statements about two 'nouns' by linking them by what may be called a 'verb'.

For example:

y is equal to x;

3 *is less than* 4;

7 *is a member of* the set of prime numbers.

The parts in italics, the 'verbs', describe *relations*.

In the three examples given above the relations are denoted by symbols; this is not always the case, for example,

8 *is a factor of* 24.

AB is perpendicular to CD.

Can you think of some other examples of relations you have used in mathematics?

Relations are not confined to statements involving numbers, lines, sets and so on. In our everyday life we use such expressions as

Alan *travels to school on the same bus as* Fred;

Heather *has the same colour eyes as* Catherine;

The headmistress *went for a holiday to* Scotland.

(*a*) Make a list of, say, six boys or girls who sit near to you in the classroom, including yourself, and also a list of the different drinks they had yesterday. If a person had a particular drink, then we indicate this by drawing a line as in Figure 1.

For example, Dick had tea and lemonade.

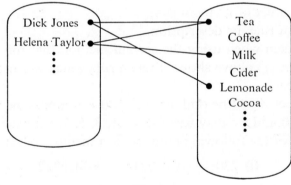

Fig. 1

Which was the most popular drink?

Who had the widest variety of drinks?

In this example the members of two sets are linked by a relation.

We can also have relations within a set.

(*b*) Consider the set of members of a particular family including aunts, uncles and first cousins but not grandparents. Each person is represented by a dot (see Figure 2).

Lines are drawn to represent 'is a parent of'. The line joining, for example, Aunt Pat to Sarah indicates:

Aunt Pat is a parent of Sarah.

28

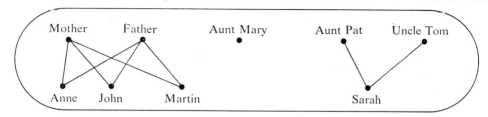

Fig. 2

Does the line also indicate Sarah is a parent of Aunt Pat?

Suggest how we might make it clear that only the first of these two statements is indicated on the diagram. Draw a diagram similar to Figure 2 for a set of members of your own family and on it indicate in different colours the following relations:

    (i) 'is a brother of';   (ii) 'is a sister of';   (iii) 'is an aunt of'.

Exercise A

1. Draw a diagram to represent the relation 'has been a pupil at' between a set of six members of your class and an appropriate set of local schools.

2. Draw a diagram to represent the relation 'is a factor of' between the set {2, 3, 4, 5, 6, 7} and the set {30, 31, 32, 33, 34, 35, 36}. Which number has no arrows joined to it? Why?

3. Draw a diagram to represent the relation 'is a factor of' in the set {2, 3, 4, 5, 6, 7, 8}.

4. Figure 3 represents the relation 'is the brother of' in a set of children. The figure is incomplete. Why?

Copy and complete the figure, as far as you can be certain.

5. In order, the eldest first, Hugh, Lucy, Sarah and Matthew, are four children in a family. Represent on a diagram the relation 'is younger than'.

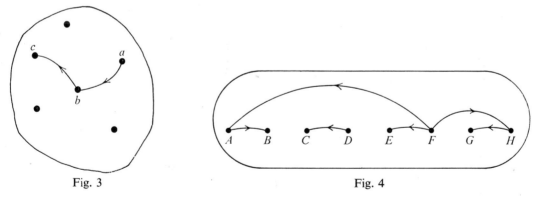

Fig. 3   Fig. 4

6. The sisters, Alison and Jane, are cousins of the four children in Question 5. Draw a diagram for the set of six children to illustrate the relation 'is a cousin of'.

7. Compare the diagrams in Questions 3, 5 and 6 and say what you can about the different ways in which the arrows are connected.

8. Figure 4 represents the relation 'beat' in a set of 8 table tennis players in a knockout tournament. The diagram is incomplete. Why? Who won the tournament and who was the runner up?

1.3   Graphs and relations

(*a*) Figure 5 represents the relation 'plays' and shows the games played by the members of a family.

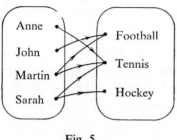

Fig. 5

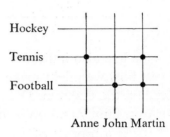

Fig. 6

A different way of representing the relation is shown (incomplete) by Figure 6. Study it carefully to see how the marked points correspond to the arrows of Figure 5.

This type of figure is called the *graph* of the relation. Copy and complete it by adding the column for Sarah. When drawing the graph of a relation it is usual, although not essential, to place the first set of elements in the relation across the page and the second set of elements up the page.

(*b*) Copy and add an arrow to Figure 7(*b*) so as to make it represent the same relation as Figure 7(*a*).

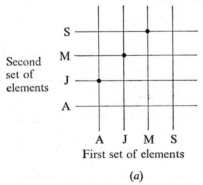

(*a*)

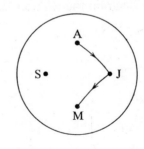

(*b*)

Fig. 7

If the relation is 'is taller than', complete your diagram by adding more arrows where appropriate.

Copy the graph and complete it by adding extra points. What do you notice about the points marked on the graph?

30

1.4   Drawing conclusions

In completing Figure 7(*b*) you probably reasoned along the following lines:

'If Anne is taller than John and John is taller than Martin then it follows that Anne must be taller than Martin'.

Drawing conclusions like this is an extremely important part of mathematics. Consider the following examples.

(i) If your birthday is 29 February then you were born in a leap year.

(ii) If $x+2 = 7$,   then $x = 5$.

We use a special shorthand, $\Rightarrow$, read '*implies*', to connect statements such as these. ($\Rightarrow$ is a relation connecting statements.)

So we write:

(iii) Your birthday is 29 February $\Rightarrow$ You were born in a leap year.

(iv) $x+2 = 7$ $\Rightarrow$ $x = 5$.

(v) Martin is Sarah's brother $\Rightarrow$ Sarah is Martin's sister.

(vi) ABC is an equilateral triangle $\Rightarrow$ $\angle ABC = 60°$.

When we use the relation '$\Rightarrow$' to connect two statements we mean that if the first statement is true then the second statement must also be true.

Write down three more examples using '$\Rightarrow$' to connect two statements.

Exercise B

1. The graph of the relation 'is a parent of' is shown in Figure 8. Express each of the elements as child, parent, or grandparent.

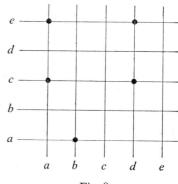

Fig. 8

2. State for each of the following whether or not '$\Rightarrow$' has been used correctly.

(*a*) You were born in a leap year $\Rightarrow$ Your birthday is February 29th.

(*b*) Peter is entitled to vote in a general election $\Rightarrow$ Peter is over 16 years old.

(*c*) Peter is over 16 years old $\Rightarrow$ Peter is entitled to vote in a general election.

(*d*) $1+2x = 12 \Rightarrow x = 4$.

(*e*) $1+2x = 12 \Rightarrow x > 4$.

(*f*) x is a sister of $y \Rightarrow y$ is a sister of x.

(*g*) x is older than y and y is older than $z \Rightarrow x$ is older than z.

(*h*) x is a cousin of $y \Rightarrow y$ is a cousin of x.

3. Consider your answer to 2(*h*). How would this property of the relation be apparent in: (*a*) the diagram with arrows; (*b*) the graph?

4. Three lines, x, y, z, are drawn on a flat surface. $\perp$ is the symbol for 'is perpendicular to'. Is it true that

(*a*) $x \perp y \Rightarrow y \perp x$.

(*b*) $x \perp y$ and $y \perp z \Rightarrow x \perp z$?

31

(*c*) Copy and complete Figure 9 for the relation 'is perpendicular to'.

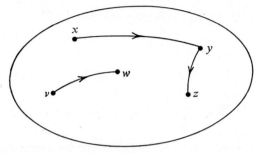

Fig. 9

5. Repeat Question 4 using the relation 'is parallel to' instead of 'is perpendicular to'.

6. Compare the diagrams drawn in Questions 4 and 5.
(*a*) How are they similar?
(*b*) How are they different?
(*c*) Investigate some more relations and state in what ways they are similar to or different from those of Questions 4 and 5.

7. (*a*) Choose ten people near you and draw the graph of the relation 'belongs to', between the set of people and the set of houses in the school.
(*b*) Draw a diagram to represent the relation 'is in the same house as' for the ten people. Comment on any properties of the relation which you notice.
(*c*) Copy and complete Figure 10 for the relation 'is in the same house as', given that *c* and *d* belong to different houses.

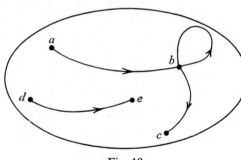

Fig. 10

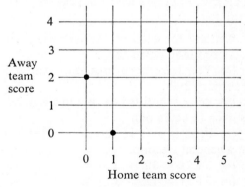

Fig. 11

8. Rovers   v. Rangers 0–2   City   v. Wanderers 1–0
   Hotspurs v. Athletic 1–0   Town v. United   3–3.

Figure 11 shows how football results can be shown on a graph, which relates the number of goals scored by the Home team to the number scored by the Away team.

Plot the results of last week's football league matches similarly (ignoring repetitions of scores), using different colours for home wins, away wins, and draws. Describe briefly the result you obtain.

2. FUNCTIONS

2.1 Mapping

(*a*) Write down the prime factors of each of the numbers 4, 6, 7, 15.

Is Figure 12 a correct illustration of the relation 'has the prime factor' between the sets {4, 6, 7, 15} and {2, 3, 5, 7}?

Sometimes, we wish to place the emphasis on *obtaining* the second set of elements from the first. If, for example, we start with 15, we obtain 3 and 5 as the prime factors; we say that the relation *maps* 15 onto 3 and 5, or that 15 is mapped onto 3 and 5 by the relation. The set of starting elements is called the *domain* of the relation and the set of finishing elements is called the *range* of the relation.

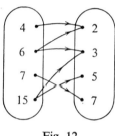

Fig. 12

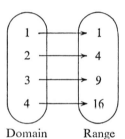

Fig. 13

(*b*) How have the members of the range been obtained from those of the domain in Figure 13?

To describe the relation we write $x \to x^2$ ('*x* is mapped onto *x* squared') with domain {1, 2, 3, 4}.

(*c*) With the same domain draw the diagram for $x \to 3x$.

2.2 Functions

Compare the relations shown in Figures 14 and 15.

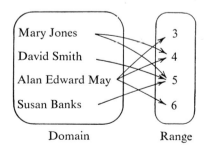

Fig. 14

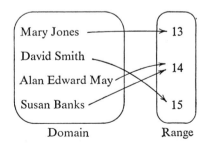

Fig. 15

(*a*) In Figure 14 a person is mapped onto the number of letters in any of his or her names. What makes Alan the most complicated person to be mapped?

(*b*) The relation shown in Figure 15 is onto age. This is obviously a less complicated diagram. There are less arrows for instance, but this relation also possesses a special property not possessed by the previous one, namely that *one and only one arrow leaves each member of the domain*. When this happens we call the relation a *function*. The idea of a function is extremely important in mathematics.

Notice that we may have more than one arrow arriving at the same place (where does this occur in Figure 15?); it is 'one arrow leaving' each member of the domain that is the criterion. Which of the relations illustrated in Figures 12 and 13 is a function?

(*c*) Draw graphs for each of the five relations considered in this section. How could you tell by looking at a graph whether or not it represented a function?

Exercise C

1. Copy and complete Figure 16 if the relation is:
(*a*) $x \rightarrow x+4$;
(*b*) $x \rightarrow 3x$.

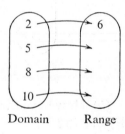

Fig. 16

2. For the function $x \rightarrow 2x+3$,
(*a*) What is 7 mapped onto?
(*b*) What number is mapped onto 29?

3. Which of the following are functions? Explain briefly why the others are not.

Domain	Relation
(*a*) {members of your form}	$x \rightarrow$ month of birthday of x
(*b*) {members of your form}	$x \rightarrow$ number of sisters of x
(*c*) {houses in the school}	$x \rightarrow$ person in your form belonging to x
(*d*) {a family of 2 boys and a girl Andrew, Bill, Kate}	$x \rightarrow$ brother of x
(*e*) $\{\frac{3}{4}, \frac{4}{3}, \frac{5}{6}, \frac{9}{7}, \frac{13}{3}\}$	$x \rightarrow$ next largest counting number
(*f*) {occupants of houses in a road}	$x \rightarrow$ next door neighbour of x
(*g*) {days in May this year}	$x \rightarrow$ time of High Tide at London Bridge on x

Draw arrow diagrams to illustrate (*d*) and (*e*).

4. Figure 17(*a*) shows the relation '$x \rightarrow$ father of x'; Figure 17(*b*) shows the relation '$x \rightarrow$ child of x'.
(*a*) What is the connection between the two relations?
(*b*) Which (if either) is a function?

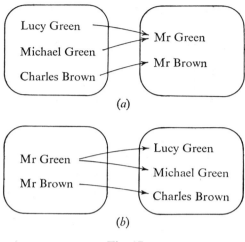

Fig. 17

5. (a) With {2, 3, 4, 5} as domain draw an arrow diagram for $x \rightarrow 2^x$, as in Figure 16.
   (b) With {0, 1, 2, 3, 4} as domain draw an arrow diagram for $x \rightarrow 4 - x$.
And draw the graph of this function.

6. Figure 18 displays the results of a race by means of the relation 'person → position'.
(a) Who won?
(b) Who deadheated?
(c) Is the relation a function?
(d) Would this relation always be a function for any race? Why?

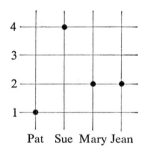

Fig. 18

3. COORDINATES AND FUNCTIONS

3.1 Coordinates

You are probably familiar with the use of numbers to describe the positions of points on a grid of lines such as is shown in Figure 19.

(a) Describe the positions of: (i) *B*; (ii) *E*; (iii) *F*; (iv) *G*. Which point is above 5 in the numbers marked across the page and opposite 2 in the numbers marked up the page? It is usual to shorten this description and just state the ordered pair of numbers (5, 2).

(*b*) Which are the points represented by: (i) (2, 1); (ii) (1, 2)?

Explain the phrase 'ordered pair'.

We call the pair of numbers the *coordinates* of the point. The first number of the pair, which refers to the numbers *across* the page, is called the *x-coordinate* whilst the second number, which refers to the numbers *up* the page, is called the *y-coordinate*. So, for instance, *I* has x-coordinate 4 and y-coordinate 2.

(*c*) Which points have: (i) $x = 5$, $y = 2$; (ii) $x = 2$, $y = 4$?

(*d*) Consider the points *H*, *B* and *E* together. Write down their coordinates. Which of the following statements is true about all three points:

(i) $x = 2$;   (ii) $x = 4$;   (iii) $y = 4$;   (iv) $y = x$?

Is it also true for other points on the line passing through *H*, *B* and *E*?

(*e*) Give equations, as above, which describe the lines:

(i) *BD*;   (ii) *AC*;   (iii) *FH*;   (iv) *FG*.

FH and *FG* are called the *base lines* and the point where they intersect, *F*, (0, 0) is called the *origin*.

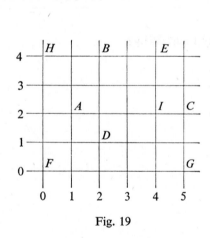

Fig. 19

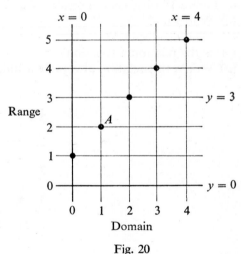

Fig. 20

3.2   Coordinates and functions

(*a*) Consider the function shown in Figure 20. Is it:

(i) $x \to 2x$;   (ii) $x \to x+1$;   (iii) $x \to x-1$?

(*b*) *A* is the point (1, 2). Write the other points of the graph similarly in co-ordinate form.

(*c*) Which of the following equations is correct for the set of points marked:

(i) $y = 2x$;   (ii) $y = x+1$;   (iii) $y = x-1$?

(*d*) With domain {0, 1, 2, 3, 4} draw the graph of the function $x \rightarrow x^2$, choosing a suitable scale for the range. What is the corresponding equation? We observe that a function, say, $x \rightarrow 2x+3$ may be represented by an equation $y = 2x+3$.

3.3　Linear functions

(*a*) Copy and complete the diagram in Figure 21, assuming the pattern is continued. It concerns a journey I made of 240 kilometres starting with a full petrol tank of 30 litres.

(*b*) The following type of table is often used to describe functions, as here 'distance travelled → petrol left'. Complete the table.

Distance travelled (km)	0	40	80	120	160	200	240
Petrol left (litres)	30	26	22	18	—	—	—

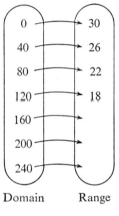

Fig. 21

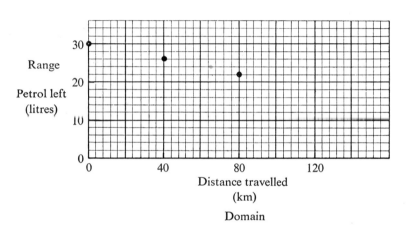

Fig. 22

Mark 7 points on your graph corresponding to the pairs of numbers (see Figure 22). If the points are joined up can a meaning be given to the in-between points?

  (i)　How much petrol was left after 135 km? (Use your graph.)
  (ii)　How far had I gone when there were only 7 litres left? (Use your graph.)
  (iii)　What is the petrol consumption in kilometres per litre?
  (iv)　How much petrol is consumed per kilometre?
  (v)　The equation of the function may be written

$$y = 30 - \frac{x}{10}.$$

What do x and y represent? Explain the appearance of the numbers 30, 10 and the subtraction sign in the equation.

Functions which give rise to straight line graphs are called *linear functions*.

Exercise D

1. Draw arrow diagrams for the following functions with the domain {1, 2, 3, 4, 5}:

(a) $x \to x+3$;   (b) $x \to 2x-1$;   (c) $x \to 4$ (a *constant* function).

2. Write down equations connecting x and y for the functions in Question 1 and represent the functions on an (x, y) graph, keeping the domain the same.

3. A butcher sells lamb at 36 new pence per kilogram and beef at 60 new pence per kilogram. On the same diagram graph the function 'mass $\to$ cost' for each type of meat.

(a) Use your graph to find the cost of 1·6 kg of lamb.

(b) What mass of beef would cost about 110 pence?

(c) A customer bought 1·3 kg of beef. How much lamb could she have bought for the same amount? (Find this graphically without noting the cost.)

4. The butcher advises his customers on cooking the beef as follows '45 minutes to the kilogram and 20 minutes extra'.

(a) For how long ought one to cook beef having a mass of:

(i) 1 kg;   (ii) 3 kg;   (iii) 0·8 kg;   (iv) x kg?

(b) If x kg requires y minutes cooking, give the equation connecting x and y.

5. A motorist calculated that the total weekly cost of running his car is given by $p = 250 + 1·5\,k$ where k stands for the number of kilometres covered in the week and p the cost in pence.

(a) Use the graph of this relation (see Figure 23) to find the weekly cost if the motorist travels:

(i) 150 km;   (ii) 230 km;   (iii) 465 km.

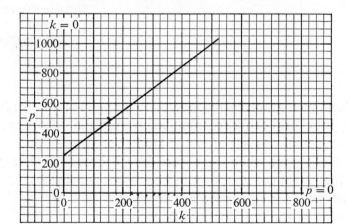

Fig. 23

(b) What is the weekly cost if the motorist does not use the car? Give possible reasons for this cost.

(c) What is the increase in cost for every kilometre covered?

(d) What is the average cost per kilometre in a week if the motorist drives:

(i) 100 km;   (ii) 300 km?

6. Alice's height at various times after she ate the cake labelled EAT ME was given by the relation
$$h = 2 + 6t,$$
where t stands for the number of seconds after eating the cake and h her height in inches (this happened before we changed over to centimetres).
   (a) By finding the values of h when t was 0, 2, 4, 6 draw a graph to illustrate the relation.
   (b) What was Alice's height $5\frac{1}{2}$ s after eating the cake?
   (c) How long did it take her to grow to a height of 24 inches?
   (d) What was her height before eating the cake and how fast did she grow?

7. A ship has z litres of oil on board after t days at sea where
$$z = 20\,000 - 1200\,t.$$
Show this information on a graph.
After how many days at sea does the ship need refuelling?

8. The price of ladders of different lengths is shown in the following table:

Number of rungs (x)	10	12	14	20
Price of ladder (y pounds)	6·30	7·40	8·50	11·80

(a) Represent this information on a graph. The cost consists of:

(i) a basic charge, and (ii) a charge per rung.

Find each of these charges.
(b) Write down an equation connecting x and y.
(c) What would be the cost of a 7 rung ladder?

9. Discuss whether any meaning can be given to the graphs drawn in Questions 6–8 if we extend them in either direction.

10. Suggest whether the following are likely to be linear functions or not.
Give reasons in each case.
(a) Area of floor space $\rightarrow$ price of house.
(b) Time $\rightarrow$ speed of a ball rolling down a hill.
(c) Counting number $n \rightarrow n$th triangle number.
(d) Number of sides $\rightarrow$ size of interior angle of a regular polygon.

3.4 Non-linear functions

Not all functions give rise to straight lines as you will have realized. Indeed some functions do not even produce continuous lines when graphed.
   (a) Consider the function

$$\text{mass of parcel} \rightarrow \text{cost of postage.}$$

If the postage rates are: not over 0·7 kg $12\frac{1}{2}$p,
                          not over 1 kg   15p,
                          not over 3 kg   22p, etc.,

what would be the postage rate for a 1·4 kg parcel?

Part of the graph is shown in Figure 24.

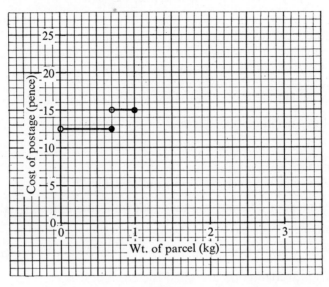

Fig. 24

Copy and complete this graph for parcels up to 3 kg. Suggest why the dots at the ends of each line are shown empty on the left and full on the right?

Is the relation

cost of postage → mass of parcel

a function?

(b) The function

number of sides → size of interior angle of a regular polygon

can easily be seen to be non-linear by drawing up a table of values.

Number of sides (x)	3	4	5	6
Size of angles ($y°$)	60	90	108	120

The number sequence 3, 4, 5, 6... increases in equal steps, but the sequence 60, 90, 108, 120... does not.

Is there an equation connecting x and y?

We can find the size of each angle in this way. Every time we take a 'walk round the block' we effectively turn through 360°. For the regular pentagon in Figure 25, the five marked angles add up to 360°. Hence each angle is $\frac{360°}{5}$. Therefore an interior angle, in degrees, is

$$180 - \frac{360}{5} = 108.$$

Now complete the general equation

$$y = 180 - \square.$$

40

Find the values of y corresponding to 7, 8, 9, 10 as values of x. Graph this function. Should the graph be a continuous curved line or just a set of points?

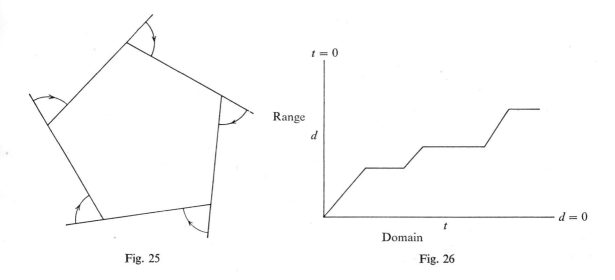

Fig. 25   Fig. 26

Exercise E

1. The graph in Figure 26 represents the function

$$\text{time } (t) \rightarrow \text{distance travelled } (d)$$

for a local train. What do the parts of the graph parallel to the base line $d = 0$ represent?

2. It is approximately 4000 km from London to New York and some of the times taken for aircraft of different speeds are shown in the table:

Approx. Speed (x km/h)	300	400	500	800	1000
Approx. Time (t hours)	13	10	8	.	`

Graph the function 'speed → time' and write the equation which approximately connects x and t.
From your graph find:
(a) the time taken for an aircraft flying at 730 km/h;
(b) the increase in speed necessary to cut the flying time from 10 to 6 hours.

3. If a car runs down a hill of slope 1 in 10, the distance s metres that it goes in t seconds is given roughly by

$$s = \tfrac{1}{2}t^2.$$

Graph this equation by taking several values of t up to, say, 24 and finding the corresponding distances. Find how long it would be before it crashed into the wall at the bottom of the hill 200 m from where it was parked.
What can you say about the speed of the car on its runaway adventure?

4. On a photographic negative 5 cm wide the area of a ship's sail is 3 cm². Make a table showing what the area would be on an enlargement of width:

(a) 10 cm;   (b) 15 cm;   (c) 20 cm;   (d) 1 cm (microfilm).

Graph your results and write down an equation relating the area, A cm², to the width of film, x cm.

Summary

A *relation* connects the members of two *sets* or members of the same set. Figure 27 represents a relation in which

$$\{2, 4, 6, 8\} \quad \text{is related to} \quad \{1, 2, 3, 4\}.$$

We can describe this as a relation which *maps x onto* $\frac{1}{2}x$ (or $x \to \frac{1}{2}x$) with *domain* $\{2, 4, 6, 8\}$ and *range* $\{1, 2, 3, 4\}$.

When *only one* arrow leaves each member of the domain we call the relation a *function*.

Relations may be represented in various ways including the use of graphs, coordinates and equations. The function $x \to \frac{1}{2}x$, sometimes represented by the equation $y = \frac{1}{2}x$.

For a function, each line up the page ($x = 0$, $x = 4$ etc.) will cross the graph only once.

A function which is represented by a straight line graph is called a *linear function* (see Figure 28).

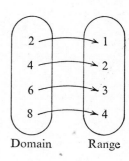

Fig. 27

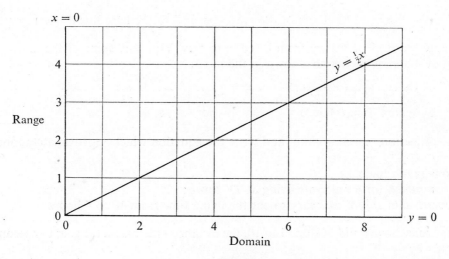

Fig. 28

3

THE SLIDE RULE

I have not kept my square, but that to come
Shall all be done by the rule.

WILLIAM SHAKESPEARE, *Antony and Cleopatra*

1. MAKING AND USING AN ADDITION SLIDE RULE

The rulers in Figure 1 are in such a position that we can use them to work out

$$3\cdot6 + 2\cdot8.$$

Could this position or setting of the rulers (we shall call them the *C* and *D scales*) be used for any other additions, for example $3\cdot6 + 5\cdot5$?

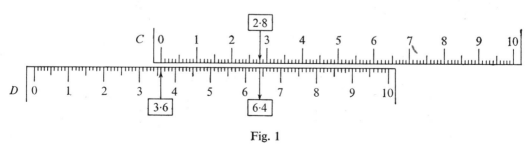

Fig. 1

4-2

Make such a slide rule for yourself from two pieces of graph paper, and use it for all the questions in Exercises A and B.

Exercise A

1. Calculate: (*a*) 2·1+2·5;   (*b*) 3·9+1·4;   (*c*) 6·2+0·4.

2. Calculate: (*a*) 3·4+1·6;   (*b*) 3·4+1·6+3·1.

3. Calculate: (*a*) 2·75+1·05;   (*b*) 2·75+1·05+0·45.

4. Make up some more additions and solve them on your slide rule. You may find some which you cannot do. Note what goes wrong and leave them until later. You can check your answer and the accuracy of your home-make rule by ordinary addition, of course. You may already have noticed that the presence of a point makes no difference to the rules!

5. Explain how you could use the rule to calculate:

(*a*) 240+350;   (*b*) 0·24+0·35.

6. What do you find if you try to calculate 3·4+7·8? Instead of the 0, try setting the 10 of the *C* scale against the 3·4 of the *D* scale. Do you get the correct answer below 7·8? If not, can you find a connection between the figure you get and the correct answer? Are you now able to do those calculations in Question 4 which you had to leave?

7. Calculate: (*a*) 7·2+5·3;   (*b*) 2·4+6·1+3·8.

1.1 Flow diagrams

The diagram in Figure 2 summarizes the method for adding two numbers on your rule. It shows the stages of the work without going into any detail and is much easier to follow than a set of instructions. We call it a *flow diagram*.

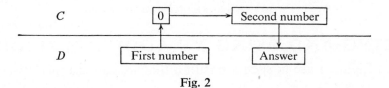

Fig. 2

Draw another flow diagram which summarizes the method when the answer is going to be greater than 10.

Figure 3 is a flow diagram for a second operation which your slide rule will do for you. Experiment with the setting shown and some simple numbers and find out what the operation is. If you cannot see, Exercise B will give you plenty of clues.

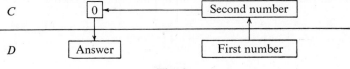

Fig. 3

44

Exercise B

1. Calculate: (*a*) 4·8 − 3·9;   (*b*) 8·4 − 0·6;   (*c*) 5·5 − 2·7.

2. Calculate: (*a*) 47–36;   (*b*) 220–195;   (*c*) 0·92–0·74.

What does unit length on your slide rule represent in each case?

3. When working on a slide rule, altering the order of the calculations (when permissible) may help to avoid coming off the scale.
Use your slide rule to calculate:

(*a*) 3·4 + 8·5 − 2·7;   (*b*) 3·4 − 2·7 + 8·5.

Do you get the same answer in each case? Is it easier to use the order in (*a*) or the order in (*b*)? Why?

Keep your slide rule carefully; you will need it again later!

2. THE LAW OF NATURAL GROWTH

In 1968 the F.A. Cup was won by West Bromwich Albion. The teams taking part in the four rounds leading up to the final are shown in Figure 4.

Fig. 4

The dotted curve shown joining the right-hand top corners of the columns is a very important one. From right to left the heights of the columns form the sequence

$$1, 2, 4, 8, 16, \ldots$$

In this example only certain points of the curve concern us. All other points have no meaning in connection with the Cup competition. This is not always the case.

(*a*) A biologist wishes to observe the way in which the number of a certain type of bacteria varies with time. In order to do this he grows a colony of the bacteria in a dish in the laboratory. Provided that the food is plentiful and the temperature is constant, he finds that the number of bacteria doubles each day.

Suppose that at one instant there are 1 000 000. Then the following table shows the size of the population at the end of each following day.

Day	0	1	2	3	4	5
Population (millions)	1	2	4	8	16	32

Presumably the population was growing before 'Day 0' of our experiment and so we can extend the table backwards. The day before the experiment began could be called 'Day −1' and so on.

Copy this table and fill in the gaps.

Day		−1	0	1	2	3	4	5	6	7
Population (millions)		$\frac{1}{2}$	1	2	4	8	16	32		

You will probably have already noticed that

$$4 = 2^2, \quad 8 = 2^3, \quad 16 = 2^4, \quad 32 = 2^5 ...,$$

that is, at the end of the xth day the population is 2^x million.

So far we have only considered the population at the end of each day. However, the growth is virtually continuous and does not leap from 1 to 2 million at the end of the first day. In this problem, unlike that of the knock-out competition, every point on the curve has meaning.

We have the function:

$$x \to 2^x.$$

The graph of this function is so important that it has been carefully drawn in Figure 5. The equation of the curve is $y = 2^x$.

Complete the following statements:

$$2^1 = \quad ; \quad 2^0 = \quad ; \quad 2^{-1} = \quad ; \quad 2^{-2} = \quad ; \quad 2^{-3} = \quad .$$

(i) What would be the size of the population half a day after the experiment began?

(ii) When would the population be 3 million?

If, when looking at Figure 5, you had given the answers to these questions as

(i) 1·5 million;   (ii) 1·5 days,

what sort of function would you have been assuming the law of growth to be? Using Figure 5, find:

(i) $2^{\frac{1}{2}}$;   (ii) $2^{1·5}$;   (iii) $2^{-1·3}$.

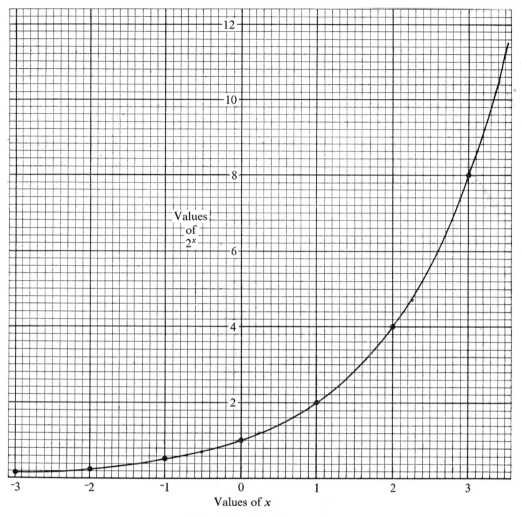

Fig. 5. The function $x \rightarrow 2^x$.

(*b*) Find out from the graph how many days it would take for the population to increase from:

(i) 1 to 3 million;   (ii) 2 to 6 million;   (iii) $1\frac{1}{2}$ to $4\frac{1}{2}$ million.

What do you find?

Do you think that the colony would always take the same time to increase its population by half? (That is, from 1 to $1\frac{1}{2}$ million, or 2 to 3 million, or 4 to 6 million.) If so, how long would it take?

Summary

The *law of natural growth* is that, if at a particular instant a population is of size P, the time taken for the population to increase in size to kP is always the same, no matter what value of P we take.

Ideas such as these are used when estimates are made of the population of the world in, say, 2000 A.D. For example, if the population in 1960 was P and in 1970 is kP, then in 2000 the population would be approximately k^4P. Of course, other factors are almost certain to affect this estimate; there might, for example, be a gradual trend towards smaller families.

Exercise C

Answer Questions 1–4 using Figure 5.

1. What are:

(*a*) $2^{2 \cdot 5}$; (*b*) $2^{1 \cdot 2}$; (*c*) $2^{1 \cdot 5}$; (*d*) $2^{1 \cdot 8}$; (*e*) $2^{0 \cdot 6}$; (*f*) $2^{-1 \cdot 5}$?

2. Find values of x which satisfy the following equations:

(*a*) $2^x = 8$;   (*b*) $2^x = 6$;   (*c*) $2^x = 2 \cdot 8$;   (*d*) $2^x = 0 \cdot 2$.

3. How many bacteria would there be:
(*a*) $2\frac{1}{4}$ days after the beginning of the experiment;
(*b*) 6 hours after the beginning of the experiment;
(*c*) 12 hours before the beginning of the experiment?

4. (*a*) How many days would it take for the population to multiply itself by 4?
   (*b*) How many times larger would the population be after 6 days than after 3 days?

5. Figure 6 is a plan view of a grand piano. The concave part is similar to the curve of growth. Can you find out why this is so? Ask your music teacher about the connection between a note on the piano and one an octave higher or an octave lower.

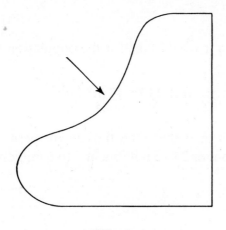

Fig. 6

Fig. 7

6. Some of you will own guitars. You will notice that there are 20 or more ridges or frets perpendicular to the strings (see Figure 7). Numbering the frets from the body end of the guitar 0–19 or more, measure the distance from each fret from the bridge of the guitar.

Draw up a table:

Fret	0	1	2	3	4	5	...
Distance from bridge (cm)							

Graph these results. Compare the curve with Figure 5. Which fret is twice as far from the bridge as the fret labelled 0? How many notes are there in an octave?

7. Radioactive materials decay in the same way as populations increase. The decay is usually measured in terms of the half-life. The half-life is the time required for half the radioactive material present at any time to disintegrate. Rates vary tremendously; the half-life of one isotope of lead is just less than $\frac{1}{2}$ hour, but that of one isotope of thorium is about 17 000 years.

Copy and complete this table for the lead isotope.

Time (min)	0	27	54	81	108
Fraction of sample left	1	0·5	0·25		

Draw a graph to find what fraction of the sample would be left after:

(a) $\frac{1}{4}$ hour;   (b) 1 hour.

If the sample had a mass of 100 g, when would the radioactive part of it have a mass less than 10 g? Would it ever all decay?

3. THE MULTIPLICATION SLIDE RULE

Advances in scientific achievement have always resulted in a demand for ways of speeding up the arithmetical work associated with problems. Galileo's invention of the telescope, for example, produced more accurate measurements of the large distances involved in astronomy. These more accurate measurements resulted in more arithmetical work; consequently it was no coincidence that, in the early seventeenth century, John Napier pioneered a new technique for calculation at the same time as Galileo was observing the movements of the moon and the planets.

We shall approach the problem in the way that Napier did.

3.1 Powers of 2

In Section 2 we looked at the relationship between the following two sequences:

$$... \;\tfrac{1}{16}\; \tfrac{1}{8}\; \tfrac{1}{4}\; \tfrac{1}{2}\; 1\; 2\; 4\; 8\; 16\; 32\; 64\; 128\; 256\; ...$$
$$\updownarrow\;\updownarrow\;\updownarrow\;\updownarrow\;\updownarrow\updownarrow\updownarrow\updownarrow\updownarrow\;\updownarrow\;\updownarrow\;\updownarrow\;\updownarrow$$
$$... \;^-4\;^-3\;^-2\;^-1\;0\;1\;2\;3\;4\;5\;6\;7\;8\;...$$

(a) Take any two numbers of the first sequence and multiply them.
For example, $4 \times 16 = 64.$

Try several pairs. What do you notice about your answers? (It may be necessary to extend the sequence.)

Write the multiplications as follows, writing alongside them the numbers of the second sequence

$$16 \leftrightarrow 4$$
$$\underline{\times 4 \leftrightarrow 2}$$
$$\underline{64 \leftrightarrow 6}$$

What can you say in each case about the numbers of the second sequence?

(b) Now try dividing two numbers of the first sequence, again writing the corresponding numbers of the second sequence alongside.

Here is an example:

$$2 \leftrightarrow 1$$
$$\underline{\div 8 \leftrightarrow 3}$$
$$\underline{\tfrac{1}{4} \leftrightarrow {}^{-}2}$$

What can you say about the numbers of the second sequence in this case?

Perhaps you think that this is very clever but do not understand why it works. Consider again the multiplication in (a) above

$$16 \leftrightarrow 4$$
$$\underline{\times 4 \leftrightarrow 2}$$
$$\underline{64 \leftrightarrow 6}$$

Now $16 = 2 \times 2 \times 2 \times 2 = 2^4,$

and $4 = 2 \times 2 \quad\quad = 2^2,$

therefore $16 \times 4 = (2 \times 2 \times 2 \times 2) \times (2 \times 2)$
$$= 2 \times 2 \times 2 \times 2 \times 2 \times 2 = 2^6 = 64.$$

That is, $2^4 \times 2^2 = 2^6.$

So writing numbers as *powers* of 2, a multiplication is converted into an addition. Similarly a division is converted into a subtraction.

Here are the numbers of the first sequence with the corresponding powers of 2 below:

$$\ldots \quad \tfrac{1}{8} \quad \tfrac{1}{4} \quad \tfrac{1}{2} \quad 1 \quad 2 \quad 4 \quad 8 \quad 16 \quad 32 \quad 64 \quad \ldots$$
$$\updownarrow \quad \updownarrow \quad \updownarrow \quad \updownarrow \quad \updownarrow \quad \updownarrow \quad \updownarrow \quad \updownarrow \quad \updownarrow \quad \updownarrow$$
$$\ldots \quad 2^{-3} \quad 2^{-2} \quad 2^{-1} \quad 2^0 \quad 2^1 \quad 2^2 \quad 2^3 \quad 2^4 \quad 2^5 \quad 2^6 \quad \ldots$$

Notice that the numbers of the original second sequence now appear as the powers of 2.

This is the basis of Napier's discovery. If all multiplications and divisions could be converted into the *far easier* operations of addition and subtraction, then the saving of time and labour would be considerable. However, so far in this chapter the only numbers which have been expressed as powers of 2 are those in the first sequence. We will get round this difficulty in the next section.

Exercise D

1. Calculate, leaving your answers as powers of 2:

(a) $2^2 \times 2^3$;   (b) $2^6 \times 2^{-1}$;   (c) $2^{-1} \times 2^{-2}$;   (d) $2^0 \times 2^{-1}$.

Write down the corresponding products without the powers.

2. The division $4 \div \frac{1}{2} = 8$ can be written in terms of powers of 2 as $2^2 \div 2^{-1} = 2^3$; notice that $2 - {}^-1 = 3$.

Write the following divisions in terms of powers of 2 and check that in each case the subtraction of the powers is correct:

(a) $16 \div 8$;   (b) $16 \div 32$;   (c) $64 \div 1$;

(d) $8 \div \frac{1}{8}$;   (e) $2 \div \frac{1}{2}$;   (f) $1 \div \frac{1}{4}$.

3. Calculate, leaving your answers as powers of 2:

(a) $2^3 \div 2^{-1}$;   (b) $2^{-1} \div 2^2$;   (c) $2^0 \div 2^{-3}$.

3.2   Converting the slide rule

In Section 1 we used a slide rule to add and subtract. We have just seen how multiplication and division can be converted into addition and subtraction. Hence by applying the function

$$x \rightarrow 2^x \quad (0 \rightarrow 1, \ 1 \rightarrow 2, \ 2 \rightarrow 4, \ 3 \rightarrow 8, \ 4 \rightarrow 16, \ 5 \rightarrow 32, \ \text{etc.}),$$

to the numbers on your addition slide rule you can convert it into a multiplication slide rule.

Mark the new figures on your graph paper rule using a different colour. Use it to check the following multiplications:

$$4 \times 4 = 16; \quad 2 \times 16 = 32; \quad 8 \times 16 = 128.$$

How do we mark intermediate points? Do we put 3 half-way between 2 and 4? Mark 3 in this position (in pencil) and see what answer the slide rule gives for 3×3.

This, then, is the wrong place for 3! Look back at Figure 5. From this you can see that when $2^x = 3$, $x = 1 \cdot 6$ approximately and we can write $2^{1 \cdot 6} = 3$ approximately.

Now in converting our addition slide rule into a multiplication slide rule we have converted the upper scale in Figure 8 into the lower.

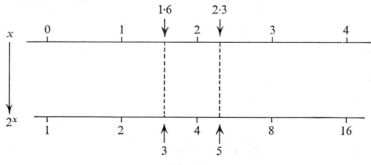

Fig. 8

51

Therefore, since $2^{1 \cdot 6} = 3$ approximately, we put the 3 opposite the 1·6 mark.

Similarly, since $2^{2 \cdot 3} = 5$, approximately, we put the 5 opposite the 2·3 mark.

By referring to Figure 5, find where to mark 6, 7, 9 and 10 on your rule. Check that you can indeed multiply and divide using the new markings.

Explain why the flow diagram for multiplication will be the same as that in Figure 2, and why the flow diagram for division will be the same as that in Figure 3.

3.3 The manufactured slide rule

The C and D scales on all manufactured slide rules are only marked from 1 to 10. This is sufficient, as we shall see later.

Most slide rules have a cursor to help you to read and position the slide. If a cursor is used, then the operations which must be made when multiplying are:

(*a*) set the cursor on the first number on the D scale;

(*b*) move 1 on the C scale under the cursor;

(*c*) set the cursor on the second number on the C scale;

(*d*) read off the answer under the cursor on the D scale.

The corresponding operations for division are:

(*a*) set the cursor over the first number on the D scale;

(*b*) move the C scale until the second number is under the cursor;

(*c*) set the cursor on 1 on the C scale;

(*d*) read off the answer under the cursor on the D scale.

Exercise E

Work out the following on your slide rule, giving the answers to 2 s.f.

1. $1 \cdot 5 \times 1 \cdot 4$.
2. $2 \cdot 5 \times 1 \cdot 8$.
3. $2 \cdot 2 \times 3 \cdot 5$.

4. $2 \cdot 3 \times 4 \cdot 3$.
5. $5 \cdot 3 \times 1 \cdot 4$.
6. $1 \cdot 7 \times 3 \cdot 7$.

7. $2 \cdot 7 \times 2 \cdot 3$.
8. $1 \cdot 9 \times 3 \cdot 1$.
9. $4 \cdot 6 \times 1 \cdot 3$.

10. $5 \cdot 9 \times 1 \cdot 1$.
11. $5 \cdot 6 \times 1 \cdot 7$.
12. $1 \cdot 1 \times 1 \cdot 8$.

13. $(1 \cdot 9)^2$.
14. $(2 \cdot 7)^2$.
15. $(3 \cdot 1)^2$.

16. $5 \cdot 4 \div 1 \cdot 2$.
17. $9 \cdot 8 \div 3 \cdot 5$.
18. $8 \cdot 5 \div 3 \cdot 4$.

19. $9 \cdot 9 \div 5 \cdot 4$.
20. $9 \cdot 1 \div 2 \cdot 4$.
21. $9 \cdot 3 \div 1 \cdot 6$.

22. $7 \cdot 3 \div 1 \cdot 7$.
23. $6 \cdot 3 \div 3 \cdot 9$.
24. $4 \cdot 6 \div 3 \cdot 8$.

25. $7 \cdot 6 \div 1 \cdot 9$.
26. $9 \cdot 5 \div 1 \cdot 7$.
27. $8 \cdot 2 \div 6 \cdot 8$.

28. $8 \cdot 9 \div 8 \cdot 1$.
29. $4 \cdot 7 \div 1 \cdot 3$.
30. $2 \cdot 9 \div 1 \cdot 7$.

3.4   Repetition of scales

Figure 9 shows a slide rule set for the multiplication

$$4 \times 5 = 20.$$

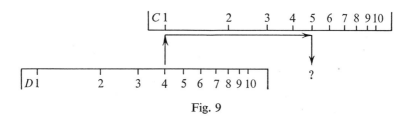

Fig. 9

Following the flow lines the answer is off the scale! Let us see what happens if we have another D scale following on immediately to the right of the one in Figure 9. This is shown in Figure 10.

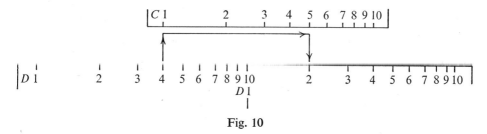

Fig. 10

The flow lines now end at 2. We must, of course, read this as 20 to obtain the correct answer.

If we only had the right-hand D scale, we could still get the correct answer by using the flow diagram of Figure 11.

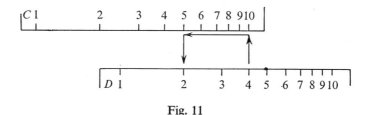

Fig. 11

The setting in Figure 11 is also the correct one for multiplications such as

$$4 \times 50 = 200; \quad 40 \times 50 = 2000; \quad 0 \cdot 4 \times 5 = 2.$$

To divide, we reverse the direction of the arrows in Figure 11. We then obtain the flow diagram for divisions such as:

$$2 \div 5 = 0 \cdot 4; \quad 20 \div 5 = 4; \quad 200 \div 50 = 4; \quad 2000 \div 50 = 40; \quad 2 \div 0 \cdot 5 = 4.$$

This demonstrates that a slide rule multiplies and divides *without* reference to the decimal point. We shall consider this again in the next section.

Summary

These are the operations for (*a*) multiplication, (*b*) division.

(*a*) (i) Cursor on first number on the *D* scale;
    (ii) 1 or 10 on the *C* scale under the cursor;
    (iii) cursor on second number on the *C* scale;
    (iv) answer on the *D* scale under the cursor.

(*b*) (i) Cursor on first number on the *D* scale;
    (ii) second number on the *C* scale under the cursor;
    (iii) cursor on 1 or 10 on the *C* scale;
    (iv) answer on the *D* scale under the cursor.

Exercise F

Use your slide rule to work out Questions 1–21 giving your answers to 2 s.f.

1. $7 \cdot 2 \times 2 \cdot 5$.
2. $8 \cdot 4 \times 2 \cdot 5$.
3. $5 \times 8 \cdot 2$.
4. $5 \cdot 6 \times 3 \cdot 4$.
5. $3 \cdot 1 \times 4 \cdot 2$.
6. $6 \cdot 7 \times 9 \cdot 7$.
7. $7 \cdot 3 \times 3 \cdot 7$.
8. $(4 \cdot 9)^2$.
9. $(8 \cdot 3)^2$.
10. $12 \div 4 \cdot 8$.
11. $19 \div 2 \cdot 5$.
12. $17 \div 6 \cdot 8$.
13. $16 \div 4 \cdot 1$.
14. $25 \div 4 \cdot 9$.
15. $31 \div 6 \cdot 6$.
16. $34 \div 46$.
17. $100 \div 11$.
18. $51 \div 88$.
19. $6 \cdot 1 \times 72$.
20. $44 \times 9 \cdot 1$.
21. $8 \cdot 2 \times 33$.

3.5 Successive Calculations

Multiply $2 \cdot 3$ by $1 \cdot 4$. Now multiply the result by $1 \cdot 8$. Your final answer ($5 \cdot 8$) is $2 \cdot 3 \times 1 \cdot 4 \times 1 \cdot 8$. If you were interested only in this final answer, was it necessary to read off the intermediate result of $2 \cdot 3 \times 1 \cdot 4$?

Using the summary above as a guide, list the operations you have used for this double multiplication. Check your list by using it to perform

(i) $2 \times 2 \times 2$;   (ii) $4 \times 5 \times 6$.

How many settings of the *C* scale would be required if four numbers were to be multiplied together?

A calculation such as $\dfrac{4 \times 3}{2}$ can, of course, be done in several ways. For instance 4 could be multiplied by 3 and the result divided by 2, or 4 could be divided by 2 and the result multiplied by 3. Perform this calculation on your slide rule by each of these methods. Which one involved only one setting of the *C* scale?

Clearly doing the division before the multiplication may be an advantage in calculations of this type.

You will probably find it best to treat calculations such as $\dfrac{6}{1\cdot5\times2}$ as two successive divisions (6 divided by 1·5 and the result divided by 2). What difficulties would arise if you did the multiplication first?

Exercise G

On your slide rule calculate, using as few settings as possible, giving the answers to 2 s.f.

1. $1\cdot7\times3\cdot8\times1\cdot1$.
2. $1\cdot3\times1\cdot6\times3\cdot2$.
3. $2\cdot6\times1\cdot9\times1\cdot7$.
4. $\dfrac{2\cdot8\times4\cdot5}{7\cdot2}$.
5. $\dfrac{7\cdot1}{1\cdot4\times2\cdot8}$.
6. $\dfrac{8\cdot3\times6\cdot4}{5\cdot5}$.
7. $\dfrac{2\cdot8}{1\cdot6\times1\cdot26}$.
8. $\dfrac{8\cdot4\times3\cdot3}{6\cdot1}$.
9. $\dfrac{2\cdot6\times1\cdot7}{1\cdot6}$.
10. $\dfrac{5\cdot3\times2\cdot8}{9\cdot4}$.

4. ESTIMATION AND ACCURACY

4.1   Estimating

As we have said, the slide rule works without reference to the decimal point.

It is quite possible that you had to think carefully in some of the divisions in Exercise F before you put the decimal point in your answer.

In more difficult cases it will be necessary to get an approximate answer first, in order to feel certain where to put the decimal point.

Write down rough answers to the following, doing all the working in your head without using a slide rule:

(i) 38×31;   (ii) $410\times8\cdot7$;   (iii) $0\cdot27\times570$;
(iv) $660\div72$;   (v) $28\div0\cdot83$.

Here are the same problems with the numbers written correct to 1 s.f. Work them out and compare the answers with your first attempts:

(i) 40×30;   (ii) 400×9;   (iii) $0\cdot3\times600$;
(iv) $700\div70$;   (v) $30\div0\cdot8$.

To work out (ii) roughly it would be good enough to say 400×10, and to work out (v) roughly it would be good enough to say $30\div1$.

Here are the answers to 1 s.f.:

(i) 1000;   (ii) 4000:   (iii) 200;   (iv) 10;   (v) 30.

These answers are *sufficient to fix the position of the decimal point*.

Now use your slide rule to find the five answers to 3 s.f.

4.2 Accuracy

One can be certain that the class will not all agree on the third significant figure in some of these five answers. For example, consider (iv) $660 \div 72$. To 3 s.f. the correct answer is 9·17 but some of you will probably have obtained 9·16 or 9·18. However, you should all agree on 1180 correct to 3 s.f. for (i).

It appears that it is easier to read to 3 s.f. at the left-hand end of the slide rule than it is at the right-hand end. However, it would be wrong to think that the slide rule is less accurate at one end than the other.

The difference between 9·17 and 9·16 (or 917 and 916) is about 1 in 1000, but the difference between say 1180 and 1170 (or 118 and 117) is only a little less than 1 in 100. So when you read to 3 s.f. at the right-hand end, you are being nearly 10 times as accurate as when you read to 3 s.f. at the left-hand end. Is it possible to even things out by estimating to 4 s.f. at the left-hand end?

Exercise H

1. Find rough answers to the following and then use the slide rule to work them out to 3 s.f.

(a) $430 \times 31 \cdot 2$; (b) $4 \cdot 17 \times 730$; (c) $0 \cdot 632 \times 425$; (d) $9 \cdot 05 \times 18 \cdot 2$;
(e) $54 \cdot 5 \times 1 \cdot 96$; (f) $7 \cdot 40 \div 36 \cdot 4$; (g) $0 \cdot 136 \div 8 \cdot 45$; (h) $24 \cdot 4 \div 0 \cdot 673$.

2. What is a man's average speed in metres per second if he runs 100 m in 10·5 s?

3. Assuming a rate of exchange of 8·86 marks to the pound, convert the cost of the following articles advertised in a German magazine into pounds.

(a) A typewriter—240 marks;
(b) A watch—136 marks;
(c) A T.V. set—521 marks.

4. Assuming a rate of exchange of 13·7 francs to the pound, convert the cost of the following articles advertised in a French magazine into pounds (to 3 s.f.)

(a) a camera—90 francs;
(b) a bottle of perfume—39 francs;
(c) a travelling case—75 francs.

4.3 Large and small numbers

Science is more and more involved with very large and very small numbers.

When Isaac Newton made his estimate of the mass of the sun he gave the figure

$$22\,000\,000\,000\,000\,000\,000\,000\,000\,000\,000 \text{ pounds.}$$

Obviously a more convenient method of writing this number is necessary. Using powers of ten it could be written 22×10^{27}.

However, if we write it so that only the 1st significant figure is to the left of the decimal point it is
$$2 \cdot 2 \times 10^{28}.$$

This is called the *standard index form* of the number.
The mass of an electron is
$$0 \cdot 000\,000\,000\,000\,000\,000\,000\,000\,000\,009 \text{ g.}$$

In standard form this number would be 9×10^{-28}.
Count the number of 0's after the decimal point and explain why there are not 28.
Expressing numbers in standard index form can help us to estimate the position of the decimal point in calculations.

Example 1

Find rough answers to:

$\qquad\qquad$ (a) $479 \times 0 \cdot 0032$; $\qquad$ (b) $0 \cdot 634 \div 261$.

(a) $479 \times 0 \cdot 0032 = 4 \cdot 79 \times 10^2 \times 3 \cdot 2 \times 10^{-3}$; $\quad$ Now $10^2 \times 10^{-3} = 10^{-1}$.
Rough answer: $5 \times 3 \times 10^{-1} = 1 \cdot 5$.
(b) $\dfrac{0 \cdot 634}{261} = \dfrac{6 \cdot 34 \times 10^{-1}}{2 \cdot 61 \times 10^2}$; $\quad$ Now $10^{-1} \div 10^2 = 10^{-3}$.
Rough answer: $\frac{6}{3} \times 10^{-3} = 2 \times 10^{-3} = 0 \cdot 002$.

Summary

(a) When dealing with large or small numbers, convert them to standard form and work out the 'powers of ten' part separately.

(b) To fix the position of the decimal point obtain a rough approximation by correcting all the numbers to 1 S.F.

(c) Endeavour to read the slide rule accurately to 3 S.F. in the region of the 5 mark. At the right-hand end there will always be some doubt as to the third figure but at the left-hand end you may be able to read the fourth significant figure.

Exercise I

1. Put the following numbers in standard index form:

(a) 4 million; $\qquad$ (b) 12 thousand; $\qquad$ (c) 243 000; $\qquad$ (d) 748×10^4;

(e) $0 \cdot 003$; $\qquad$ (f) $\dfrac{1}{100\,000}$; $\qquad$ (g) $45 \div 10^4$.

2. Put the following numbers in standard index form to 2 S.F.:

(a) $3 \cdot 141\,592\,653\,5$;
(b) the distance of the moon from the earth, 384 000 km;
(c) the number of people vaccinated against poliomyelitis in Manchester in 1960, 22 294;
(d) the coefficient of linear expansion of copper per degC, $0 \cdot 000\,0187$.

3. Arrange these sets of numbers in order of size, largest first, by writing them in standard index form:

(a) $42 \cdot 6 \times 10^3$, 420×10^2, $0 \cdot 067 \times 10^5$,   41 600;

(b) $0 \cdot 023\,250 \times 10^{-2}$, $2 \cdot 37 \div 10^2$, $0 \cdot 000\,29 \times 10^3$.

Why is it easier to arrange numbers in order of size when they are in standard index form?

4. Work out the following to 3 s.f. by converting the numbers to standard index form:

(a) 2070×361;   (b) $0 \cdot 0051 \times 57\,300$;   (c) $(460)^2$;

(d) $1090 \div 0 \cdot 12$;   (e) $0 \cdot 96 \div 101$;   (f) $(0 \cdot 062)^2$.

5. (a) The population of Manchester in 1962 was 659 000 and the area 51 square kilometres. Find the average number of persons per square kilometre.

(b) The number of live births recorded in Manchester in 1962 was 13 571. Find to 3 s.f. the number of live births per thousand of the population.

6. In an experiment, a mild steel rod was found to expand in length by $0 \cdot 0146$ cm when heated from 15 to 100 degC. What would be the amount that it expanded when its temperature increased by 1 degC? (Assume the expansion is linear.)

7. Radio signals travel at $3 \cdot 0 \times 10^5$ kilometres per second and sound travels at 330 metres per second. The microphone in a studio is $1 \cdot 0$ m from the weather man giving the forecast and a ship's navigator is listening through earphones 840 km from the transmitter.

Find, to 2 s.f., the time the weather man's voice takes:

(a) to reach the microphone;

(b) to travel from the transmitter to the navigator.

Why would it have been foolish in this case to ask for the answers to be given to 3 s.f.?

8. The needle of a record player moves $8 \cdot 9$ cm when playing a particular L.P. record. If the record makes 830 revolutions estimate to 2 s.f. the width of the groove.

5. SQUARES AND SQUARE ROOTS

Set the cursor to 3 on the D scale and read the A scale. What do you find? Repeat, setting 4, 5 and $1 \cdot 5$ on the D scale.

What is the function

$$\text{number on } D \text{ scale} \rightarrow \text{number on } A \text{ scale?}$$

What is the inverse of this function? Verify your answer by using your rule to find $\sqrt{16}$, $\sqrt{4}$ and $\sqrt{81}$.

The squares of numbers from 1 to 10, and the square roots of numbers from 1 to 100 can be read direct. For numbers outside these ranges the slide rule can only provide the digits—the position of the decimal point will have to be found by working out an approximate answer.

Example 2

(a) Calculate $45 \cdot 3^2$.

Approximating, $45 \cdot 3^2 \approx 50^2 = 2500$. ($\approx$ means 'approximately equal to'.)

Slide rule digits are 205.

Hence, $45 \cdot 3^2 = 2050$, to slide rule accuracy.

(b) Calculate $0 \cdot 216^2$.

Approximating, $0 \cdot 216^2 \approx 0 \cdot 2^2 = 0 \cdot 04$.

Slide rule digits are 467.

Hence $0 \cdot 216^2 = 0 \cdot 0467$, to slide rule accuracy.

Squares can obviously be found by multiplying the number by itself using the C and D scales, as well as by reading from the D to A scale direct. Which of these two methods will be (i) the quicker, (ii) the more accurate?

Exercise J

Use your slide rule to calculate Questions 1–20.

1. 34^2.   2. 19^2.   3. $5 \cdot 4^2$.   4. 186^2.

5. 860^2.   6. $26 \cdot 5^2$.   7. $5 \cdot 15^2$.   8. 1230^2.

9. $0 \cdot 52^2$.   10. $0 \cdot 91^2$.   11. $0 \cdot 22^2$.   12. $0 \cdot 084^2$.

13. $0 \cdot 0031^2$.   14. $0 \cdot 0032^2$.   15. 105^2.   16. $0 \cdot 105^2$.

17. 8090^2.   18. $0 \cdot 602^2$.   19. $0 \cdot 62^2$.   20. $0 \cdot 000\,77^2$.

5.1 Square roots

Copy and complete the following table:

Number	Square root	Number of digits to left of D.P. in number	Number of digits to left of D.P. in square root
I	I		
10			
100	10	3	2
1000			
10 000	100	5	
100 000			
1 000 000	1000		4

Is there a connection between the last two columns? Divide each of the numbers in the third column by 2. This gives the number of pairs. Can you now be more precise about the connection?

A convenient way of finding the number of digits in the square root of a number is therefore to pair off its digits from the decimal point to the left. The number of pairs (including the half pair if there is one) then gives the number of digits in the square root.

5-2

Example 3

Using a slide rule, calculate (*a*) $\sqrt{920}$, (*b*) $\sqrt{920000}$.

(*a*) Pairing off from the decimal point to the left

$$\sqrt{9}\,|\,20$$
$$\approx 3\,|\,0.$$

(only 1 s.f. is required at this stage).

Looking below 9·2 on the *A* scale one finds 3·03 on the *D* scale. Looking below 92 on the *A* scale one finds 9·6 on the *D* scale. From the approximation it is plainly the former that is needed.

The slide rule figure needed is therefore 3·03, so $\sqrt{920} = 30 \cdot 3$ (to slide rule accuracy).

(*b*) Similarly

$$\sqrt{92}\,|\,00\,|\,00$$
$$\approx 9\,|\,\ 0\,|\,0.$$

The slide rule figure needed in this case is therefore 9·6, so $\sqrt{920000} = 960$ (to slide rule accuracy).

A similar discussion of the square roots of numbers less than 1 will convince you that each digit of the square root corresponds to a pair of digits in the number itself. In this case it is necessary to pair off from the decimal point *to the right*.

Example 4

Using a slide rule, calculate $\sqrt{0 \cdot 0092}$.

Pairing off from the decimal point to the right

$$\sqrt{0 \cdot}\,|\,00\,|\,92$$
$$\approx 0 \cdot\,|\,\ 0\,|\,9.$$

The slide rule figure needed is therefore 9·6, so $\sqrt{0 \cdot 0092} = 0 \cdot 096$ (to slide rule accuracy).

Exercise K

Use your slide rule to calculate the square roots of the following.

1. 275.	2. 27·5.	3. 0·0275.	4. 0·0000275.
5. 4040.	6. 4·04.	7. 8920000.	8. 89200.
9. 0·00684	10. 1085.	11. 0·0382.	12. 0·00208.
13. 56155.	14. 27·185.	15. 300·5.	16. 23·4².
17. $(4 \cdot 1 \times 10^4)$.	18. $(7 \cdot 9 \times 10^{-4})$.	19. $(3 \cdot 48 \times 10^{12})$.	20. $(8 \cdot 3 \times 10^{-5})$.

Summary

Squares are found by slide rule by reading from scale *D* to scale *A*.

Square roots are found by reading from scale *A* to scale *D*.

In both cases the position of the decimal point must be found by estimation. In the case of square roots care has also to be taken to choose between 72 and 7·2 on the *A* scale when calculating (say) $\sqrt{720}$. The simplest method is to pair off the digits from the decimal point.

5.2   Some standard formulae

Many types of problem involve substituting in a formula and performing a calculation. Here are some familiar formulae which we shall need in the following exercise:

Circumference of a circle (radius *r*) $C = 2\pi r$.

Area enclosed by circle $A = \pi r^2$.

Sector: region bounded by two radii and an arc (top view of a slice of round cake).

Area of sector is $\dfrac{\theta}{360} \times$ area enclosed by circle ($\theta°$ is the angle of the sector).

Area of curved surface of *cylinder* (height *h*), $A = 2\pi rh$.

Volume of cylinder $V - \pi r^2 h$.

Pythagoras's Theorem: The sum of the areas of the squares drawn on the shorter sides of a right-angled triangle is equal to the area of the square drawn on the longest side. If *a*, *b* and *c* are the lengths of the sides, then

$$a^2 + b^2 = c^2.$$

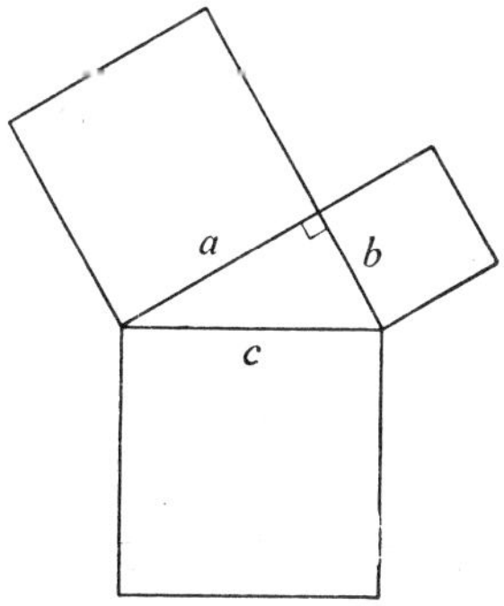

Fig. 12

Exercise L

1. Use your slide rule to calculate the circumferences of circles with the following radii:

(*a*) 6·1 cm;   (*b*) 29·4 cm;   (*c*) 18 m;   (*d*) 0·045 mm.

2. Use a slide rule to find the diameter of circles with the following circumferences:

(*a*) 88 m;   (*b*) $4\frac{1}{2}$ km;   (*c*) 5100 m.

Sketch a slide rule setting that enables you to read the answers to all three parts in one setting.

3. A tricycle has wheels whose *diameter* including the tyres is 52 cm. What is their circumference? How far does a wheel go forward in 80 revolutions?

4. The radius of the cylinder on the winch at the top of a well is 11 cm. How many times must the handle be turned to draw up a bucket of water through 6 m?

5. How long a piece of string would you need to enclose an area of 1 m²? Try to guess the answer, and then check your answer by calculation.

6. Use your slide rule to find the radius of a circle of area;

(*a*) 40 cm²;   (*b*) 845 m².

7. Find the length of arc cut off:

(*a*) from a circle of radius 35 cm by two radii at 60°;
(*b*) from a circle of radius 210 cm by two radii at 135°;
(*c*) from a circle of radius 1·4 cm by two radii at 300°.

8. Find the area of a sector:

(*a*) of 48° from a circle of radius 4·55 cm;
(*b*) of 240° from a circle of radius 2·03 cm.

9. A box for Brie cheese is made in the shape of a 30° sector from a circle of radius 20 cm. What is the area of its top? The box is 2 cm deep. Find the volume of the box.

10. Find the area of the curved surfaces of the cylinders with the following dimensions:

(*a*) radius 4 cm, height 3 cm; (*b*) radius 3 cm, height 4 cm.
Did you expect them to be equal? Say why.

11. Find the volumes of the cylinders in Question 10. Did you expect them to be equal? Explain your findings.

12. Find the area of the curved surface and the volume of a cylinder of height 2·45 cm and base radius 0·55 cm.

13. The volume of a cylinder is 80 cm³ and the area of its base is 16 cm².
Find its height and the radius of its base, to slide rule accuracy.

14. In the triangle *ABC*, *AB* = 3 cm, *BC* = 5 cm and $\angle ABC = 90°$. Find *AC*.

15. In the triangle *LMN*, *LM* = 2 cm, *LN* = 3 cm and $\angle LMN = 90°$. Find *MN*.

16. Find the length of the diagonal of:

(*a*) the square *ABCD* in which *AB* = 3 cm;
(*b*) the rectangle *EFGH* in which *GH* = 20 m and *HE* = 24 m.

17. Find the length of a side of:

(*a*) a rhombus whose diagonals measure 8 cm and 6 cm;
(*b*) a square whose diagonal measures 16 cm.

4
OPERATIONS

He marched them up to the top of the hill,
And he marched them down again.

ANONYMOUS, *The Noble Duke of York*

1. OPERATIONS ON SETS

Mathematics can be said to start with sets and numbers. In this chapter we are mainly concerned with the ways in which these can be combined by what we call *operations*.

Examples of standard operations on numbers are addition and division. Because each operation produces a result from *two* elements they are properly called *binary operations*. Name some other binary operations.

We shall first, however, investigate some binary operations on sets.

1.1 Venn diagrams

The pupils in your class can be divided up into sets in many different ways. Each set will have its own particular feature. For example, members of your class who wear glasses form one set, while those with birthdays in March form another.

(a) Suppose all the pupils in your class whose birthdays are in March go to the front of the room and stand together. Let this set be called *M*. Now let *S*, the set of pupils in your class whose birthdays are in September, also stand together at the front of the room. If the teacher draws a curve on the floor around each set, what can you say about the two curves? Make a sketch.

Fig. 1

(b) Figure 1 shows the members of a class whose parents own a caravan—set *C*. Let *G* be the set of children in the class who wear glasses and suppose *G* and *C* go to the front of the classroom. If we draw curves around *G* and *C* what can we say about the two curves?

(c) Figure 2 shows the result when the members of a class who have blue eyes, *B*, and the members who play chess, *P*, go to the front of the class. How many members of this class have blue eyes and play chess?

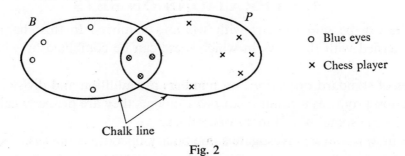

Fig. 2

(d) Suggest pairs of sets, whose members are pupils in your class, which would lead to the diagrams shown in Figure 3.

Diagrams like these are called *Venn diagrams* after John Venn (1834–1923), a Cambridge mathematician.

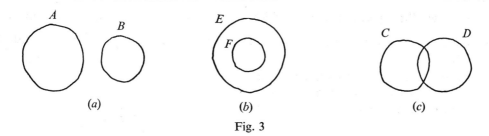

Fig. 3

(e) If $G = \{2, 4, 6, 8, 10\}$ and $H = \{1, 2, 3, 4, 5\}$, draw a Venn diagram which would show the connection between these sets. Shade in the region which represents the members common to both sets.

We call the set whose members are common to the sets G and H the *intersection* of G and H and denote it by $G \cap H$.

The symbol, which looks like a 'U' upside down, is sometimes called *cap* so that $G \cap H$ is read as '*G* cap *H*'.

In the above example, $G \cap H = \{2, 4\}$. '$\cap$' is a binary operation on the two sets G and H. Compare $G \cap H = \{2, 4\}$ with $3 + 5 = 8$.

Figure 4 is a Venn diagram representing the sets:

$A = \{$Austin cars$\}$ and $B = \{$red cars$\}$. The shaded region represents $A \cap B$ which is the set of red Austin cars.

It often happens that two sets have no members in common, so their intersection will have no members.

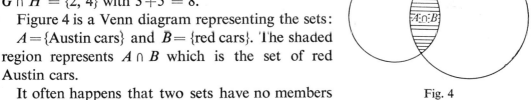

Fig. 4

When this happens, the sets are said to be *disjoint*, and their intersection is called the *empty* set, written $\varnothing$ (the Danish letter 'oe'). For example, if $P = \{$pupils$\}$ and $T = \{$teachers$\}$, then P and T are disjoint and $P \cap T = \varnothing$.

Exercise A

1. Illustrate the relation between the following pairs of sets by drawing Venn diagrams:

(a) $A = \{1, 2, 3, 4\}$, $B = \{5, 6, 7, 8\}$;

(b) $A = \{$the vowels$\}$, $B = \{$the first five letters of the alphabet$\}$;

(c) $A = \{1, 2, 3, 4, 5\}$, $B = \{1, 5, 3\}$;

(d) $A = \{$letters in your surname$\}$, $B = \{$letters in your Christian names$\}$;

(e) $A = \{$children in your class who wear glasses$\}$,
     $B = \{$children in your class who do not wear glasses$\}$;

(f) $A = \{$types of birds who nest in England$\}$, $B = \{$thrush, robin, wren, rook$\}$;

(g) $A = \{\triangle, \bigcirc, \square, *\}$, $B = \{\square, *, \triangle, \dagger\}$;

(h) $A = \{3, 7, 11, 5, 9\}$, $B = \{$odd numbers between 2 and 12$\}$.

2. List the members of the sets $A \cap B$ for the pairs of sets given in Question 1.

65

3. Write down three pairs of sets for which Figure 5 would be a suitable Venn diagram.

4. When the members of a set *B* are all members of another set *A*, then *B* is called a *subset* of *A*.

(*a*) Draw a Venn diagram to illustrate this.
(*b*) If *B* is a subset of *A*, what is $A \cap B$?
(*c*) Give three examples of subsets.

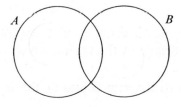

Fig. 5

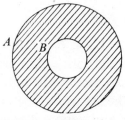

Fig. 6

5. Copy and complete the following:

(*a*) $\{a, \quad\} \cap \{b, c\} = \{c\}$;
(*b*) $\{7, 9, \quad\} \cap \{5, \quad, 2, 3\} = \{9, 3\}$;
(*c*) $\{d, \quad, \quad, b, t\} \cap \{p, \quad, d, \quad\} = \{a, \quad, e\}$.

6. If *A* is a complete set of playing cards and *B* is the subset of *A* consisting of the picture cards, then *A* and *B* can be represented by the Venn diagram shown in Figure 6.

(*a*) Describe in words, the set of cards represented by the shaded region.
(*b*) Name two sets of cards *C* and *D* for which

$$C \cap D = \{\text{the four of Hearts}\}.$$

7. Let *P* = {even numbers greater than 7} and *Q* = {counting numbers less than 25}. Give two members of the set $P \cap Q$.

8. In a class of 30 pupils, there are 19 who play tennis and 16 who play hockey.

(*a*) Draw a Venn diagram to illustrate the sets *T* and *H*, where *T* = {tennis players in the form} and *H* = {hockey players in the form}.

(*b*) If everyone in the class plays at least one of these games, how many members has the set $T \cap H$?

9. A paper boy delivers 27 copies of the *Daily Mirror* and 22 copies of the *Guardian* in a street of 40 houses.

What is (*a*) the smallest number of houses, (*b*) the largest numbers of houses, which could have had two papers delivered? (Assume no house receives more than 2 papers.)

1.2 Universal sets

If $\qquad A = \{\text{children born in August}\}$

and $\qquad B = \{\text{children who come to school on a bicycle}\}$,

describe the set $A \cap B$. How many members has this intersection set? Before we can answer this question we will have to decide what complete set of children we are thinking about. For instance, we might have to consider all the children in one class, all the children in one school, all the children who enjoy dancing, or many other possible sets. In each case the complete set under consideration is called the *universal set*. It is denoted by the letter $\mathcal{E}$, written in curly script. It is the initial letter of the French word for a set, which is 'Ensemble'.

(*a*) Let $E = \{$even numbers$\}$. The members of E will depend on the universal set under consideration. Write down the members of E in the following cases:

 (i) $\mathcal{E} = \{5, 6, 7\}$;

 (ii) $\mathcal{E} = \{$counting numbers between 11 and 19$\}$;

 (iii) $\mathcal{E} = \{$whole numbers$\}$.

(*b*) In most cases in which we talk of sets it is necessary to be quite clear before starting what set we are taking as $\mathcal{E}$. All the sets involved will then be subsets of $\mathcal{E}$. It is useful to have a special symbol for the relation 'is a subset of' and we write '$\subset$'. The sentence

$$\text{'}A \subset \mathcal{E} \quad \text{and} \quad B \subset \mathcal{E}\text{'},$$

is read

'A is a subset of the Universal Set and B is a subset of the Universal Set'. It is possible to write the symbol the other way round. '$\mathcal{E} \supset A$' has the same meaning as '$A \subset \mathcal{E}$'. How would you read '$\mathcal{E} \supset A$'?

1.3 Complementary sets

(*a*) Let $\mathcal{E} = \{$children in your class$\}$ and $G = \{$children wearing glasses$\}$. All the members of $\mathcal{E}$ *not* in G also form a subset of $\mathcal{E}$. We call this set the *complement* of G in $\mathcal{E}$ and denote it by G' (read 'G dash'). What can you say about

$$G \cap G'?$$

What is the complement of G'?

(*b*) Figure 7 shows a Venn diagram of a universal set $\mathcal{E}$ and two subsets one inside and one outside the curve. If we denote these subsets by A and B, what can you say about A, B, $\mathcal{E}$ and $\varnothing$. Does it matter which subset is labelled A?

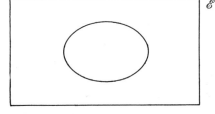

Fig. 7

Exercise B

1. $\mathcal{E} = \{1, 2, 3, 4, 5, 6\}$.

(*a*) Write down A' if $A = \{2, 4, 6\}$.

(*b*) Write down B if $B' = \{$numbers less than 3$\}$.

(*c*) Write down $(P \cap Q)'$ if $P = \{2, 3\}$ and $Q = \{5, 2\}$.

2. Let $\mathcal{E} = \{$positive even numbers less than 10$\}$ and $A = \{$multiples of 4$\}$. Write down A'.

3. If $P = \{1, 3\}$ and $P' = \{2, 7, 8\}$ write down $\mathscr{E}$.

4. Name three possible sets $\mathscr{E}$ having $\{1, 3\}$ and $\{2, 7, 8\}$ as subsets.

5. (a) Let $\mathscr{E} = \{\text{letters of the English alphabet}\}$
and
$A = \{\text{letters used in the sentence 'The quick brown fox jumps over the lazy dog.'}\}$
Write down A'. We still say A is a subset of $\mathscr{E}$ and write $A \subset \mathscr{E}$. In what way is this a special kind of subset?

(b) Let $\mathscr{E} = \{\text{the five vowels}\}$. Consider the subset B, where $B = \{\text{vowels in the word 'hymn'}\}$.
In what way is this a special kind of subset?

6. A, B and C are sets such that $A \subset B$ and $B \subset C$. Give an example of three sets which satisfy these relations indicating which are A, B and C. Is it always, sometimes or never true that $A \subset C$?

7. Is it possible to have b, A and $\mathscr{E}$ such that $b \in A$ and $b \in A'$?

8. If $\mathscr{E} = \{\text{whole numbers}\}$ and $O = \{\text{odd numbers}\}$ then we can write

$$5 \in O,$$
$$8 \in O',$$
$$\tfrac{3}{4} \notin \mathscr{E} \quad (\notin \text{ indicates 'is not a member of').}$$

In the following cases state whether the element in the first column belongs to the set in the second column, to its complement, or to neither. Write your answers using the symbol $\in$.

	Element	Set (A)	Universal set ($\mathscr{E}$)
(a)	Penny	Silver coins	British coins
(b)	Square	Figures with 4 sides	Polygons
(c)	Mauve	Primary colours	Colours
(d)	$\sqrt{3}$	Even numbers	Whole numbers

1.4 Union of sets

Let $\mathscr{E} = \{\text{new cars for sale at this year's Motor Show}\}$, $A = \{\text{Austin cars}\}$ and $B = \{\text{blue cars}\}$. The shaded region in Figure 8 represents the cars which are possible buys for a man who says 'I am looking for a blue car, but I wouldn't mind having an Austin, whatever its colour'. The region represents the members of $\mathscr{E}$ that are Austins *or* blue *or both*. This subset of $\mathscr{E}$ is called

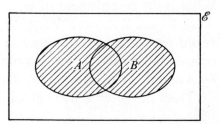

Fig. 8

the *union* of the two sets A and B, and is denoted by $A \cup B$. A simple way of reading this is '*A cup B*'.

Exercise C

1. Given that $A = \{a, b, c, d\}$ and $B = \{a, c, e\}$, list the members of (a) $A \cap B$; (b) $A \cup B$.

2. Given that $E = \{2, 4, 6, 8\}$ and $Y = \{2, 2^2, 2^3\}$, list the members of (a) $E \cap Y$; (b) $E \cup Y$. Write down a relation satisfied by E and Y.

3. Let $\mathcal{E}$ = {letters in the word CUSHION}, S — {letters in the word SHUN} $V = \{I, O, U\}$. List the members of: (a) V'; (b) $V \cup S$; (c) $V' \cup S$; (d) $V \cup S'$.

4. Examine the following assertions and say whether they are true or false:

(a) $x \in A \Rightarrow x \in A \cup B$;
(b) $x \in A \Rightarrow x \in A \cap B$;
(c) $A \subset B$ and $x \in A \Rightarrow x \in B$;
(d) $A \subset B$ and $x \in B \Rightarrow x \in A$.

5. Simplify, for any A, $\mathcal{E}$:

(a) $A \cap A$;   (b) $A \cup A'$.

6. Is it always, sometimes or never true that $(A \cap B) \subset (A \cup B)$?

7. What can you say about sets P and Q if:

$\qquad$ (a) $P \cap Q = P$;   (b) $P \cup Q = Q$?

Are these statements equivalent to each other?

8. (a) List all the subsets of {Alan, Betty, Charles}. (Remember to count the empty set and original set as subsets.)

$\qquad$ (b) How many subsets has a set containing:

$\qquad\qquad$ (i) one member;   (ii) two members;   (iii) three members?

What do you notice about your answers? Guess the number of subsets of a set with four members and check your answer. Can you extend the pattern? Try to give a reason for your answer.

9. Simplify:

(a) $A \cup A$;   (b) $A \cap A$;   (c) $A \cup \mathcal{E}$;
(d) $A \cap \mathcal{E}$;   (e) $A \cup \varnothing$;   (f) $A \cap \varnothing$;
(g) $\varnothing \cup \mathcal{E}$;   (h) $\varnothing \cap \mathcal{E}$.

10. What can you say about the sets L and M in Figure 9?
$\qquad$ Simplify: (a) $L \cap M$;   (b) $M \cup L$.

Fig. 9

11. Copy the following statements and state which are always true, which are sometimes true (when?) and which are never true. If any are meaningless, then say so.

(a) $P \cup Q = Q \cup P$;   (b) $P \cap Q = Q \cap P'$;   (c) $P \cup P = 2P$;
(d) $P \cap P = P^2$;   (e) $P \cup Q = P \cap Q$;   (f) $(P \cap Q)' = P' \cup Q'$.

12. Let $\mathcal{E}$ = {birds alive at this moment}; C = {canaries}; S = {singing birds}; W = {well-fed birds}. Change the following statements of relations between sets into English sentences. For example (a) tells us that 'the set of all canaries is a subset of the set of singing birds', or, more briefly (and in more day-to-day language) 'all canaries sing'. (The truth of the sentences does not matter!)

(a) $C \subset S$;   (b) $S \subset C$;   (c) $C \subset S'$;
(d) $S = C \cap W$;   (e) $S \cap W \cap C' = \varnothing$.

Write the following sentences in symbols.

(f) Some canaries are well-fed.
(g) Some singing canaries are not well-fed.
(h) All canaries that do not sing are ill-fed

69

Summary *Notation*

1. *Sets*
A set is a general name for a collection of things or numbers. There must be a way of deciding whether or not any particular item belongs to the set. This may be done by making a list of the objects or by giving a statement which describes them.

2. *Members*
The members, or elements, of a set are the individual objects of the set.
'The element x is a member of set A'. $x \in A$
'The set whose members are...'. $\{...\}$

3. *Equal sets*
Two sets are said to be equal if they have the same members. For example if
$A = \{$the vowels$\}$ and $B = \{a, e, i, o, u\}$
then $A = B$.

4. *Subsets*
A is a subset of B if every member of A is $A \subset B$ also a member of B.

5. *Union*
The union of sets A and B is the set of all $A \cup B$ elements which are members of set A or set B (or both).

6. *Intersection*
The intersection of sets A and B is the set $A \cap B$ of all elements which are members of both set A and set B.

7. *Universal sets*
The universal set is the set of all elements $\mathscr{E}$ under consideration.

8. *Empty set*
The empty set is the set with no members. $\varnothing$

9. *Complementary sets*
The complement of set A in $\mathscr{E}$ is the set of A' all members of $\mathscr{E}$ which are not members of A.

10. *Venn diagrams*

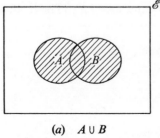

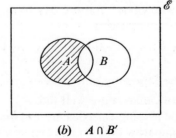

(a) $A \cup B$ Fig. 10 (b) $A \cap B'$

2. PROPERTIES OF OPERATIONS

2.1 Closure

(*a*) Copy and complete the following ($\square$ is used to indicate a missing number):

Operation

(i) Addition	$3+4 = \square$, $4+3 = \square$, $6+2 = \square$, $2+6 = \square$;
(ii) Division	$3\div4 = \square$, $4\div3 = \square$, $6\div2 = \square$, $2\div6 = \square$;
(iii) Raising to the power	$3^4 = \square$, $4^3 = \square$, $6^2 = \square$, $2^6 = \square$.

(*b*) In each of the above calculations the initial two numbers are counting numbers, that is members of N where $N = \{1, 2, 3, 4, ...\}$.

Note which answers are also members of N.

Obviously for addition no matter what two initial counting numbers are chosen the result will also be a counting number. Is the same statement true for the other two operations: (i) division; (ii) raising to the power?

(*c*) We describe this by saying that the set N is *closed* under the operation of addition, but *not* under division because, for instance,

$$2\div6 = \tfrac{1}{3},$$

and $\tfrac{1}{3}$ is not a member of N.

Is N closed under: (i) raising to the power; (ii) subtraction; (iii) multiplication?

(*d*) Consider $S = \{\text{square numbers}\}$. Copy and complete:

$$9+16 = \square; \qquad 36+64 = \square; \qquad 25+81 = \square.$$

Is S closed under addition?

2.2 Commutativity

The twelve questions in Section 2.1 (*a*) are arranged in pairs, for instance $6\div2$ and $2\div6$. For which pairs has this interchanging of the numbers altered the result?

With addition, we can obviously interchange the two numbers without altering the result, for instance $20+4$ and $4+20$ both result in 24.

For division, however,

$$20\div4 = 5, \quad \text{whereas} \quad 4\div20 = \tfrac{1}{5}.$$

We say that   addition of numbers is *commutative,*

and   division of numbers is *non-commutative.*

Does $A \cap B = B \cap A$? Is intersection commutative?

Are the following operations commutative:

(i) raising to the power;   (ii) subtraction;   (iii) multiplication;   (iv) union?

2.3   Operation tables

(*a*) Let us use E to stand for 'any even number' and O to stand for 'any odd number'. By $E+O$ we denote 'any even number added to any odd number'. What can you say about the answer? Is $\{E, O\}$ closed under the operation of addition? To confirm your answer, copy and complete the following table:

		Second number	
$+$		E	O
First number $\begin{cases} E \\ O \end{cases}$			O

This table is called an *operation table*. An ordinary multiplication table is an example of an operation table.

(*b*) Even and odd are really words which describe numbers according to their relationship to the number 2.

Members of E are exactly divisible by 2.

Members of O leave a remainder of 1 when divided by 2.

In a similar way we can classify integers by the way in which they are related to 3, though we will have no special names for them. Figure 11 shows how this can be done.

This is a spiral divided into three sections.

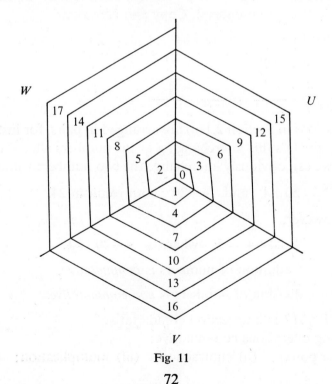

Fig. 11

72

U denotes 'integers exactly divisible by 3'.

V denotes 'integers that when divided by 3 leave a remainder of 1'.

W denotes 'integers that when divided by 3 leave a remainder of 2'.

Is $\{U, V, W\}$ closed under the operation of addition? Copy and complete the operation table for U, V, W.

+	U	V	W
U			
V		U	
W			

Exercise D

1. (a) Name an operation under which the set of even numbers is closed but the set of odd numbers is not closed.

(b) Name an operation under which the sets in (a) are both closed.

(c) Make an operation table similar to that in Section 2.3 (a) for $\{E, O\}$ under multiplication. Is $\{E, O\}$ closed under multiplication?

2. We may use the expression 'add two numbers together'. Which of the following expressions would be acceptable:

(a) Subtract two numbers together.

(b) Multiply two numbers together.

(c) Divide two numbers together?

3. What is a commuter?

4. Taking as elements the sets $\varnothing$, A, B, P where $A = \{a\}$, $B = \{b\}$ and $P = \{a, b\}$, copy and complete the operation table.

		Second element		
∩	∅	A	B	P
∅		∅		
A			A	
B				
P				

First element

(a) Is $\{\varnothing, A, B, P\}$ closed under intersection?

(b) Sketch in lightly two lines running diagonally across the table. Which one could be called a line of symmetry? Why?

(c) What property of the operation is connected with the symmetry of the table?

5. The Venn diagram in Figure 12 is shaded to indicate the set $A \triangledown B$, that is, the set of elements which are members of A but not members of B.

Taking

$$A = \{1, 3, 5, 7, 9, 11\} \quad \text{and} \quad B = \{3, 6, 9, 12\},$$

(a) list $A \triangledown B$;

(b) list $B \triangledown A$;

(c) is $A \triangledown B = B \triangledown A$?

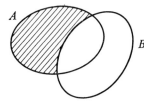

Fig. 12

6. We can do arithmetic on a 'clockface' with the set of numbers {0, 1, 2, 3, 4, 5} (see Figure 13). The numbers are combined in the following manner

$$2+3 = 5.$$

This means 'start at 0, turn through 2 and then turn through 3 in a clockwise direction'.

Similarly $0+2 = 2$ and $3+4 = 1$.

(a) Copy and complete the following combination table. The examples already given are included in the table.

(b) Is the operation + commutative?

(c) What symmetrical feature of the numbers in the table helps you to decide?

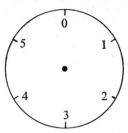

Fig. 13

		Second number					
+		0	1	2	3	4	5
First number	0	—	—	2	—	—	—
	1	—	—	—	—	—	—
	2	—	—	—	5	—	—
	3	—	—	—	—	1	—
	4	—	—	—	—	—	—
	5	—	—	—	—	—	—

7. Using the clockface in Question 6 we could also combine numbers in the following manner

$$5-3 = 2.$$

This means 'start at 0, turn through 5 clockwise and then turn through 3 in an anticlockwise direction'. Similarly,

$$1-4 = 3.$$

		Second number					
−		0	1	2	3	4	5
First number	0	—	—	—	—	—	—
	1	—	—	—	3	—	—
	2	—	—	—	—	—	—
	3	—	—	—	—	—	—
	4	—	—	—	—	—	—
	5	—	—	2	—	—	—

(a) Copy and complete the table shown.

(b) Is the operation − commutative?

(c) Is {0, 1, 2, 3, 4, 5} closed under −?

8. (a) Form an addition table for clock arithmetic with {0, 1, 2} (see Figure 14).

(b) Compare it with the table you made for {U, V, W} in Section 2.3(b). Why does the same pattern arise?

(c) Make tables in each case for the operation of multiplication and again compare the patterns that arise.

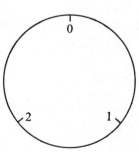

Fig. 14

2.4   Combination

(*a*) Comment on the results of the instructions in (i) and (ii).

(i) 'Put on your shoes after you have put on your socks'.
'Paint the door after you have stripped it'.
'Rub yourself with a towel after you have had a bath'.

(ii) 'Put on your socks after you have put on your shoes'.
'Strip the door after you have painted it'.
'Have a bath after you have rubbed yourself with a towel'.

You will have realized that when two instructions are combined in this way the order in which they are to be carried out may make a big difference to the result.

(*b*) Suppose we denote the instructions

$$\text{'Take one step forward' by } \mathbf{F},$$

and   'About turn' by $\mathbf{A}$

and then write the combined instruction 'Take one step forward *after* you have done an about turn'

as $\mathbf{F} \circ \mathbf{A}.$

What would we mean by $\mathbf{A} \circ \mathbf{F}$?
Is $\mathbf{F} \circ \mathbf{A} = \mathbf{A} \circ \mathbf{F}$? (Try them both.)

We have here an important binary operation, which we shall call *combination*, for which the elements are instructions. The symbol ○ may be read *combined with* but you may prefer to read it as *following* as a reminder that

$$\mathbf{X} \circ \mathbf{Y},$$

means that instruction $\mathbf{X}$ is carried out *following* instruction $\mathbf{Y}$.

(*c*) Consider the instructions

'turn right for a quarter turn' denoted by $\mathbf{R}$,
'turn left for a quarter turn' denoted by $\mathbf{L}$,
'about turn' denoted by $\mathbf{A}$.

Give a friend facing you the combined instruction 'Turn right after you have done an about turn' that is

$$\mathbf{R} \circ \mathbf{A}.$$

What single instruction would have achieved the same result?
So we write

$$\mathbf{R} \circ \mathbf{A} = \mathbf{L}.$$

Copy and complete where possible:
(i) $\mathbf{L} \circ \mathbf{A} = \square$;   (ii) $\mathbf{L} \circ \mathbf{L} = \square$;   (iii) $\mathbf{R} \circ \mathbf{L} = \square$.
Is $\{\mathbf{R}, \mathbf{L}, \mathbf{A}\}$ closed under the operation of combination?

2.5 Identity elements

(a) If we now add a fourth instruction

'stay where you are' denoted by S

to the three instructions in Section 2.4(c) above, we can now express the result of the combined operation R ∘ L, writing

$$R \circ L = S.$$

Give two other combinations which result in S.

What are: (i) S ∘ L; (ii) A ∘ S; (iii) S ∘ R?

You may have noticed that when any element is combined with S then the same element results. For instance R ∘ S = R. We say that S is the *identity* element. It does not alter anything.

(b) We are now in a position to complete the operation table.

Second element

∘	S	L	A	R
S				
L				
A				
R			L	

First element

The entry R ∘ A = L is shown. As usual its position is across from the first element (R) and under the second element (A). In working out the result (L) we had in mind though that instruction R was carried out following instruction A.

Copy and complete the table.

Describe the row and column opposite the identity elements S.

(c) What is the identity element when ordinary numbers are: (i) added; (ii) multiplied?

2.6 Inverse elements

(a) Examine the operation table you have just completed and note where the identity element appears inside the table.

Which pair of elements combine to yield S?

We call a pair such as this an *inverse pair*. So L is the inverse of R and R is the inverse of L.

Why would we describe A as *self-inverse*?

Would it be correct to describe S as self-inverse?

(b) Copy and complete the table below for the operation of multiplication on the clockface. (See Figure 15.)

×	0	I	2	3	4
0	0	0	0	0	0
I	0	I	2		
2	0		4		3
3	0	3			
4	0				

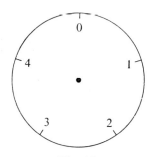

Fig. 15

(i)   Which element is the identity?

(ii) Which elements form inverse pairs?

(iii) Which elements are self-inverse?

(iv) Which element has no inverse?

Exercise E

1. A girl counted the plumstones on her plate at dinner time with the rhyme

cottage, palace, mansion, pigsty, cottage, palace,...

A plate with 3 stones on it would give 'mansion', and one with 6 stones on it would give 'palace'.

If the stones on the two plates were added there would be 9 stones, giving 'cottage'.

If we write ⊕ to stand for addition, then

mansion ⊕ palace = cottage.

Copy and complete the following table. Which element is the identity?

⊕	Cottage	Palace	Mansion	Pigsty
Cottage				
Palace				
Mansion		Cottage		
Pigsty				

2. Make up operation tables for a clock arithmetic with {0, 1, 2, 3} for the operations + and ×.

In each case state the identity element.

Rearrange the first table in the order 1, 2, 3, 0 and compare the pattern with that of the table in Question 1. What do you notice? Why does a similar pattern arise?

3. Place two pennies side by side on the table.

They can be 'altered' in four ways:

(a) leave them as they are, **I**;   (b) turn the right one over, **R**;

(c) turn the left one over, **L**;   (d) turn them both over, **B**.

What single alteration is equivalent to performing alteration **L** following alteration **R**?

Is {**I**, **R**, **L**, **B**} closed under this operation of combination?

Construct the operation table.

Could this table be made to have a similar pattern to those in Questions 1 and 2 by altering the order of the letters **I**, **R**, **L** and **B**? Explain.

4. By inspection find the identity element of the following table for the operation o on the set $\{K, L, M\}$:

o	K	L	M
K	L	M	K
L	M	K	L
M	K	L	M

Notice that we need not know anything about K, L, M or about the operation o in order to be able to spot the identity.

5. Clock arithmetic with the set $\{0, 1, 2, 3, 4, 5\}$ is also referred to as arithmetic modulo 6 or mod 6; one with the set $\{0, 1, 2, 3\}$ is mod 4.

(*a*) Find the identity elements for:
    (i) $\{1, 5\}$ under the operation $\times$ mod 6;
    (ii) $\{0, 2, 4\}$ under the operation $\times$ mod 6.

(*b*) Compare the patterns of combination tables for:
    (i) $\{1, 2, 3, 4\}$ under $\times$ mod 5;
    (ii) $\{1, 3, 7, 9\}$ under $\times$ mod 10;
    (iii) $\{0, 1, 2, 3\}$ under $+$ mod 4.

What rearrangements of the numbers are necessary in order to produce similar patterns?

6. Have all the operation tables constructed so far a line of symmetry from top left to bottom right (along the leading diagonal)? What does this imply about the operations involved? Construct a table for $\{0, 1, 2, 3,\}$ for the operation of subtraction mod 4. Is this set closed under subtraction? Explain why the leading diagonal is not a line of symmetry. Discuss whether an identity element exists in this case.

Summary

A set P is said to be *closed* under an operation o if,

whenever $a \in P$ and $b \in P$, it follows that $a \circ b \in P$.

A table showing the results of an operation on pairs of members of a set is called an *operation table*, for example,

+	E	O
E	E	O
O	O	E

The table for a certain operation on a set may sometimes have the same pattern as the table for a second operation on that set, or on a different set.

The *identity element* of a set P under an operation o is a member e of P such that

whenever $a \in P$ then $a \circ e = e \circ a = a.$

If $a \circ b = b \circ a = e,$

then a and b are called an *inverse pair*.

If,   whenever $a \in P$ and $b \in P$, it follows that

$$a \circ b = b \circ a,$$

then the operation is said to be *commutative*.

3. NEGATIVE NUMBERS

3.1   Inverse numbers

(*a*) Figure 16 shows a clockface which we may use for arithmetic mod 20.

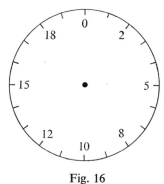

Fig. 16   Fig. 17

Copy and complete the following additions (mod 20):

(i) $2+7 = \square$;   (ii) $18+7 = \square$;   (iii) $2+13 = \square$,   (iv) $18+13 = \square$.

(*b*) Does modular arithmetic give the same results as ordinary arithmetic

               (i) always;   (ii) never;   (iii) sometimes?

(*c*) What is the identity element for addition (mod 20)?

What are the inverses of (i) 2, (ii) 7 under addition (mod 20)?

(*d*) If we use a star symbol (∗) to denote 'the inverse of' we may replace, for instance, 14 by 6* because $14+6=0$. Figure 17 indicates the result of replacing all the numbers from 11 to 19 by starred numbers.

Using Figure 17 copy and complete

(i) $2+7 = \square$;   (ii) $2^*+7 = \square$;   (iii) $2+7^* = \square$; (iv) $2^*+7^* = \square$.

Compare these with the four questions in (*a*) above.

We might explain the working of $2+7$ by saying

     'Start at 0, turn through 2 clockwise and then through 7 clockwise'.

The result is 9 clockwise. So $2+7 = 9$.

Explain the others in a similar way.

3.2   Negative numbers

The numbers we have used in this chapter so far have been mainly counting numbers but you should be familiar with the picture of a number line (see Figure 18) showing what we call the *positive* and *negative integers,* and zero.

79

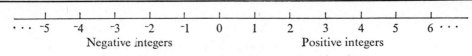

Fig. 18

(a) What is the identity element for addition?

(b) Copy and complete: (i) $^-4+4 = \square$; (ii) $6+\square = 0$; (iii) $\square+^-3 = 0$.

So we may regard the complete set of integers as being

    (i) our original counting numbers (the positive integers),

    (ii) the identity under additon (zero), and

    (iii) the set of inverse numbers under addition (the negative integers).

3.3 Addition

Copy and complete the following:

(i) $2+7 = \square$; (ii) $^-2+7 = \square$; (iii) $2+^-7 = \square$; (iv) $^-2+^-7 = \square$.

Compare your results with those of Section 3.1 (d).

We can see the similarity pictorially by bending the number line (see Figure 18) so as to suggest a giant clockface (see Figure 19). How does this resemble Figure 17? What important difference is there?

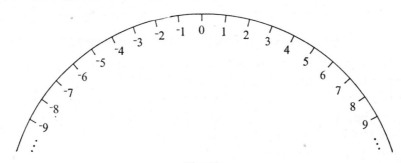

Fig. 19

In this way we may think of the addition of positive and negative integers using the idea of clockwise and anticlockwise turns.

Can you suggest two words to replace clockwise and anticlockwise that could equally well apply to the number line in Figure 18?

3.4 Multiplication

(a) Whenever new kinds of numbers are considered it is important to be able to add, subtract, multiply and divide, using them. We have looked at addition and now turn to multiplication.

Look again at Figures 16 and 17.

Copy and use Figure 16 to complete:

(i) $2 \times 3 = \square$;   (ii) $18 \times 3 = \square$;   (iii) $2 \times 17 = \square$;   (iv) $18 \times 17 = \square$.

Copy and use Figure 17 and the above results to complete:

(v) $2 \times 3 = \square$;   (vi) $2* \times 3 = \square$;   (vii) $2 \times 3* = \square$;   (viii) $2* \times 3* = \square$.

What similarities and differences do you notice about these last four answers?

(b) Figure 20 shows a similar clockface although only a few numbers have been marked.

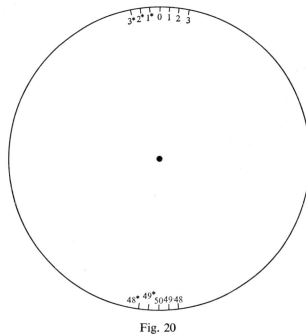

Fig. 20

What is the modulus involved?

What number has been replaced by (i) 49*;   (ii) 1*?

Using this clockface we could work out, say,

$3 \times 7*$ as follows:

$7*$ is the inverse of 7 which is 93;

now $3 \times 93 = 79$ (mod 100),

but   79 is the inverse of 21 which is written 21*,

so $3 \times 7* = 21*$.

(c) Copy and complete:

(i) $4 \times 5 = \square$;   (ii) $4* \times 5 = \square$;   (iii) $4 \times 5* = \square$;   (iv) $4* \times 5* = \square$.

Compare your answers with those above. Make some observations about the sort of answers you have obtained.

In which cases was the answer a starred number?

In which cases was the answer *not* a starred number?

(*d*) We know that in ordinary arithmetic if two positive numbers are multiplied together then the answer is also positive. Use the connection between the starred numbers and the negative integers to say what sort of answer you would expect when

      (i) a *positive number* and a *negative number* are multiplied,

     (ii) two *negative numbers* are multiplied.

(*e*) We may conveniently summarize these properties in a table,

×	Pos.	Neg.
Pos.	Pos.	Neg.
Neg.	Neg.	Pos.

(*f*) Make a multiplication table, like the one below, showing the results of multiplying together all possible pairs of integers from $^-4$ to 4 inclusive.

×	⋯	⁻4	⁻3	⁻2	⁻1	0	1	2	3	4	⋯
⋮											
⁻4											
⁻3											
⁻2											
⁻1											
0											
1											
2											
3											
4									16		
⋮											

Exercise F

State the value of y in each of the equations in Questions 1–20.

1. $^-4+6 = y.$ 2. $4+^-6 = y.$ 3. $^-4+^-6 = y.$ 4. $7+y = 7.$ 5. $y+^-3 = 1.$

6. $^-5+y = ^-2.$ 7. $^-4+y = ^-7.$ 8. $6+y = 5.$ 9. $y+3 = ^-5.$ 10. $^-4+y = 9.$

11. $6×^-4 = y.$ 12. $^-5×7 = y.$ 13. $^-4×^-7 = y.$ 14. $^-5×y = 25.$ 15. $3×y = ^-18.$

16. $y×^-6 = 12.$ 17. $^-7×^-7 = y.$ 18. $y×^-4 = 16.$ 19. $y×y = 36.$ 20. $8×y = ^-8.$

21.

Fig. 21

(a) Give some examples (e.g. temperature scale) where it would be sensible to have the integers shown on a number line (see Figure 21).
(b) Where on the line would you find numbers greater than 4?
(c) Where would you find numbers less than ⁻2?
(d) Which is the warmer 2 °C or ⁻3 °C.
(e) Arrange in order of size (smallest first),

$$2, {}^-3, {}^-5, 7, 0, 6, {}^-8, 10.$$

22. One way of picturing integers is shown in Figure 22.

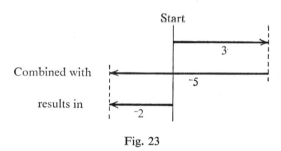

Fig. 22

So 3 is thought of as 'shift 3 to the *right*' and ⁻5 is thought of as 'shift 5 to the *left*'.
Put together these shifts give

Start

Combined with

results in

Fig. 23

So '3 to the right followed by 5 to the left results in 2 to the left' can be written $3 + {}^-5 = {}^-2$.
Make up some more examples like this.

3.5 Division

(a) The two operations, division and subtraction can be dealt with quite simply by using their close relationship to multiplication and addition respectively. Let us take division first.
Consider

$$40 \div 8 = \square \quad \text{and} \quad 40 = 8 \times \square.$$

What do you notice about the missing numbers?
Write down some similar pairs of calculations using other numbers instead of 40 and 8. Comment on the results.
(b) Find the missing number in $91 = 7 \times \square$, and hence give the result of the division $91 \div 7$.

83

We can perform the division

$$a \div b = \square,$$

by considering the multiplication $a = b \times \square$.

(c) (i) What is the missing number in $^-12 = 4 \times \square$?

(ii) Give the result of $^-12 \div 4 = \square$.

What are: (iii) $^-12 \div ^-4$; (iv) $12 \div ^-4$?

(d) When is the result of a division (i) negative? (ii) positive?

Is $\{... ^-3, ^-2, ^-1, 0, 1, 2, 3, ...\}$ closed under division?

What kind of numbers are needed as answers to divisions like $12 \div 9$?

The extension of the number system to include fractions will be considered in some detail in *Book* 4. We shall assume from now on that we are working in the fuller system unless stated otherwise.

3.6 Subtraction

(a) If you give £1 ($= 100$ pence) for an article costing 68 pence how much change would you get? Write down the calculation. The shopkeeper might very well be heard to say   '68, 70, 80, 90, £1. Thank you. Next please'.

What is he doing? Has he performed a subtraction? Did you?

Compare the missing numbers in

$$53 - 18 = \square \quad \text{and} \quad 53 = 18 + \square.$$

Make up a few similar pairs of calculations. What do you notice?

We can summarize by saying that to solve

$$a - b = \square,$$

we may find the missing number in $a = b + \square$.

(b) Find the missing number in $^-3 = 2 + \square$ and hence solve $^-3 - 2 = \square$.

(c) What are: (i) $6 - ^-4$;   (ii) $2 - ^-8$;   (iii) $3 - 5$?

(d) We have seen above that if $a - b = c$ then $a = b + c$ and it is equally obvious that if $a = b + c$ then $a - b = c$.

We can express this by writing

$$a - b = c \iff a = b + c,$$

meaning that each statement implies the other. They are equivalent statements.

Exercise G

State the value of x in each of the equations.

1. $8 \div ^-2 = x$.
2. $^-7 \div ^-1 = x$.
3. $0 \div ^-2 = x$.
4. $10 \div x = ^-2$.
5. $x \div ^-3 = 6$.
6. $^-2 - 5 = x$.
7. $^-3 - ^-8 = x$.
8. $^-10 - ^-4 = x$.
9. $6 - x = ^-5$.
10. $x - ^-2 = 2$.
11. $x + 3 + ^-7 = ^-2$.
12. $^-5 + x + 4 = 6$.

13.

Fig. 24

The thick black line (see Figure 24) represents the set of numbers between ⁻3 and 2, the circles indicating exclusion of the end points. We shorten the description to $\{x: ^-3 < x < 2\}$. The colon is read 'such that' and the statement ⁻3 < x < 2 gives the rule of membership of the set.

(a) Draw similar diagrams to represent the sets:

$$P = \{x: 2 < x < 5\};$$

$$Q = \{x: ^-6 < x < 3\};$$

$$R = \{x: ^-4 < x < ^-1\cdot6\}.$$

(b) Name some members of: (i) $P \cap Q$; (ii) $Q \cap R$; (iii) $R \cap P$.

14. Give 4 members of $\{(x, y): x - y = 2\}$ having negative y numbers.

15. Find the range corresponding to the given domains for each of the following functions:

(a) $x \rightarrow 1 + 2x$, $\{^-2, ^-1, 0, 1, 2\}$;
(b) $x \rightarrow 3 - \frac{1}{2}x$, $\{^-4, ^-2, 0, 2, 4\}$;
(c) $x \rightarrow 4 - x^2$, $\{^-3, ^-2, ^-1, 0, 1, 2, 3\}$.

Represent each function graphically. In each case show the effect of making the domain continuous between the given values by joining the points of the graph by a straight line or a curve.

3.7 Subtraction revisited

(a) First of all we will look at subtraction in arithmetic mod 9. (See Figure 25.)

Copy and complete:

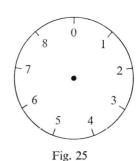

$$8 - 5 = \square = 8 + \triangle,$$
$$6 - 5 = \square = 6 + \triangle,$$
$$5 - 5 = \square = 5 + \triangle,$$
$$3 - 5 = \square = 3 + \triangle,$$
$$1 - 5 = \square = 1 + \triangle,$$
$$0 - 5 = \square = 0 + \triangle.$$

Fig. 25

What is equivalent to turning 5 anticlockwise?
What addition is equivalent to: (i) subtracting 7 (mod 9); (ii) subtracting 3 (mod 9).
How are the pairs of numbers related?
What is the missing word in the following statement:
'in modular arithmetic we can subtract a number by adding the ... of that number'?

(b) Working with the complete set of integers, perform the subtractions below in the usual way and then do the additions alongside.

85

Subtraction	Addition of inverse
$7-2 = \square$	$7+{}^-2 = \square$
${}^-3-2 = \square$	$3+{}^-2 = \square$
$4-9 = \square$	$4+{}^-9 = \square$
${}^-2-9 = \square$	${}^-2+{}^-9 = \square$
$6-{}^-4 = \square$	$6+4 = \square$
${}^-6-{}^-4 = \square$	${}^-6+4 = \square$
$3-{}^-8 = \square$	$3+8 = \square$
${}^-2-{}^-7 = \square$	${}^-2+7 = \square$

Were the results the same for each pair?

We can replace the operation 'subtract' by 'add the inverse'.

Example

Solve the equation $6x-5 = {}^-14$.

First method. Solution:

$$6x-5 = {}^-14,$$
$$\Rightarrow 6x+{}^-5 = {}^-14, \qquad \text{(replacing } -5 \text{ by } +{}^-5\text{)}$$
$$\Rightarrow \quad 6x = {}^-9, \qquad \text{(because } {}^-9+{}^-5 = {}^-14\text{)}$$
$$\Rightarrow \quad x = {}^-1{\cdot}5.$$

Alternatively we could use the fact that

$$a-b = c \quad \text{and} \quad a = b+c \text{ are equivalent statements.}$$

This leads to

Second method. Solution: $6x-5 = {}^-14,$

$$\Rightarrow 6x = 5+{}^-14,$$
$$\Rightarrow 6x = {}^-9,$$
$$\Rightarrow \quad x = {}^-1{\cdot}5.$$

Exercise H

Solve the following equations:

1. $2x+7 = 5$.

2. $6x+17 = 5$.

3. $5x-4 = 21$.

4. $8x+4 = {}^-16$.

5. $24-5x = 4$.

6. $23-2x = 37$.

7. $\frac{1}{2}x-3 = 4$.

8. $\frac{3}{2}x+7 = 13$.

9. $\frac{1}{3}x+8 = 5$.

10. $4-\frac{1}{4}x = {}^-1$.

11. $\frac{2}{3}x+11 = 6$.

12. $4x-11 = {}^-5$.

REVISION EXERCISES

SLIDE RULE SESSION NO. 1

Calculate the following, giving all answers to 3 s.f.

1. $943 \div 0.026$.
2. 41×650.
3. 2.43×16.2.
4. $374 \div 12.7$.
5. $\dfrac{1.82 \times 0.64}{52.7}$.
6. $\dfrac{2.47}{21.4 \times 0.98}$.
7. $\dfrac{0.153 \times 7.62}{42.1 \times 89.1}$.
8. $\sqrt{(57.2)}$.
9. 3.95^2.
10. $\sqrt{(4.25)}$.

SLIDE RULE SESSION NO. 2

Calculate the following, giving all answers to 3 s.f.

1. $(0.0155)^2$.
2. 22.5×19.6.
3. $0.159 \div 0.0725$.
4. 1.64×12.6.
5. 705×0.0915.
6. $3.05 \div 695$.
7. $158.5 \div 0.124$.
8. $\dfrac{13.3 \times 495}{67.5 \times 0.722}$.
9. 43.5^2.
10. $\sqrt{(43.5)}$.

A

1. Find the angles marked a, b, c, d, e, in Figure 1.

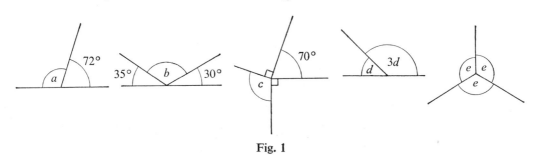

Fig. 1

2. $10^3 + 2 \times 10^2 + 3 \times 10 + 7$ can be written 1237_{10} for short. Write these expressions in short ways:

(*a*) $2^4 + 2^2 + 2 + 1$; (*b*) $2 \times 6^3 + 3 \times 6^2 + 5 \times 6 + 2$.

3. Write the following numbers in standard index form, correct to 3 s.f.:

(*a*) 0.001234; (*b*) 432.1; (*c*) 78987; (*d*) 1000889; (*e*) 0.080080008.

4. Basil Brayne handed in the following answers to a slide rule test. Say whether the answers are right (to 3 s.f.) or wrong; and when they are wrong correct them.

(*a*) $27.9 \times 1.72 = 4.80$; (*b*) $16.5 \times 3.02 = 49.8$;
(*c*) $1.18 \times 22.1 = 26.7$; (*d*) $23.4 \times 16.1 \times 227 = 85500$;
(*e*) $17.5 \times 8.97 = 157$.

5. Fred (a mathematical fly) is sitting at the vertex A of a cube (see Figure 2). He decides to go for a walk along the edges so as to visit each vertex once (and only once), and then to arrive back at the starting point A. If each edge of the cube is 10 cm long, what is the distance he must travel? Sketch the cube, and show one of Fred's possible routes by marking the edges with arrowheads. Would it be possible for him to walk along each of the edges once and only once?

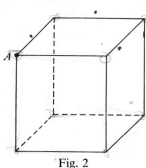

Fig. 2

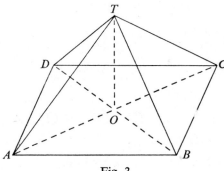

Fig. 3

6. Figure 3 shows an oblique projection of a square based pyramid, all the edges of which are equal in length.

(a) What shape are the faces?
(b) What is the shape $ABCD$ *in the drawing*? What shape is it really?
(c) Name a line equal in length to TA which appears to be different in length on the drawing.
(d) If each edge is 6 cm, find the height, TO, of the pyramid.

7. A boy was asked to solve the equation
$$\tfrac{1}{5} = 2x+6.$$
He began by multiplying both sides by 5, and wrote
$$1 = 10x+6.$$
Find the solution he obtained, and explain why it was wrong. What is the correct value of x?

8. If $\mathscr{E} = \{$natural numbers less than 17$\}$, $E = \{$even numbers less than 17$\}$, $T = \{$multiples of 3 less than 17$\}$, $F = \{$multiples of 4 less than 17$\}$, list the members of the sets:

(a) E';   (b) $E \cap T$;   (c) $F \cup T$;   (d) $F' \cap T'$;   (e) $(F \cup T)'$.

9. Solve the following equations:

(a) $2x-1 = {}^-3$;   (b) $\tfrac{1}{4}x-3 = 5$;
(c) $1-\tfrac{2}{3}x = 7$;   (d) $4-5x = 11$.

10. The cost in pounds (C) of hiring a car is given by the formula
$$C = 11+\frac{2n}{75},$$
where n is the number of kilometres travelled.

(a) What is the total cost if the car is driven only 1500 km?
(b) Does the cost per kilometre become larger or smaller as the distance increases?
(c) What is the average cost per kilometre if the car travels 3000 km?
(d) If the total cost is £31, how far has the car travelled?

B

1. Calculate the angles marked *p*, *q* and *r*.

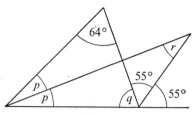

Fig. 4

2. Find the whole numbers (differing only by 1) between which the following must lie:

(*a*) $\sqrt{7}$;   (*b*) $\sqrt{31\cdot5}$;   (*c*) $\sqrt{70}$;   (*d*) $\sqrt{99\cdot9}$;   (*e*) $\sqrt{1\cdot53}$.

(For example, since the square root of 6 must lie between 2 and 3, we can write $2 < \sqrt{6} < 3$.)

3. Give the next three terms of each of these sequences:

(*a*) 4, 2, 0, ⁻2, ...;   (*b*) 1, ⁻3, ⁻7, ...;
(*c*) 6, 5, 3, ⁻1, ...;   (*d*) 1, ⁻2, 4, ⁻8, ...;
(*e*) ⁻11, ⁻10, ⁻8, ⁻5, ...

4. Find the value of $pq+r$ when:

(*a*) $p = 1, q = 1, r = ^-1$;   (*b*) $p = ^-7, q = 0, r = ^-2$;
(*c*) $p = ^-7, q = 1, r = ^-3$;   (*d*) $p = ^-7, q = ^-1, r = 3$;
(*e*) $p = ^-7, q = ^-1, r = ^-3$.

5. In a village of 176 houses it is discovered that the only newspapers that anyone buys are the *Daily Telegraph, Daily Mail* and the *Sketch*. The newsagent delivers at least one paper to every house, but never 2 copies of the same paper to one house. Altogether they deliver 40 copies of the *Sketch*, 71 *Telegraphs* and 98 *Mails*. Including those houses that take all three papers, 12 take the *Sketch* and the *Mail*, 13 take the *Mail* and the *Telegraph*, and 15 take the *Telegraph* and the *Sketch*. Draw a Venn diagram and hence find how many take all three.

6. Find the range of each of the following functions, where the domain in each case is $\{^-2, ^-1, 0, 1, 2\}$:

(*a*) $x \rightarrow 1+x$;   (*b*) $x \rightarrow (1+x)^2$;   (*c*) $x \rightarrow (1+x)^3$.

7. With which of the following plane shapes is it possible to tessellate planes:

(*a*) isosceles triangle;   (*b*) rectangle;   (*c*) irregular quadrilateral;
(*d*) regular pentagon;   (*e*) regular hexagon?

Draw a diagram to show how a plane can be tessellated with a combination of squares and regular octagons.

8. (*a*) List the members of the sets:

        (i) $A = \{$multiples of 6 less than 50$\}$;
        (ii) $B = \{$multiples of 8 less than 50$\}$;
        (iii) $A \cap B$.

(*b*) What is the L.C.M. of 6 and 8? What is the L.C.M. of 8 and 12?

9 All quadrilaterals that are members of the set R have property D. Is it necessarily true that quadrilaterals that have the property D are members of the set R? Give an example to illustrate your answer.

10. Does an identity element exist for the operation o on the set $\{a, b, c, d\}$ as shown in the table?

If so, name it and name the inverse pairs in the set for this operation.

Is the operation commutative?

o	a	b	c	d
a	b	a	d	c
b	a	b	c	d
c	d	c	a	b
d	c	d	b	a

C

1. A pyramid has a base 2 cm square, and its vertex is 2 cm above the centre of the base. By accurate drawing, find the length of a sloping edge of the pyramid.

2. Say which of the following are true and which are false:

(a) $0 \cdot 1 \times 0 \cdot 1 = 0 \cdot 1$; (b) $0 \cdot 1 \div 0 \cdot 01 = 10$; (c) $\dfrac{100 \times 0 \cdot 02}{0 \cdot 04} = 200$;

(d) 5^2 is greater than 5; (e) $0 \cdot 5^2$ is greater than $0 \cdot 5$.

3. Find the area of the Meccano piece shown in Figure 5, given that the radii of the large arcs are 2·5 and 3·5 cm, that the ends are semi-circles and the holes have diameters of 0·28 cm. (Take π to be 3·14.)

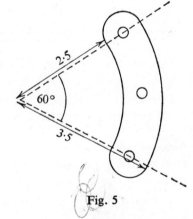

Fig. 5

4. Draw diagrams similar to Figure 27 (p. 42) for the following functions, with domain $\{1, 2, 3, 4, 5\}$:

(a) $x \to 3x - 5$; (b) $x \to \tfrac{1}{2}x^2$.

5. Find x in each case if:

(a) $2 + x = 5$; (b) $2 + x = {}^-5$; (c) $x + {}^-6 = {}^-4$;
(d) $x + {}^-6 = {}^-8$; (e) $3 + x = 0$.

Draw diagrams similar to Figure 23 (p. 83) to illustrate your answers to (a) and (c).

6. If possible sketch plane figures with the following specifications, dotting in lines of symmetry and showing centres of rotational symmetry by crosses. Where the specifications are impossible, say so.

(a) One line of symmetry, no rotational symmetry.
(b) No line of symmetry, quarter-turn rotational symmetry.
(c) Two different centres of rotational symmetry.
(d) Three lines of symmetry, no rotational symmetry.
(e) An infinite number of lines of symmetry.

7. A school pays for its electricity by paying a fixed charge of £2000 and then at a rate of £4 for each 1000 units consumed.

(a) Draw a graph to illustrate this.

(*b*) If the total charge is £*C* and the number of units consumed is *U*, write down the relation between *C* and *U*.

(*c*) If 40000 units are consumed find the cost.

(*d*) If the total bill is £2804, how many units were consumed?

8. If $p = {}^-1, q = {}^-2, r = {}^-3$, find the value of:

(*a*) $pq - r$; (*b*) $p - q - r$; (*c*) pqr;
(*d*) $p + rq$; (*e*) $(p)^q$; (*f*) $(q)^r$.

9. Teletown is in a fringe area. *A* = {houses receiving BBC 1 clearly}, *B* = {houses receiving BBC 2 clearly} and *C* = {houses receiving ITV clearly}. Describe the sets *A* ∩ *B*, *A'* ∩ *B'*, *A* ∪ *B*, *A* ∪ *B* ∪ *C*. If *A*, *B*, *C* ⊂ $\mathscr{E}$, $n(\mathscr{E}) = 1000$, (where $n(\mathscr{E})$ means the number of elements in $\mathscr{E}$), $n(A) = 500$ and $n(B) = 400$, give the greatest and least values of $n(A \cup B)$ and $n(A \cap B)$. If $n(A \cap C) = 200$, what (if anything) can you say about $n(C)$?

10. For $S = \{1, 2, 3, 4, 5, 6\}$ sketch graphs similar to Figure 7 (*a*) in Chapter 2, to represent the relations:

(*a*) *p* is a factor of *q*; (*b*) *p* is a prime factor of *q*;
(*c*) $p = q - 2$; (*d*) $p \sim q$ is even.

(*Note*: ~ represents 'the positive difference between'.)

D

1. Figure 6 shows the net for a certain solid.

(*a*) Name it.

(*b*) State its number of vertices (*V*), edges (*E*), faces (*F*).

(*c*) $F + V = ?$

(*d*) $E + 2 = ?$

(*e*) Does $F + V = E + 2$?

(*f*) Do any of the lengths have to be equal?

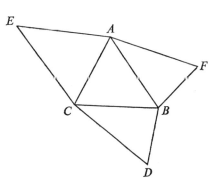

Fig. 6

2. Taking the value of π to be 3·14, use your slide rule to find the volume and the total surface area of a solid cylinder of height 4·45 cm and radius 1·85 cm.

3. Taking the set {1, 2, 3, 4, 5} as domain, give the ranges of the following functions:

(*a*) $x \to 2x$; (*b*) $x \to 3x - 7$; (*c*) $x \to 4x + 5$; (*d*) $x \to x^3$; (*e*) $x \to x^2 + x$.

4. What signs must be inserted in order to make the following statements correct?

(*a*) $5 - 4 - 1 = 5 - (4 \quad 1)$; (*b*) $5 - (3 - 2) = 5 - 3 \quad 2$;
(*c*) $(9 - 2) - (6 \quad 3) = 4$; (*d*) $6 \quad (2 \quad 1) = 3$;
(*e*) $(6 \quad 1) - (5 \quad 3) = 5$.

5. What can be said about *x* in each of the following cases:

(*a*) $2x - 1 = 49$; (*b*) $3x^2 + 1 = 76$ (two answers);
(*c*) $2^x = 8$; (*d*) $2(2x + 9) = 4(x + 4\frac{1}{2})$;
(*e*) $1/x = \frac{2}{3}$

6. Figure 7 shows an oblique projection of a cuboid.

(*a*) Name a length equal to *AC*, and appearing to be equal.

91

(b) Name a length equal to *AC*, but appearing to be shorter.
(c) *EFGH* appears to be a parallelogram. What shape is it really?
(d) Name an angle which is a right-angle, but appears greater.
(e) Name an angle which is a right-angle, but appears smaller.

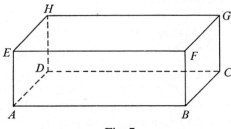

Fig. 7

7. The distances one can see from various heights are given in the following table.

Height (x m)	1	10	30	50	100	150
Distance seen (y km)	3.3	10·6	20·8	27·2	39·2	45·6

(a) Graph the function 'height → distance you can see.'
(b) How far can you see from 80 m up?
(c) Under what conditions do you suppose the table is reliable?

8. Graph the function $x \to x^2$ for $x = ^-3, ^-2\frac{1}{2}, ^-2, ..., 2, 2\frac{1}{2}, 3$. Join the points with a smooth curve and hence estimate the value of $\sqrt{5}$.

9. Figure 8 represents a tetrahedron, all of whose edges are of equal length.

(a) Describe any planes of symmetry it has.
(b) Describe its axes of rotational symmetry.
(c) What is the order of rotational symmetry in each case?

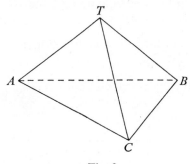

Fig. 8

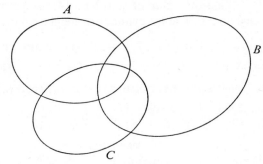

Fig. 9

10. On separate diagrams, of the type shown in Figure 9, indicate by shading:
(a) $(A \cup B) \cap C$;   (b) $(A \cap B) \cup C$;
(c) $(A \cup C) \cap (B \cup C)$;   (d) $(A \cap C) \cup (B \cap C)$.
Are any of the above equivalent?

E

1. Copy Figure 10 (*b*) to (*f*). Mark lines of symmetry with dotted lines and centres of rotational symmetry with small circles. As an example the lines of symmetry of the rectangle in Figure 10(*a*) are already marked.

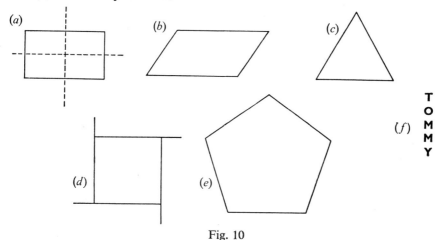

Fig. 10

2. Find the third angles of the triangles in which two angles are:
(*a*) 100°, 20°;   (*b*) 60°, 60°;   (*c*) 45°, 90°;   (*d*) 17°, 23°.
What can be said about the lengths of the sides of the triangle in (*b*)?

3. Pythagoras Park at Mathsville is, of course, in the shape of a right-angled triangle, the two shortest boundaries being 85 m and 200 m long. Cars are parked all round the outside of the triangle. The town council make all the parking spaces of exactly the same length, and 17 of them fit exactly along the shortest side. How many spaces will fit along the longest side?

4. Find x if:
(*a*) $x+3+2x+4 = 28$;   (*b*) $(2 \times 3)+x = 10$;
(*c*) $1-x = x-1$;   (*d*) $x \times 2 \times x = 72$.

5. All 24 members of Form 2Z like either cornflakes or porridge or both. It is known that 18 like cornflakes (or both) and that 17 like porridge (or both). What is the smallest number who could like both? Illustrate with a Venn diagram. What is the largest possible number?

6. Here are the playing times for various lengths of recording tape moving at 20 cm per second.

Tape length (x m)	192	288	384	576	768
Playing time (y min)	16	24	32	48	64

(*a*) Graph the function 'tape length → playing time'.
(*b*) Is this a linear function?
(*c*) If you wanted to record a piece of music lasting 20 min, what would be the shortest length of tape which would do?
(*d*) How long would 504 m of tape play?

7. Give an example to show that you understand what is meant when an operation is said to be commutative. Is subtraction commutative? Is division? If xMy means $\frac{1}{2}(x+y)$ [for example, $2M4 = 3$], work out $12M20$ and $20M12$. Is the operation M commutative? Work out $(3M5)M7$ and $3M(5M7)$. Are the brackets needed?

8. Find x when:

(a) $2^x = 8$;   (b) $8 \div 2^x = 16$;   (c) $4 \times 2^x = 2^5$;   (d) $8 \times 2^x = 2^{6.5}$.

9. In a certain district 70% of the families own a TV set, and 30% of the families own cars. Which of the following statements are *not necessarily* true, but *could* be true?

(a) All families have either a TV set or a car.
(b) Most car owners have a TV set.
(c) Fewer than half the owners of TV sets have cars.
(d) If the number of car owners increased by 50% of the present number, there would be more car owners than TV owners.

10. The triangles in Figure 11 are all equilateral, and $AB = 4$ cm.

(a) Name the solid of which $ABCDEF$ is a net.
(b) With which point does C coincide when the net is folded up?
(c) Calculate the total surface area of the solid. (The answer may be given in a form which includes a square root.)

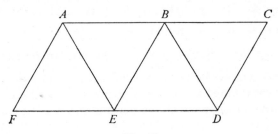

Fig. 11

F

1. Copy and complete the patterns shown in Figure 12. The dotted lines are lines of symmetry, and the small circles are centres of rotational symmetry.

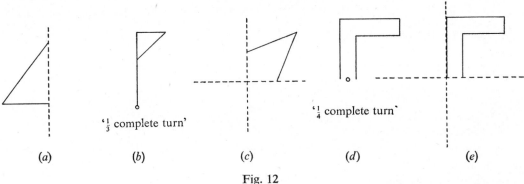

'$\frac{1}{3}$ complete turn'   '$\frac{1}{4}$ complete turn'

(a)   (b)   (c)   (d)   (e)

Fig. 12

2. It is 7·5 cm from the 1 to the 2 on the C scale of my slide rule. How far is it between the 1 and the 4? What number is 15 cm from the 2? If I put the 1 of the C scale opposite the 3 of the D scale, what number is opposite the 2·5 of the C scale? (Try to answer this question without looking at a slide rule.)

3. The function $x \rightarrow 2x+1$ is to be represented on an (x, y) graph.

(a) Write down the equation relating x and y.
(b) Draw the graph, with the domain $\{x: 0 < x < 5\}$ and a scale of 1 cm to 1 unit.

4. In this question the operation † means 'square the first number and multiply by the second'; for example, $3 † 4 = 3^2 \times 4 = 36$.

(a) Work out:
(i) $2 † 3$;   (ii) $3 † 2$;   (iii) $(2 † 3) † 4$;   (iv) $2 † (3 † 4)$.
(b) Is the operation † commutative?

In (b) explain what you mean.

5. If $r = 1$, $s = 2$, $t = {}^-3$, work out:

(a) $(r+s)+t$;   (b) $r+(s+t)$;   (c) $(r-s)-t$;   (d) $r-(s-t)$;   (e) $r+(s-t)$.

6. With which of the following is it possible to fill space completely?

(a) Equal spheres;   (b) equal cubes;   (c) equal regular tetrahedra;
(d) equal regular hexagonal prisms (like unsharpened hexagonal pencils);
(e) equal cylinders.

7. The curved surface of a conical tent is a sector of a circle, radius 3 m, angle 240°. Calculate:

(a) the radius of the base;
(b) the area of the base of the tent;
(c) the height of the tent;
(d) the volume, given that the volume of a cone is $\frac{1}{3} \times$ area of base × height.

[Take $\pi = 3\cdot14$.]

8. Is the shaded area in Figure 13:

(a) $(X \cup Y) \cap Z$;   (b) $(X \cap Y) \cup Z$;
(c) $(X \cap Y') \cap Z$;   or   (d) $(X \cap Y)' \cap Z$?

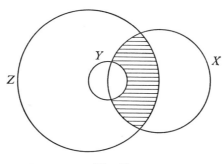

Fig. 13

95

9. $\mathscr{E} = \{10, 11, 12, 13, 14, 15, 16, 17, 18, 19\}$; list the members of the following subsets of $\mathscr{E}$:

(a) {prime numbers};   (b) {multiples of 3};

(c) {numbers such that $(x-13)^2 = 4$}.

List the members of any two subsets A and B of $\mathscr{E}$ such that $A \cap B = \varnothing$ and $A \cup B = \mathscr{E}$.

10. $ABCDV$ is a pyramid. $ABCD$, 8 cm square, is its base; and V is 6 cm vertically above the centre of the base.

(a) Draw an oblique projection of the pyramid.

(b) Calculate the length of BV.

(c) State the angle between the planes AVC and BVD.

(d) Is it true that BV and AD are skew lines?

(e) State how you would calculate the angle between the planes BCV and $ABCD$, but do not actually carry out the calculation.

(f) P is a point 27 cm from B and 19 cm from A. Is there a single plane containing the three points P, B and A?

5

DISPLACEMENTS

Let's all move one place on.
LEWIS CARROLL, *Alice in Wonderland*

1. TRANSLATIONS

When an object moves from one position to another *without being turned* we say it has undergone a *translation*.

The drawings in Figure 1 show a polygonal man, $ABCD$, and his images $A_1B_1C_1D_1$ and $A_2B_2C_2D_2$ after two translations. Consider the translation which maps $ABCD$ onto $A_1B_1C_1D_1$.

(*a*) What can you say about the distances AA_1, BB_1, CC_1 and DD_1? Can you say the same for all distances between pairs of corresponding points?

(*b*) Would the translation from the first position to the second position be fully described if you knew only the distance moved? If your answer to this question is 'no', what else would you need to know?

(*c*) If you are told the distance and direction of A_2 from A, can you find the position of B_2, C_2 and D_2?

When an object is translated, each point of the object undergoes the same change of position because it moves the same distance in the same direction as every other

97

point. When you see a platoon of soldiers drilling, a formation team dancing, or a pair of ice skaters figure-skating and see them moving as a single body it is because each person involved moves in exactly the same way.

The answers to the above questions tell us that a translation is fully described when the direction and distance of the motion is known. In mathematics a quantity which describes a change in position is called a *displacement*.

The change in position of a point can be clearly illustrated by arrows as shown in Figure 1 (for they possess both length and direction) and because of this it is convenient to represent displacements by arrows.

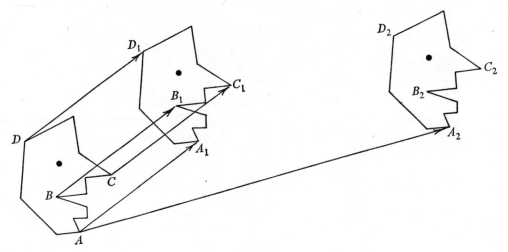

Fig. 1

To distinguish between the line segment AA_1 joining A and A_1 and the displacement from A to A_1 a new notation is needed. We shall use the notation $\mathbf{AA_1}$ for the displacement. You will see that the printer has printed the symbols in bold letters. Bold letters cannot be shown clearly in ordinary handwriting so you will use a wiggly underline—like this AA_1—to indicate a displacement.

The translation could be fully described by any one of these displacements. They are all equivalent, so we can write

$$\mathbf{AA_1} = \mathbf{BB_1} = \mathbf{CC_1} = \mathbf{DD_1}.$$

Because all equivalent displacements represent the same translation, it is often convenient to have one symbol, usually a small letter, for the translation.

The translation which maps $ABCD$ into $A_1B_1C_1D_1$ could be named $\mathbf{a}$ (written a) and then

$$\mathbf{a} = \mathbf{AA_1} = \mathbf{BB_1} = \mathbf{CC_1} = \mathbf{DD_1}.$$

(*d*) Which pairs of arrows in Figure 2 represent equivalent displacements?

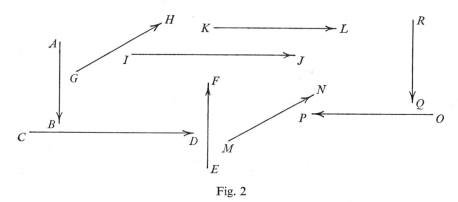

Fig. 2

(*e*) An object could be given a translation **p** followed by a translation **q**, and if a single translation **r** had the same effect we could write **p** ⊕ **q** = **r**.

(*f*) Figure 3 shows two arrows representing **a** and **b**. What is **a** ⊕ **b**? Two translations which cancel each other out are said to form an inverse pair and we write **b** = ⁻**a**.

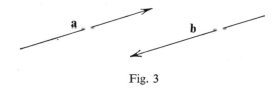

Fig. 3

When **a** is followed by ⁻**a** the resulting translation is the zero translation **0**:

$$\mathbf{a} \oplus {}^-\mathbf{a} = \mathbf{0}.$$

This equation is very similar to the equation relating the integers ⁺3 and ⁻3 where

$$^+3 + {}^-3 = 0.$$

For this reason we will drop the ⊕ and use an ordinary + sign when translations are to be 'added'. But remember that we are *not* dealing with numbers here.

How do we always indicate that the elements are *not* numbers?

Exercise A

Questions 1–7 refer to Figure 4 overleaf.

1. The parallelogram *S* is given a translation equivalent to:

(*a*) **DK**;   (*b*) **BK**;   (*c*) **PF**;   (*d*) **FJ**.

Where will it be after each translation?

2. Give other displacements which represent the same translation as:

(*a*) **AC**;   (*b*) **RA**;   (*c*) **CK**;   (*d*) **DG**.

3. In Figure 4 name a displacement which will describe the translation of the parallelogram S to the position:

(a) *RBAS*;   (b) *CFGB*;   (c) *FJIG*;   (d) *DEFC*.

4. In Figure 4 (a) If S undergoes a translation equivalent to **RG** followed by a translation equivalent to **CD**, where will it be?

   (b) What single translation is equivalent to **RG+CD**?
   (c) Where would S be if it were translated first by **CD** and then by **RG**?
   (d) What can you say about **RG+CD** and **CD+RG**?

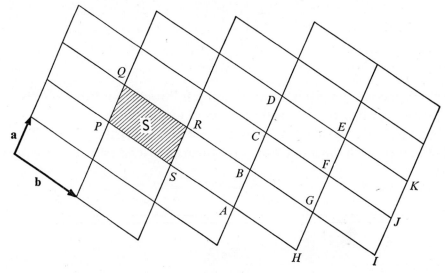

Fig. 4

5. In Figure 4 name single displacements which will describe the same translations as the following combinations:

(a) **BC+CF**;   (b) **RG+IK**;   (c) **SA+BC+EK**;   (d) **RC+BH**.

6. Using the notation of Figure 4, (a) Suggest a reasonable meaning for 2**a**.
   (b) Name displacements which describe:

   (i) **a**;   (ii) 2**a**;   (iii) 3**a**;   (iv) ⁻**a**.

   (c) If S is given a translation **a+2b** where will it be?
   (d) Name single displacements which describe the same translations as:

   (i) 2**a+2b**;   (ii) 3**a+b**;   (iii) 2**a+3b**;   (iv) **b+⁻a**;   (v) 2**a+⁻b**.

7. In Figure 4 which of the following are true:

(a) **BE** = 2**a+b**;   (b) **a+2b** = 2**b+a**;   (c) **HB** = **b+⁻a**;
(d) **DJ** = 2**b+⁻a**;   (e) **KS** = 3(⁻**a+⁻b**);   (f) **AC** = **EG**;
(g) 2**a+2b** = 2(**a+b**)?

8. In Figure 5, (a) If **AP**, **BQ** and **CR** are equivalent displacements, what can you say about the triangles ABC and PQR?

   (b) Will area $APQRC$−area ABC = area $APQRC$−area PQR?

100

(*c*) What relation can you find between the areas of the parallelograms *ABQP*, *CBQR* and *APRC*?

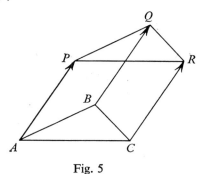

Fig. 5

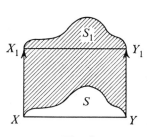

Fig. 6

9. A shape *S* (see Figure 6) is translated to S_1 so that $XX_1\,Y_1Y$ forms a rectangle.

(*a*) What can you say about the areas of S_1 and *S*?

(*b*) If $XY = 4$ cm and $XX_1 = 3$ cm, what is the area of:

(i) the rectangle XX_1Y_1Y;   (ii) the shaded region?

10. (*a*) How do you move a ruler when drawing parallel lines?

(*b*) Explain how an engineer or draughtsman makes use of translation when he uses a T-square and a set-square to draw parallel lines (see Figure 7).

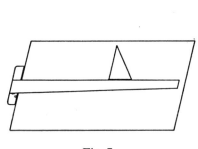

Fig. 7

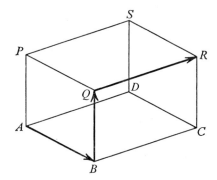

Fig. 8

11. Fred, the mathematical fly, travels from one corner of a room *A* to the opposite corner *R*, by walking along the edge *AB*, *BQ* and *QR* (see Figure 8). His change of position **AR** is equivalent to **AB + BQ + QR**.

(*a*) Using displacements, describe alternative routes he could have taken from *A* to *R*, not passing through the same point more than once, if he travelled along only:

(i) 3 edges;   (ii) 5 different edges;   (iii) 7 different edges.

(*b*) Can he reach *R* from *A* by walking along an even number of edges?

(*c*) Name displacements equivalent to:

(i) **AB**;   (ii) **RB**;   (iii) **AQ**;   (iv) **CA**.

12. On a training flight a navigator described each leg of his route by giving its length in kilometres and its bearing in degrees as follows:

AB = (150, 050°);   **BC** = (85, 340°);   **CD** = (325, 205°);   **DE** = (190, 330°).

(Bearings are measured in a clockwise direction from North.)

(*a*) If the plane changed course at *E* to fly back to base *A*, in which direction did it fly?

(*b*) If the plane had been recalled to base when it was at *D* and travelled back to *A* at 650 kilometres per hour, how long would it have taken to fly from *D* to *A*?

(*c*) Describe **BE**.

2. NAVIGATION

Figure 9 shows a map used in an imaginary game of 'Pirates'.

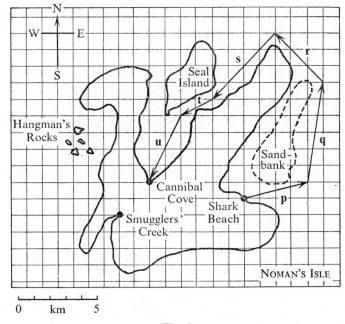

Fig. 9

A boat leaves Shark Beach and follows the route shown by the arrows to Cannibal Cove.

(*a*) How far east does the boat travel on leg **p**?

(*b*) Describe leg **q** by giving the boat's easterly and northerly changes in position.

(*c*) Make use of negative numbers to describe:

(i) the easterly change in position produced by **r**;

(ii) the northerly change in position produced by **s**.

Since each leg of the course is defined when easterly and northerly changes of position are known, these can be given instead of the direction and length of each leg. The first leg, for example, is defined by saying that it is a change of position of 4 km east and 1 km north and we shall write this as

$$\mathbf{p} = \begin{pmatrix} 4 \\ 1 \end{pmatrix}.$$

Similarly $\mathbf{q} = \begin{pmatrix} 1 \\ 6 \end{pmatrix}$ and $\mathbf{r} = \begin{pmatrix} -3 \\ 3 \end{pmatrix}.$

This way of representing a displacement is called a *vector*.

(*d*) Write **s**, **t** and **u** as vectors.

(*e*) Do the vectors $\begin{pmatrix} 3 \\ 1 \end{pmatrix}$ and $\begin{pmatrix} 1 \\ 3 \end{pmatrix}$ represent the same change in position?

(*f*) What vector represents the displacement equivalent to

$$\mathbf{p} + \mathbf{q}?$$

Vectors are often used in connection with coordinates, and the upper element then represents the *increase* in the *x*-coordinate which is termed the *x-component*, and the lower element represents the increase in the *y*-coordinate (the *y-component*).

Exercise B

1. In Figure 10, A has coordinates $(1, 2)$ and C has coordinates $(7, 4)$. Represent AC as a vector.

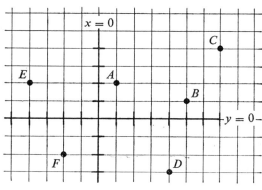

Fig. 10

2. A, B, C, D, E and F are shown in Figure 10. Write the following as vectors:

$$BC; \quad DB; \quad AB; \quad AF; \quad EC; \quad AE.$$

3. The points A, B, C, D, E and F in Figure 10 are given a translation represented by $\begin{pmatrix} 3 \\ 2 \end{pmatrix}$. What are the coordinates of the corresponding image points?

4. The vectors

$$\begin{pmatrix} 1 \\ 7 \end{pmatrix}, \begin{pmatrix} 2 \\ 2 \end{pmatrix}, \begin{pmatrix} 3 \\ 0 \end{pmatrix}, \begin{pmatrix} -1 \\ -3 \end{pmatrix}, \begin{pmatrix} -3 \\ -2 \end{pmatrix}, \begin{pmatrix} -2 \\ -4 \end{pmatrix},$$

when taken in order, describe a route from Cannibal Cove (see Figure 9). Draw the route on tracing paper and explain briefly in words where it goes.

5. Mark on graph paper three points P, Q and R such that

$$\mathbf{PQ} = \begin{pmatrix} 5 \\ 3 \end{pmatrix} \quad \text{and} \quad \mathbf{QR} = \begin{pmatrix} 2 \\ 7 \end{pmatrix}.$$

(a) What vector represents $\mathbf{PR}$?

(b) How could you have answered (a) without a drawing?

(c) If S is a fourth point such that

$$\mathbf{PS} = \begin{pmatrix} 2 \\ 7 \end{pmatrix} \quad \text{what is} \quad \mathbf{SR}?$$

What name do we give to the quadrilateral $PQRS$?

In Questions 6–8 the vectors $\mathbf{u}$, $\mathbf{v}$, *and* $\mathbf{w}$ *are defined as follows*:

$$\mathbf{u} = \begin{pmatrix} 2 \\ 3 \end{pmatrix}; \quad \mathbf{v} = \begin{pmatrix} -1 \\ -2 \end{pmatrix}; \quad \mathbf{w} = \begin{pmatrix} -3 \\ 1 \end{pmatrix}.$$

6. Without using a diagram give the vectors equivalent to:

(a) $^-\mathbf{u}$, $^-\mathbf{v}$, $^-\mathbf{w}$; (b) $\mathbf{u}+\mathbf{v}$, $2\mathbf{u}+\mathbf{w}$, $\mathbf{u}+^-\mathbf{v}$;

(c) $2\mathbf{w}+3\mathbf{v}$, $^-\mathbf{u}+\mathbf{w}$, $3(^-\mathbf{v})+\mathbf{u}$; (d) $4\mathbf{u}+^-\mathbf{v}+2\mathbf{w}$, $5(^-\mathbf{v})+^-\mathbf{w}+2\mathbf{u}$.

7. Find the vector $\mathbf{x}$ which satisfies:

(a) $\mathbf{x} = \mathbf{u}+\mathbf{v}$; (b) $\mathbf{x}+^-\mathbf{v} = \mathbf{u}$; (c) $\mathbf{x}+\mathbf{w} = \mathbf{v}$;

(d) $\mathbf{u}+\mathbf{x} = \mathbf{v}+\mathbf{w}$; (e) $2\mathbf{x} = \mathbf{u}+\mathbf{v}+\mathbf{w}$; (f) $\mathbf{u}+\mathbf{v} = \mathbf{x}+^-\mathbf{w}$.

8. By considering combinations of $\mathbf{u}$, $\mathbf{v}$ and $\mathbf{w}$, illustrate that:

(a) $\mathbf{u}+\mathbf{v} = \mathbf{v}+\mathbf{u}$; (b) $(\mathbf{u}+\mathbf{v})+\mathbf{w} = \mathbf{u}+(\mathbf{v}+\mathbf{w})$.

Summary

When an object S moves in such a way that every point of S moves the same distance in the same direction, it has undergone a *translation*.

1. A *displacement* $\mathbf{AB}$ is the quantity which describes a change in position from a point A to a point B. It can be defined by giving the length of the line segment AB

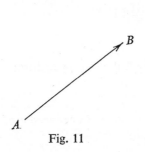

Fig. 11

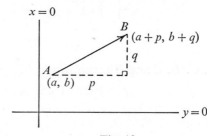

Fig. 12

and the direction of B from A (see Figure 11). We are only interested in the end-points A and B here and not in the nature of the movement as we were when dealing with a *translation*.

104

2. When points are described by Cartesian coordinates a displacement can best be described by the change in the coordinates. In Figure 12, $\mathbf{AB} = \begin{pmatrix} p \\ q \end{pmatrix}$, and $\begin{pmatrix} p \\ q \end{pmatrix}$ is called a vector; p and q are called the x- and y-*components* of the vector, respectively.

3. Displacements which describe the same translation are said to be *equivalent*.

$$\mathbf{a} = \mathbf{b} = \mathbf{c} \quad \text{(see Figure 13)}.$$

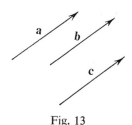

Fig. 13

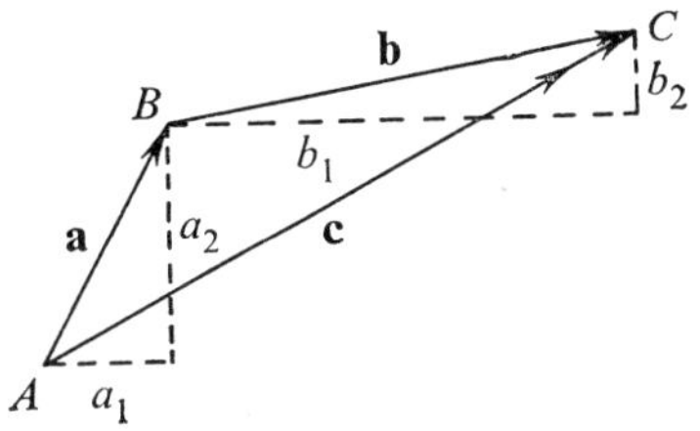

Fig. 14

4. When a displacement $\mathbf{AC}$ is a combination of two displacements

$$\mathbf{AB} \quad \text{and} \quad \mathbf{BC} \quad \text{(see Figure 14)}$$

then $$\mathbf{AC} = \mathbf{AB} + \mathbf{BC} \quad \text{or} \quad \mathbf{c} = \mathbf{a} + \mathbf{b},$$

or $$\mathbf{c} = \begin{pmatrix} a_1 \\ a_2 \end{pmatrix} + \begin{pmatrix} b_1 \\ b_2 \end{pmatrix} = \begin{pmatrix} a_1 + b_1 \\ a_2 + b_2 \end{pmatrix}.$$

5. When two displacements are equal in length but opposite in direction, they form an inverse pair (see Figure 15).

$$\mathbf{a} + {}^{-}\mathbf{a} = \mathbf{0},$$

where $\mathbf{0}$ is the displacement with no length, the *zero displacement*.

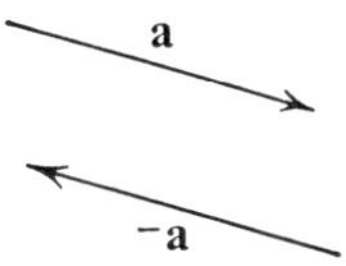

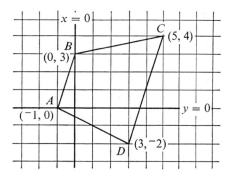

Fig. 15

6. If $$\mathbf{a} = \begin{pmatrix} a_1 \\ a_2 \end{pmatrix} \quad \text{then} \quad \mathbf{a} + \mathbf{a} = 2\mathbf{a} = \begin{pmatrix} a_1 \\ a_2 \end{pmatrix} + \begin{pmatrix} a_1 \\ a_2 \end{pmatrix} = \begin{pmatrix} 2a_1 \\ 2a_2 \end{pmatrix} = 2\begin{pmatrix} a_1 \\ a_2 \end{pmatrix}.$$

3. VECTOR GEOMETRY

We have seen in previous sections how a vector can be used to describe a change in position, or the position of one point in relation to another. Because of this, vectors are a useful tool for demonstrating certain geometric properties. In this section we will be particularly concerned with the property of parallelism of vectors.

A quadrilateral (see Figure 16) has vertices

$A(^{-}1, 0), \quad B(0, 3), \quad C(5, 4) \quad \text{and} \quad D(3, ^{-}2).$

(*a*) Find the vectors representing $\mathbf{AB}$ and $\mathbf{DC}$.

Fig. 16

(b) What do you notice about the components of these two vectors? Can you say anything about the length and direction of **AB** compared with those of **DC**?

(c) (i) Represent the following vectors as displacements on graph paper, starting each from a different point:

$$\binom{2}{3}; \quad \binom{4}{6}; \quad 3\binom{2}{3}; \quad \binom{3}{4\frac{1}{2}}; \quad \binom{-2}{-3}; \quad \binom{-6}{-9}.$$

(ii) Each of the vectors in (i) can be expressed as $k\binom{2}{3}$, for some number k. Find k in each case.

What can you say about the displacement when k is negative?

When two displacements are parallel then one is a certain multiple of the other. This multiple tells us how many times longer one displacement is than the other.

The arrows in Figure 17 show a set of parallel displacements.

We see that

$$\mathbf{b} = 3\mathbf{a}, \quad \mathbf{c} = {}^-\mathbf{a}, \quad \mathbf{d} = 2\mathbf{a},$$

$$\mathbf{e} = \tfrac{3}{2}\mathbf{a}, \quad \mathbf{f} = 3({}^-\mathbf{a}) = {}^-3\mathbf{a}.$$

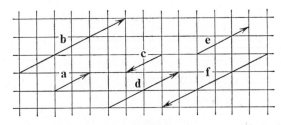

Fig. 17

(d) Consider the following example.

ABC is a triangle in which Z and Y are the mid-points of AB and CA respectively. It looks possible that YZ is parallel to CB. What do you think about the relation between their lengths? Can we prove something here using displacements?

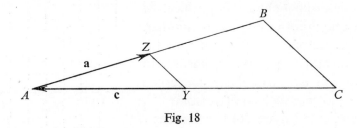

Fig. 18

Suppose we express the relative positions of points in Figure 18 in terms of the two convenient displacements **a** and **c**.

Then $\mathbf{AB} = 2\mathbf{a}$ and $\mathbf{CA} = 2\mathbf{c}$.

Hence $\mathbf{CB} = \mathbf{CA} + \mathbf{AB}$

$= 2\mathbf{c} + 2\mathbf{a}$

$= 2(\mathbf{c} + \mathbf{a})$.

However, $\mathbf{YZ} = \mathbf{YA} + \mathbf{AZ}$

$= \mathbf{c} + \mathbf{a}$.

Therefore, $\mathbf{CB} = 2\mathbf{YZ}$ which shows that YZ is parallel to CB and is half its length.

Exercise C

1. The positions of three vertices of a parallelogram are known (see Figure 19).

(*a*) Find the coordinates of the fourth vertex, *C*, by finding the vector representing $\mathbf{AB}$ and using the fact that $\mathbf{DC} = \mathbf{AB}$.

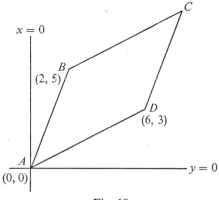

Fig. 19

(*b*) Find the coordinates of:
    (i) a point *E* given that $\mathbf{BE} = \mathbf{AC}$ and (ii) a point *F* given that $\mathbf{AF} = \mathbf{BD}$.
(*c*) Show that *FE* is parallel to *AB* and that $\mathbf{FE} = 3\mathbf{AB}$.

2. (*a*) Let

$$\mathbf{a} = \begin{pmatrix} -1 \\ 2 \end{pmatrix}, \quad \mathbf{b} = \begin{pmatrix} 3 \\ 1 \end{pmatrix} \quad \text{and} \quad \mathbf{c} = \begin{pmatrix} 2 \\ -1 \end{pmatrix}.$$

Plot on graph paper (a scale of 1 cm to a unit is suitable) the points *A*, *B*, *C*, *D*, ..., *L* given by the following equations (*O* is the origin):

$\mathbf{OA} = \mathbf{a}$.	$\mathbf{OB} = \mathbf{b}$.	$\mathbf{OC} = \mathbf{c}$.	$\mathbf{OD} = \mathbf{a} + \mathbf{b}$.
$\mathbf{OE} = \mathbf{b} + \mathbf{c}$.	$\mathbf{OF} = \mathbf{c} + \mathbf{a}$.	$\mathbf{OG} = 2\mathbf{a}$.	$\mathbf{OH} = {}^-\mathbf{b}$.
$\mathbf{OI} = \mathbf{a} + \mathbf{b} + \mathbf{c}$.	$\mathbf{OJ} = \mathbf{c} + {}^-\mathbf{b}$.	$\mathbf{OK} = 3\mathbf{c} + \mathbf{a}$.	$\mathbf{OL} = {}^-2\mathbf{a} + \mathbf{b}$.

(*b*) *AFJH* is a parallelogram because $\mathbf{AF} = \mathbf{HJ}$. Find other sets of four points on your diagram which form the vertices of parallelograms.

(*c*) Is *GK* parallel to *HJ*? Give reasons for your answer.

3. The vertices of a hexagon are given as:

$$A(1, 3), \quad B(4, 1), \quad C(2, {}^-3), \quad D(0, {}^-2), \quad E({}^-3, 0) \quad \text{and} \quad F({}^-2, 2).$$

(*a*) Show that: (i) *CB* is parallel to *EF* and (ii) **CB** = 2**EF**.

(*b*) Is *ABDE* a parallelogram?

4. Figure 20 shows the edges of a box. **AB** = **x**, **AD** = **y** and **AE** = **z**.

(*a*) Express the following in terms of **x**, **y** and **z**:

$$\textbf{AF}; \quad \textbf{BG}; \quad \textbf{FH}; \quad \textbf{AG}; \quad \textbf{EC}; \quad \textbf{BH}; \quad \textbf{FD}.$$

(*b*) If *P* is the mid-point of *FB* and *Q* is the mid-point of *HD*, show that *EPCQ* is a parallelogram.

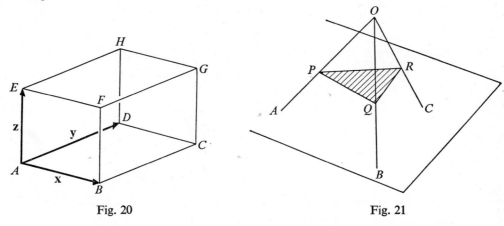

Fig. 20   Fig. 21

5. A tripod with unequal legs *OA*, *OB* and *OC* is standing on a horizontal floor (see Figure 21). The legs are held rigid by a small triangular table *PQR* attached to the mid-point of each leg.

By taking **OA**, **OB** and **OC** to be 2**a**, 2**b** and 2**c**, respectively, show that the table *PQR* is horizontal.

4. WAVES

4.1

The title of this section probably makes you think of the seaside and the waves of the sea, and these are certainly related to what we shall be considering. However modern physics has shown that types of wave motion are studied in such apparently different topics as heat, light, electricity, sound and even matter itself.

In Figure 22, a boy is moving one end of a rope up and down while the other end is fixed to a tree—a child watches him from a swing.

Fig. 22

The movement of the rope clearly looks like a wave, but what is the connection between the swing and a wave?

Imagine yourself swinging on a park swing.

(*a*) At what moments will you be moving forward fastest?

(*b*) How does your forward speed change as you swing? Is it ever negative?

Figure 23 shows the result of graphing the function 'time → forward speed'. The sketches above the graph show the appropriate positions of the swing. The pendulum of a clock has a similar motion and we call the time for one oscillation (that is, a *complete* swing forward and back) the *periodic time* of the motion.

(*c*) What is the periodic time of the swing in Figure 23?

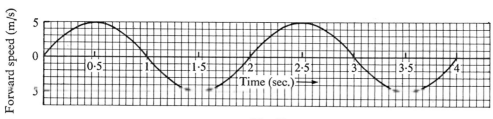

Fig. 23

Many motions when analysed produce similar graphs, for example:

(i) the up and down motion of a weight suspended from a piece of elastic or a spring, see Figure 24(*a*);

(ii) the vibration of a tuning fork, see Figure 24(*b*).

(iii) the vibration of a violin or piano string, see Figure 24(*c*).

(The simple pendulum of course produces similar results when a pen is attached. For more complicated arrangements, see *Mathematical Models* (2nd Edition), by Cundy and Rollett, pages 242–53.)

The motions of the pendulum and the tuning fork have greatly different periodic times—for a clock pendulum it might be 1 second, and for a 'middle C' tuning fork it is $\frac{1}{261}$ seconds. The number of complete oscillations per second is called the *frequency* of the motion. What is the frequency of 'middle C'?

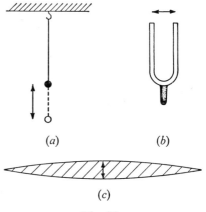

Fig. 24

We can hear notes having frequencies between about 30 and 30000 oscillations a second; the greater the frequency, the higher the note. The notes of a bat are about the highest that humans can hear and you will probably find that some of your friends can hear bats while others cannot.

There are other common phenomena which produce such waves. One example is shown in Figure 25; the graph results from plotting the height of the water at the entrance to a harbour at different times between 9 a.m. and 8 p.m.

(*d*) At what times of day was it: (i) high tide; (ii) low tide?

(*e*) A ship has a depth of 7 m below the water line. During what times would it go aground?

(*f*) What was the average level of the water?

(*g*) When was the water falling most rapidly?

Make up similar questions about this harbour mouth and answer them.

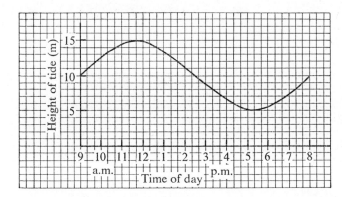

Fig. 25

Exercise D

1. The up and down motion of a piston in a car engine and the needle in a sewing machine are examples of oscillations whose variation with time when plotted on a graph lead to curves such as have just been drawn. Suggest other examples.

2. An electric motor is marked 230 volts A.C. 50 cycles. Find out what is meant by A.C. and 50 cycles.

3. Many radio sets are labelled V.H.F. Find out what these letters stand for.

4. Look at a gramophone disc through a magnifying glass or find a photograph showing a magnification of a disc. What do you find?

5. The strings which give the note 'middle C' on a piano vibrate with a frequency of 261. The C an octave higher has double the frequency, the frequency of the C above that is doubled again and so on.

(*a*) What is the frequency of the note which is:
(i) 3 octaves above middle C; (ii) 2 octaves below middle C?

(*b*) Sketch the graph of 'time → displacement' for the centre of the middle C wire as it vibrates from side to side. On the same graph show how another wire one octave higher vibrates.

6. The table below shows the maximum height, in metres, of each morning tide at Falmouth for January, 1962.

	1st week	2nd week	3rd week	4th week	5th week
Sunday	—	5.40	4·61	5·15	4·58
Monday	4·36	5·51	4·39	5·15	4·40
Tuesday	4·21	5·51	4·45	5·12	4·24
Wednesday	4·42	5·40	4·64	5·06	4·18
Thursday	4·73	5·25	4·82	4·97	—
Friday	5·0	5·04	5·0	4·85	—
Saturday	5·25	4·79	5·09	4·73	—

(a) Graph the function 'day of the month → height of the morning tide'. (It is best to start the height scale from about 4 m as no height is less than this.)

(b) Do points between those plotted have any meaning?

(c) From your graph find the probable height of the afternoon tide on the following days of the month;

      (i) 4th;   (ii) 14th;   (iii) 20th;   (iv) 29th.

7. When a car hits a bump the suspension of the car makes the passenger in the back seat bounce up and down. The graphs in Figure 26 show the effects of two different kinds of suspension. One scale shows the height of the back seat above the road surface when the car hits a bump while the other scale measures the distance travelled along the road.

Explain which suspension you would prefer, giving reasons.

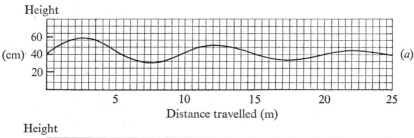

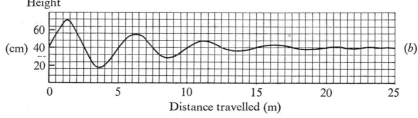

Fig. 26

4.2 Sine and cosine

We shall now go on to look at a mathematical model for these situations. Figure 27 shows a unit displacement **OA** which is fixed at one end O, the origin of a Cartesian coordinate system but free to rotate about O. We shall choose the line $y = 0$ as a starting direction. We will call this the *central direction*.

With OA drawn in the position in Figure 27, B is the point $(0·5, 0)$. OA is of length 1; what is the length of BA? The theorem of Pythagoras tells us

$$BA^2 = OA^2 - OB^2,$$

from which $BA = 0·87$.

We can now write $\mathbf{OA} = \begin{pmatrix} 0·5 \\ 0·87 \end{pmatrix}$.

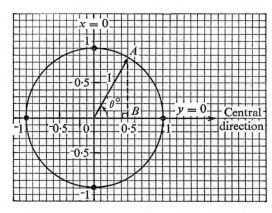

Fig. 27

We shall follow separately the changes of the upper element (x-component) and lower element (y-component) of this displacement as the angle $\theta°$ changes. $\theta°$ is measured in an anticlockwise sense from $y = 0$. What is $\theta°$ in Figure 27?

What is the maximum value of the y-component of the displacement $\mathbf{OA}$?

Between what two values does this y-component oscillate?

(a) *The y-component of the rotating unit displacement.*

Draw two lines to represent the domain (angles from $0°$ upwards) and range (members from ⁻1 to 1) as in Figure 28 for the relation $\theta° \to S(\theta°)$.

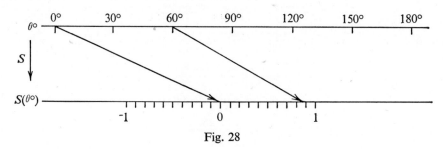

Fig. 28

Now draw carefully a larger scale drawing of Figure 27 with radius of length, say, 10 cm as the unit. Draw OA in various positions at $20°$ intervals.

Put in arrows, corresponding to the various positions of OA, such as those shown in Figure 28 ($0° \to 0$ and $60° \to 0·87$) which concern the y-component of the displacement.

What is the image of (i) 90°; (ii) 120°?

Can you extend the domain to values of θ greater than 180°?

Continue the process up to and beyond 360° by rotating OA still further. Do you ever need to go beyond 1 for $S(\theta°)$? What other limit is there for $S(\theta°)$? Is this relation a function?

(b) *The x-component of the rotating unit displacement.*

Repeat (a) for the x-component drawing diagrams such as Figures 27 and 28 again. Let OA rotate from $\theta° = 0°$ (where the x-component is 1) to $\theta° = 360°$. Call this new relation $\theta° \rightarrow C(\theta°)$.

Is it a function?

Is there any relationship between the way in which $\theta° \rightarrow S(\theta°)$ and $\theta° \rightarrow C(\theta°)$ behave, and what was discussed in Section 4.1?

We can see this more clearly if we represent the two relations on a coordinate graph.

Figure 29 shows points with $\theta°$-coordinates 30° and 60°.

Which function does Figure 29 show, $\theta° \rightarrow S(\theta°)$ or $\theta° \rightarrow C(\theta°)$?

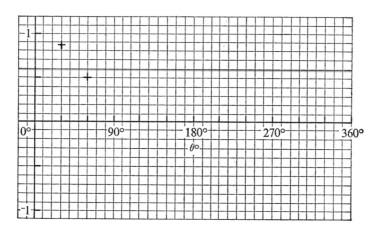

Fig. 29

Copy and complete the figure for a domain

$$0° \leqslant \theta° \leqslant 360° \text{ (that is, } \theta° \text{ from } 0° \text{ to } 360° \text{ inclusive).}$$

Would it be correct to join the points by a smooth curve? If you find this difficult to decide, try to answer these questions.

In Figure 28 should 37·25° have an arrow going to $S(37·25°)$?

Should 37·252° have an arrow to $S(37·252°)$?

Is the relationship you have drawn a *continuous* function?

We call the function $\theta° \rightarrow S(\theta°)$ the *sine* function, and $\theta° \rightarrow C(\theta°)$ the *cosine* function.

The *sine* of $\theta°$, written $\sin \theta°$, gives the *sideways* displacement (**BA** in Figure 27) of the *unit* displacement **OA** from the central direction.

113

The *cosine* of $\theta°$, written $\cos \theta°$, gives the *central* displacement of the *unit* displacement **OA** measured in the central direction (**OB** in Figure 27).

These statements are true whatever the angle and can be taken as the definitions of the sine and cosine functions.

Figure 30 shows both the functions drawn on the same grid.

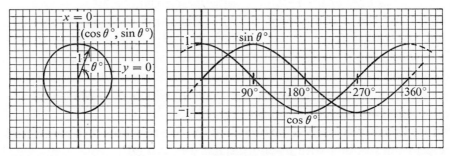

Fig. 30

Notice that the two curves are exactly the same shape, a translation from left to right mapping the cosine function onto the sine function.

Figure 31 shows the sine function drawn so that readings can be taken from it.

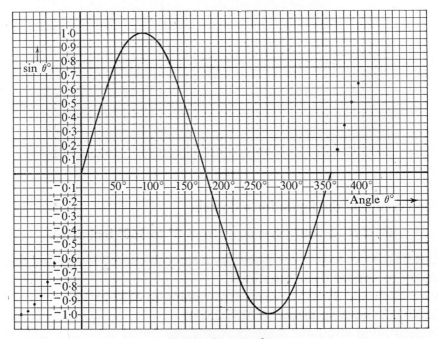

Fig. 31. $\theta° \rightarrow \sin \theta°$.

Using Figure 31 find three angles whose sine is 0·5. For how many angles between 0° and 360° does the sine function map onto the number 0·5? 0·87? If you choose any number between $^-1$ and 1 how many angles give this number under the sine function?

114

Exercise E

1. Use the graph in Figure 31 to find:

(a) the values of sin 25°, sin 200°, sin 315°, sin 72°;

(b) the angles whose sines are 0·75, 0·29, ⁻0·97.

2. (a) If θ is between 0 and 30, what can you say about sin θ°?

(b) Give the range of values for sin θ° when θ is between 180 and 270.

(c) What can you say about the value of θ if sin θ° is between 0 and 1?

3. Graph the function, $\theta° \to \cos \theta°$, for the domain from 0° to 360° on the same scale as Figure 31.

4. Use the graph you drew for Question 3 to find;

(a) the values of cos 15°, cos 67°, cos 135°, cos 284°;

(b) the angles whose cosines are 0·50, 0·85, 0·34, ⁻0·90.

5. (a) Assuming that x is between 0 and 90, solve the equations:

(i) $\sin x° = \cos 60°$;   (ii) $\sin x° = \cos 20°$;

(iii) $\cos x° = \sin 40°$;   (iv) $\cos 80° = \sin x°$;

(v) $\sin x' = \cos 23°$.

(b) What can you say about x if $\sin x° = \cos x°$?

5. MENSURATION

We now move on to consider how these functions can be used in dealing with problems which you may have solved in the past by scale drawings.

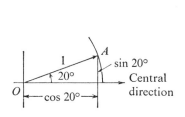
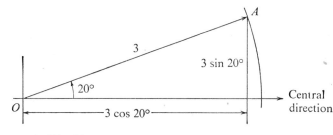

Fig. 32

(a) If the displacement is 3 units at an angle of 20° to a chosen central direction, it is 3 times the size of a unit displacement in the same direction, that is, in vector form, $\begin{pmatrix} 3 \cos 20° \\ 3 \sin 20° \end{pmatrix}$. (See Figure 32.)

115

Check that the following formulae for lengths in Figure 33 are correct.

$$e = 7 \cdot 8 \sin 60° \text{ cm.}$$
$$f = 7 \cdot 8 \cos 60° \text{ cm.}$$
$$k = 46 \cos 120° \text{ km.}$$
$$l = 46 \sin 120° \text{ km.}$$

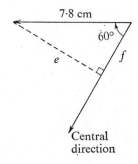 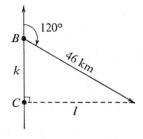

Fig. 33

Hence calculate e, f, k, and l to an accuracy of 2 s.f. with your slide rule. Is k negative? Suggest why this is and explain how you could calculate BC without using negatives at all.

How would you find $x°$ in Figure 34(*a*)?

Does Figure 34(*b*) help you?

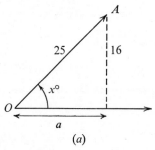 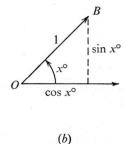

(*a*) (*b*)

Fig. 34

Since the triangle in Figure 34 (*a*) is an enlargement of that in Figure 34 (*b*) then $OA = 25\ OB$ and hence $16 = 25 \sin x°$

$$\Rightarrow \sin x° = \tfrac{16}{25} = 0 \cdot 64.$$
$$\Rightarrow x° \approx 40° \quad \text{(using Figure 31).}$$

What is the length of a?

Exercise F

In this exercise, use your slide rule to multiply whenever appropriate, and give all your answers to an accuracy of 2 s.f. Use 3-figure tables of sine and cosine functions in conjunction with Figure 31.

Bearings are measured clockwise from North; for example, 270° is due West.

1. (*a*) The cosine table gives cos 70° = 0·342. What is cos 110°?
   (*b*) The sine table gives sin 14° = 0·242. What is cos 76°?

2. Calculate the lengths denoted by small letters in Figure 35.

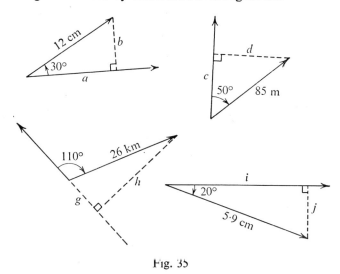

Fig. 35

3. Find the angles denoted by $x°$, $y°$, and $z°$ in Figure 36 and use these to calculate the lengths *a*, *b* and *c*.

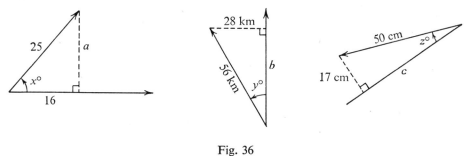

Fig. 36

4. A cliff lift is 120 m long and makes an angle of 30° with the vertical. How high does it rise?

5. Two paddle steamers left an east coast holiday resort and while one travelled 4 km on a bearing of 030° the other travelled 6 km on a bearing of 130°.
   (*a*) Find which steamer had travelled further eastwards and by how much.
   (*b*) How much further north is the first steamer than the second?

6. A mountain railway at its steepest section rises vertically through 68 m in 200 m of track length.
   (*a*) Through what height does a train rise in travelling 1 m along the track?
   (*b*) At what angle is the track to the horizontal?

7. The course of a ship after leaving port is as follows:

20 km on a bearing of 040°,

then   60 km on a bearing of 080°,

then 110 km on a bearing of 300°.

(*a*) Calculate:

    (i) the distance moved northwards on each leg of the course;

    (ii) the distance moved eastwards on each leg of the course:

(*b*) How far:

    (i) north,   (ii) east of the port is the ship at the end of the course?

Express each leg of the course as a vector.

8. Figure 37 shows a drawing of a barn roof.

(*a*) What is the width of the barn?

(*b*) What is the area of:

    (i) the two sloping faces of the roof;

    (ii) the floor?

(*c*) What would be the area of the floor if the pitch of the roof were increased to 60° (the roof area remaining the same)?

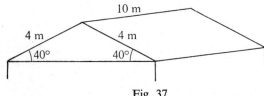

Fig. 37

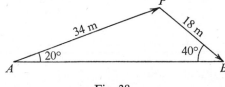

Fig. 38

9. In a cross-country race, a boy has to get from *A* to *B* but takes the path *APB* (see Figure 38) to avoid a muddy patch. How much further does he run by doing this?

10. Find values of *x* which satisfy the following equations:

(*a*) $9 = x \sin 30°$;   (*b*) $4 = 7 \sin x°$;   (*c*) $x = 5 \sin 68°$;

(*d*) $6\cdot4 = x \cos 50°$;   (*e*) $7\cdot5 = 15 \cos x°$;   (*f*) $x = 5\cdot8 \cos 70°$.

11. A small nail is picked up by the tyre of a bicycle wheel. If the diameter of the wheel is 0·66 m how high above the road is the nail when the wheel has turned through an angle of:

(*a*) 45°;   (*b*) 210°;   (*c*) 480°?

12. A ship's radar picks up the trace of an iceberg 5 km away on a bearing of 220°.

(*a*) How far is the ship

    (i) east,

    (ii) north of the iceberg?

(*b*) If the iceberg is stationary and the ship is sailing on a bearing of 230°, how near does the ship get to the iceberg?

Summary

When a displacement **AB** is made at an angle of $\theta°$ to a direction called the central direction, the displacement **AN** parallel to the central direction is called the central displacement. Similarly, the displacement **NB** is called the sideways displacement. These displacements depend on θ and the length AB.

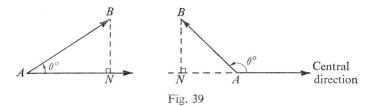

Fig. 39

The length of the sideways and central displacements of a unit displacement making an angle of $\theta°$ with the central direction are denoted by $\sin \theta°$ and $\cos \theta°$.

If R is the length of a displacement making an angle of $\theta°$ with some specified direction, then the central and sideways distances are respectively

$$R \cos \theta° \quad \text{and} \quad R \sin \theta°.$$

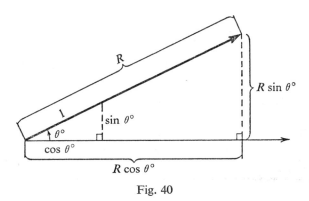

Fig. 40

The values of $\cos \theta°$ and $\sin \theta°$ can be found for any angle $\theta°$, by considering the Cartesian coordinates of the end-point of a unit displacement from the origin making an angle of $\theta°$ with the line $y = 0$. The way in which $\sin \theta°$ and $\cos \theta°$ change as θ changes can be illustrated graphically, and one obtains a curve known as a sine wave (or cosine wave). Curves of this type occur frequently in physics and engineering.

6

TRANSFORMATIONS I

1. REFLECTION

If you place a mirror beside a picture or drawing, the edge of the mirror acts as a line of symmetry between the picture and its reflection. Figure 1 shows an amusing example of this.

Fig. 1

We call the figure being reflected the *object* and its reflection the *mirror-image*. How do the object and mirror-image differ?

Of course, it is possible to have mirror-images without an actual mirror. A right-hand glove looks like a left-hand glove when it is reflected in a mirror and because of this we say that one glove is the mirror-image of the other. Give other examples of pairs of mirror-images which appear in everyday life.

Figure 2 shows the models of two molecules of lactic acid which are mirror-images of one another. In most ways these molecules behave identically but there are significant differences. For example, one kind may be consumed by certain micro-organisms, but not the other.

120

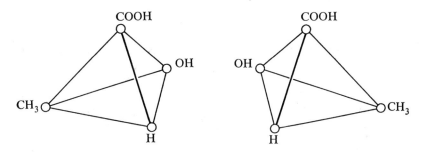

Fig. 2. Molecules of lactic acid.

In finding the mirror-image of an object it is convenient to use the property that the line segment PP' joining a point P of an object to its image P' is bisected at right-angles by the mirror (see Figure 3). The mirror line is the *mediator* of PP'.

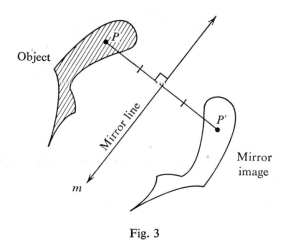

Fig. 3

The molecules of lactic acid are solid objects, but for the remainder of this section most of the questions will be about two-dimensional shapes.

To make our figures easier to understand we shall mark the line of a mirror by a double-ended arrow as in Figure 3 and we shall always assume that mirrors are two sided.

Exercise A

1. Copy Figure 4 and draw the images of the objects after reflection in the mirrors shown.

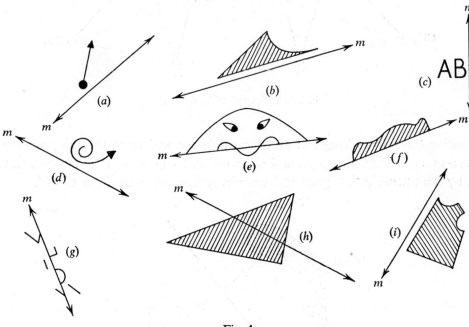

Fig. 4

2. Draw an object and a mirror line. Exchange your drawing with a neighbour's, and draw the mirror-image.

3. Write your initials in block capitals and draw mirror lines m_1 and m_2 as in Figure 5.

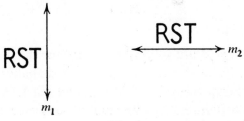

Fig. 5

(*a*) Draw the mirror-images in each case, preferably using different colours for the object and the image.

(*b*) Hold your drawing of the object and image up to a mirror. What do you notice?

(*c*) Write your initials using a pen and blot them before the ink dries. What do you notice about the image on the blotting paper?

(*d*) Write your initials on thin paper and look at them from the reverse side of the paper. What do you notice?

4. Figure 6 is the mirror-image of a boy's surname. What is the name?

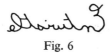

Fig. 6

5. The line with equation $x = 3$ is taken as the mirror line. The triangle PQR has its vertices at the points with coordinates (⁻1, ⁻1), (4, 4) and (3, ⁻3). What are the coordinates of the vertices of the mirror-image?

6. (a) When an object is reflected, are any (i) points, (ii) lines, left unchanged?
(b) A line of an object is at 24° to the line of the mirror. What can you say about the angle between the corresponding image line and the mirror? Draw a diagram to illustrate your answer.
Where do the two lines, produced if necessary, intersect?
(c) The angle between an object line and its corresponding image line is 60°. Describe the position of the mirror line.

7. Draw a line l and mark a point P not on l. Use compasses to draw circles which have their centres on l and pass through P. The circles all intersect at P. What other point do they have in common?
How could this question help you to construct a mirror-image of a point P?

8. Fred, the mathematical fly, walks around the circle in Figure 7 in a clockwise direction. Draw the path of his image in the mirror and indicate its direction.

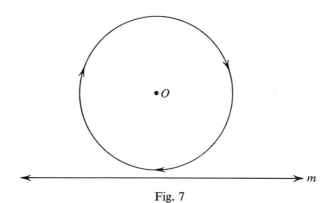

Fig. 7

9. Draw a circle and mark on its circumference three points A, B, and C not equally spaced. Take any other point P on the circumference of the circle and construct its mirror-image P_1 by taking the line through BC as the mirror line. (Question 7 gives a neat way of doing this.) Now construct:
(a) P_2, the image of P after reflection in the mirror line CA;
(b) P_3, the image of P after reflection in the mirror line AB.
What do you notice bout P_1, P_2 and P_3? Repeat the drawing with a new circle and a different set of points A, B, C and P. Does the same relation hold between the new set of images?

Summary

(a) If P' is the reflection in m of the point P, m is the mediator of PP'. P' is called the image of P in m. P will then be the image of P'.

Points of m are their own images.

(Q and Q' are coincident.)

(b) Lines parallel to m have images which are also parallel to m.

(c) Lines perpendicular to m are their own images, but directions along these lines are reversed by the reflection.

(d) m bisects the angle between any line and its image.

(e) Reflection preserves length and angle but does not preserve the direction of rotation.

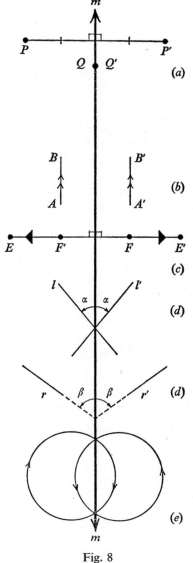

Fig. 8

2. ROTATION

In Chapter 1 you learned how to find the order of rotational symmetry of a figure by measuring the angle of the rotations which would map the figure onto itself. In this section we shall consider what happens to a shape when it is rotated about a fixed axis through a certain angle. To help you to discover these properties it will be useful to have a sheet of tracing paper or transparent plastic.

124

On a sheet of paper P draw 3 lines in different colours through a point O. Trace the lines onto a transparent sheet P' and use a pin or your compass point to make an axis at O perpendicular to the plane of the paper (see Figure 9(a)). Keep the paper P still and rotate the top sheet P' through a quarter-turn about O', the point corresponding to O.

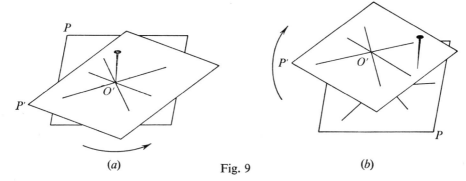

(a)   Fig. 9   (b)

(a) Compare each line with its original position.

Through what angle has each line turned?

Did any point(s) remain fixed in this rotation?

(b) Arrange the tracing to coincide with the original again and put your pin in any other place other than O (see Figure 9(b)). Rotate the sheet through a quarter-turn about this new axis. What angle has each line turned through now?

(c) Repeat the above experiment using different centres of rotation and rotating through angles other than quarter-turns. Use your protractor to measure the angles between lines and their images in each case.

What conclusion do you draw?

(d) Will P' be mapped onto the same position by a quarter-turn in an anticlockwise direction as it will by a quarter-turn in a clockwise direction?

(e) For what angle of rotation does the direction in which the rotation is made not matter?

Now that you are familiar with negative numbers we can use these to describe the direction of rotation. In science, engineering and mathematics it is the convention that an *anticlockwise* turn *is positive* while a *clockwise* turn *is negative*.

(For example, in 20 minutes the minute hand of a clock turns through $^-120°$.)

(f) What positive turn will have the same effect as a turn of $^-100°$?

Exercise B

1. (a) Cut out a triangle from a piece of card (see Figure 10) and, using a pin or a compass point as an axis of rotation, make separate drawings of the triangle to show the effect of turning it from its original position through angles of $60°$, $^-30°$, $105°$ and $180°$, about the different axes as shown in Figure 10.

125

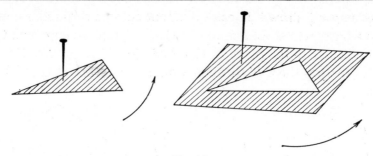

Fig. 10

(b) Measure the angles between the corresponding sides of the triangle in each case. What do you deduce?

Does it make any difference to your answer whether the axis of rotation is inside or outside the triangle? Can you suggest any centre(s) of rotation which might produce a different result?

2. (a) Make free-hand drawings on graph paper to show what happens to the line segment AB when it is rotated through a right-angle about the points C_1, C_2 and C_3 shown in Figure 11.

(b) Is $A'B'$ perpendicular to AB in each case?

(c) What can you say about the distance of the image of AB from C_3 under any rotation with C_3 as centre?

(d) If AB is turned through 180° about C_3, what can you say about the direction of $A'B'$?

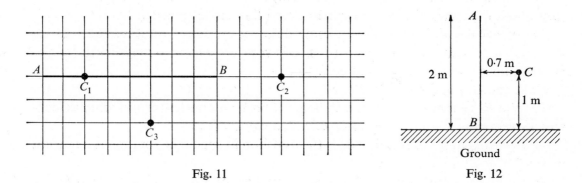

Fig. 11   Fig. 12

3. An 'up-and-over' garage door AB (seen edge-on in Figure 12) moves about a horizontal axis through C. If the door is 2 m high and C is 0·7 m from the door and 1 m above the ground, draw a sketch showing the position of the door when raised.

What height is the tallest vehicle which could use the garage?

4. On a piece of graph paper mark the points: $A(2, 1)$, $B(3, 4)$, $C(-2, 3)$, $D(-3, -1)$, $E(2, -2)$ and $F(0, 0·6)$. Trace these points and the lines $x = 0$ and $y = 0$. Put a pin through the origin and rotate the tracing paper through 180°.

(a) What are the coordinates of A', B', ..., F', the images of A, B, ..., F?

(b) If P has coordinates (h, k), what will be the coordinates of P'?

(c) Repeat (a) and (b) but rotate through an angle of 90°.

126

5. **R** is the operation of turning through $120°$ about the point A with coordinates $(2, 1)$. Use graph paper and tracing paper to find the image of the triangle LMN where L, M and N have coordinates $(4, 2)$, $(^-1, ^-2)$ and $(3, 0)$, respectively.

(*a*) What point on the graph paper is mapped onto the point $(3, 4)$ by **R**?

(*b*) Without measurement give the angle between the segments LM and $L'M'$.

(*c*) What is the path traced out by L in moving to L'?

6. (*a*) What is the path traced out by a point on a record-player turn-table as a record is played?

(*b*) Cut out any shape from a piece of paper. Put a pin through it at some point A and make a small hole for a pencil point to go through at another point, P. Rotate the cut-out and mark the path of P with the pencil (see Figure 13).

(*c*) What is the path of P?

(*d*) If P_1 and P_2 are two points on the path of P, what can you say about AP_1 and AP_2?

(*e*) If Q is any other point on the paper what can you say about the path of Q when the cut-out is rotated?

(*f*) What angle does PQ turn through when the cut-out is rotated through $137°$?

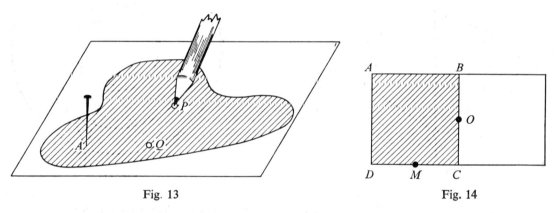

Fig. 13   Fig. 14

7. Copy Figure 14. If the shaded square is rotated about O through $180°$ it will map onto the unshaded square.

(*a*) Draw in the path of: (i) C, (ii) D, (iii) M.

(*b*) What other centres of rotation could be used to map one square onto the other? Give the centres and the angles of rotation.

(*c*) What other transformations would map one square onto the other?

2.1   Centres of rotation

You have discovered what happens to a figure when it is rotated about a fixed axis through a given angle. We now want to find a way of obtaining the angle of rotation and the position of the centre of rotation when the initial and final positions of a figure are known.

Fig. 15

Suppose that A' (see Figure 15) is the image of A after a rotation. O, the point midway between A and A', is one of the centres of rotation which could be used to map A onto A'.

(a) What would the angle of rotation be in this case?

(b) Find centres of other rotations which would map A onto A'.

How many possible centres are there?

How can you describe the set of all possible centres?

(c) If O_1 is one of the possible centres you find in (b), what can you say about the lengths of O_1A and O_1A'?

You will have seen from answering the above questions that any point P on the mediator of AA' is a possible centre of rotation (see Figure 16).

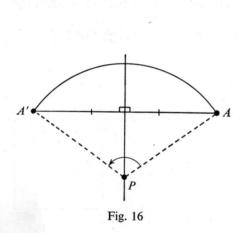

Fig. 16

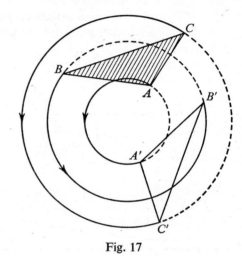

Fig. 17

Figure 17 shows the initial and final positions of a triangle under a rotation about a fixed centre (the circles indicate the paths taken by A, B and C).

The problem is to find the centre of rotation, O.

(d) Make a copy of triangles ABC and $A'B'C'$ by pricking through. On what line must O lie if:

(i) A is mapped onto A';

(ii) B is mapped onto B'?

Draw these lines carefully.

(e) The point P at which these lines intersect will be a possible centre of a rotation mapping A onto A' and B onto B'. Will the angle required to rotate A onto A' be the same as that required to rotate B onto B'?

(f) What can you say about the mediator of CC' if P is the centre of a rotation which maps C onto C'? Draw in the mediator of CC' and measure $\angle CPC'$.

In finding centres of rotation you can make use of the above work by drawing the mediators of object points and their images. As all the mediators pass through O, the centre of rotation, it is sufficient to find the intersection of any two mediators.

128

A mediator is easily drawn accurately by using compasses to draw two equal circles, one with a centre at the object point and one with its centre at the image point (see Figure 18). The radius chosen must be large enough for the circles to intersect: the line through the points of intersection is the mediator of AA'.

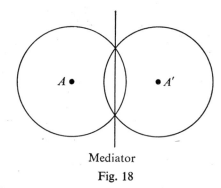

Mediator

Fig. 18

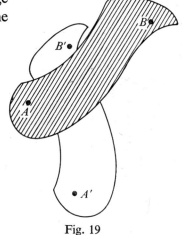

Fig. 19

Exercise C

1. Figure 19 shows a shape (shaded) and its image (unshaded) after a rotation.

(*a*) Make a tracing of the figure and, by drawing the mediators of AA' and BB', find the centre of rotation O.

(*b*) Which is the smallest (i) positive angle of rotation, (ii) negative angle of rotation, which will map A onto A' and B onto B'?

2. Each of the parts of Figure 20 shows an object (shaded) and its image after a rotation.

(*a*) Estimate the positive angle of rotation in each case.

(*b*) Copy the diagrams accurately by pricking through and mark in the centres of rotation. Only draw the mediators when you cannot find the centre without their help.

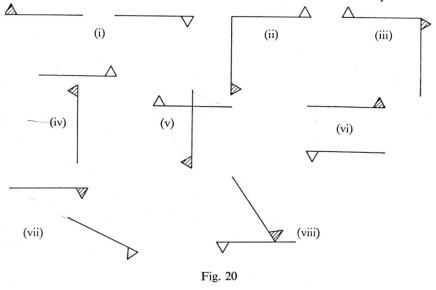

Fig. 20

129

3. Work with your neighbour, each draw two outlines of an object. Exchange drawings. Find the centre of rotation. When is it impossible to find the centre of rotation?

4. In Figure 21 are four congruent shapes.

(a) Why is it impossible to map S onto S_1 by a rotation in the plane of the paper?

(b) Can you rotate S_2 onto S_3? If not, why not?

(c) Make a copy of Figure 21 and find the centre and angle of rotation to: (i) map S onto S_2; (ii) map S onto S_3.

(Note that it is only necessary to copy the positions of 2 points from each of S, S_2 and S_3 and this can be quickly done by pricking through the page with a compass point.)

(d) Describe transformations which will map:

    (i) S onto S_1;

    (ii) S_2 onto S_3.

Fig. 21

5. Figure 22 gives a bird's eye view of the position, at two different moments, of a lorry being driven around a roundabout.

(a) Find the centre of the roundabout.

(b) If the length of the lorry is 7 m, what is the approximate radius of the roundabout?

Fig. 22

6. The vertices of two squares $ABCD$ and $PQRS$ are given by the two sets of coordinates: $\{(-3, 1), (1, 1), (1, -3), (-3, -3)\}$ and $\{(-1, 3), (3, 3), (3, -1), (-1, -1)\}$.

Draw the squares on graph paper.

(a) Find the centre and angle of rotation to map:

    (i) $ABCD$ onto $QRSP$;

130

    (ii) *ABCD* onto *SPQR*;

    (iii) *ABCD* onto *RSPQ*.

(*b*) (i) Draw in the mirror line of a reflection which would map one square onto the other. What is its equation?

    (ii) What is the image of *C* under this reflection?

(*c*) What motion would map *ABCD* directly onto *PQRS*?

7. The points *H*(2, 4) and *K*(⁻1, 1) are mapped onto the points *H'*(0, ⁻1) and *K'*(3, ⁻4) by a rotation.

(*a*) What are the coordinates of the centre of rotation?

(*b*) What is the image of the point (0, 0)?

(*c*) Which point is mapped onto (⁻3, ⁻4)?

Summary

1. Rotation takes place in a plane about an axis through a given centre *O* and through a given angle $\theta°$. The new figure is called the image of the old figure.

2. A rotation is positive when anticlockwise. A positive rotation through an angle $\theta°$ is equivalent to a negative rotation through an angle $360° - \theta°$ about the same centre (see Figure 23 (*a*)).

3. The centre is the only fixed point under a rotation. Every other point moves on the arc of a circle about this centre.

4. Every line in a figure turns through the same angle — the angle of rotation (see Figure 23 (*b*)).

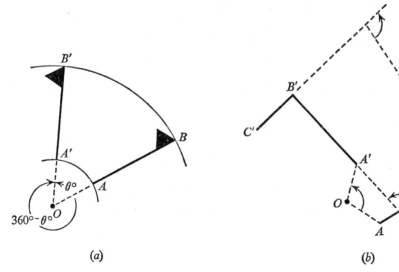

(*a*)   (*b*)

Fig. 23

3. COMBINED TRANSFORMATIONS

You have now met three kinds of geometric transformation: translation, reflection and rotation. These three transformations all have the property that they will move a figure from one position to another without changing its shape or size. For this reason they are called *isometries* (Greek: *isos*-equal, *metron*-measure).

There is, however, one important difference between the image of a figure after reflection and the image after either translation or rotation. What is it? (If you are in doubt, try the effect of each of the three transformations on a letter F.)

We now consider the effect of combining two or more isometries. Look at Figure 24. It is evident that the simplest transformation that will map triangle 1 ($\triangle$1 for short) onto $\triangle$2 is a 180° rotation or *half-turn* about the point (5, 3). Is there any other simple transformation that will do the same job? What happens if *two* transformations are combined?

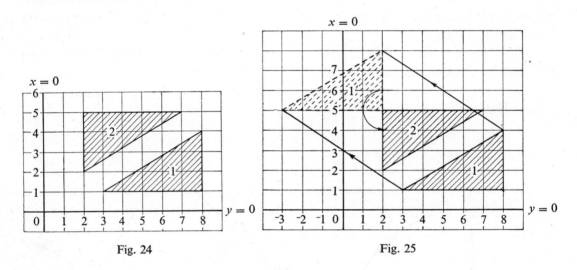

Fig. 24   Fig. 25

It is plain that there are many *pairs* of transformations that will map $\triangle$1 onto $\triangle$2. For instance, the translation described by the vector $\begin{pmatrix} -6 \\ 4 \end{pmatrix}$, followed by a half-turn about the point (2, 5) will do this (see Figure 25). What translation will be needed to complete the transformation if we start with a half-turn about the point (3, 1)? If we start with a translation of $\begin{pmatrix} -6 \\ 1 \end{pmatrix}$, is there a rotation which will then map the image of $\triangle$1 onto $\triangle$2? If so, describe it. Discuss whether the same pairs of transformations will work if applied the other way round (that is, is their combination commutative?).

132

Exercise D

1. What is the difference between a transformation and a translation?

Question 2 refers to Figure 26.

2. (*a*) What single transformation will map: (i) $\triangle 3$ onto $\triangle 7$; (ii) $\triangle 3$ onto $\triangle 8$?

(*b*) $\triangle 3$ is given the translation $\begin{pmatrix} 10 \\ 0 \end{pmatrix}$. Describe the second transformation which will map it (i) onto $\triangle 7$; (ii) onto $\triangle 8$.

(*c*) $\triangle 4$ is reflected in the line $x = 0$. State with reasons whether its image can be translated onto (i) $\triangle 8$; (ii) $\triangle 7$.

(*d*) Describe a pair of translations which will map $\triangle 7$ onto $\triangle 4$. Try to find two further pairs. How many such pairs are there? Do you need the diagram to help you describe them?

(*e*) $\triangle 8$ can be mapped onto $\triangle 3$ by means of a pair of rotations. If the first one maps $\triangle 8$ onto $\triangle 7$, describe each rotation. Will the same rotations, performed in the reverse order, have the same effect?

(*f*) What translation must be given to $\triangle 4$ so that its image coincides with the image of $\triangle 7$ under the translation $\begin{pmatrix} 2 \\ 6 \end{pmatrix}$?

Use your answer to find a pair of translations under which $\triangle 7 \rightarrow \triangle 4$.

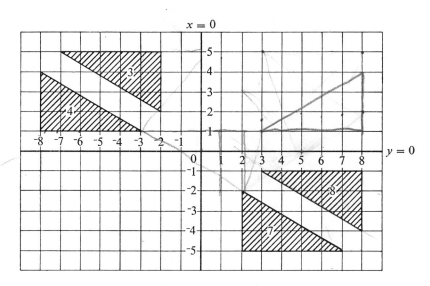

Fig. 26

D3. Is it *ever* possible for:

(*a*) a rotation followed by a reflection to be equivalent to a translation;

(*b*) a translation followed by a rotation to be equivalent to a reflection;

(*c*) a reflection followed by another reflection to be equivalent to a rotation?

Question 4 *refers to Figure* 27.

4. (*a*) What single transformation, if any, will map:

    (i) △2 onto △7;   (ii) △2 onto △1;
    (iii) △2 onto △8?

(*b*) △2 is given a rotation of ⁻90° about *O*. Describe the second transformation needed to map its image onto △7.

(*c*) △8 is given a half-turn about (6, 1). Describe if possible a second transformation which will map its image: (i) onto △7; (ii) onto △2.

(*d*) △2 is reflected in its longest side. Discuss which further triangles its image can now be mapped onto by a single rotation and how you would find the centre of such a rotation.

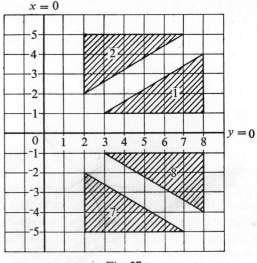

Fig. 27

Summary

An isometry maps a figure onto another figure that is identical as far as lengths and angles are concerned. Such figures are called *congruent*. The congruence may be

        ·   (*a*) direct,   or   (*b*) opposite (see Figure 28).

         (*a*) Direct

         (*b*) Opposite

Fig. 28

The isometries we have met are:

$$\left.\begin{array}{r}\text{translations}\\ \text{rotations}\end{array}\right\} \to \text{direct isometries,}$$

        reflections   → opposite isometries.

Given two congruent figures there is always a sequence of rotations, reflections and translations which will map one onto the other.

3.1   The algebra of transformations

It is very useful to have a shorthand way of denoting transformations. We shall use capital letters in bold type. Of course, we have to make sure that we have defined their meaning precisely. Draw the lines $x = 0$, and $y = 0$ on squared paper.

Draw the flag F with its shaft joining the points (2, 2) and (2, 4) (see Figure 29) and with the flag pointing to the right, as shown. You will find it useful to have a tracing of F on a piece of tracing paper or transparent plastic sheet.

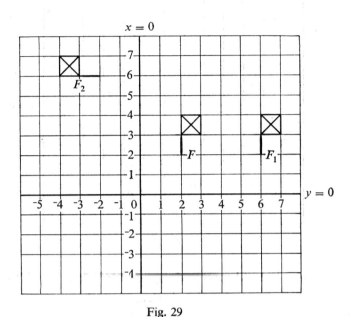

Fig. 29

Let us denote a rotation of 90° about (0, 0) by **P**. (In hand-written work we would denote it by $\underset{\sim}{\text{P}}$.)

The image of F under this will be denoted by **P**(F). Mark **P**(F) on your squared paper.

Let us denote the translation $\binom{4}{0}$ by **T**. If we now carry out the transformation **T** on **P**(F), we denote the resulting image by **TP**(F)—an abbreviation for **T**(**P**(F)). Mark this on your squared paper. What single transformation maps F onto **TP**(F)?

Copy F_1 and F_2, two images of F shown in Figure 29, onto your paper. (Neither of them should coincide with **P**(F) or **TP**(F).) Describe the single transformations that map F onto: (a) F_1; (b) F_2. How could you map F_1 onto F_2 by a single transformation?

Sometimes we are interested only in the transformation and not in the particular figure which is being transformed. We then write the letter only, without anything in brackets after it. Describe again the single transformation that has the same result as the combined transformation **TP**. Call this **R**. Discuss the meaning of the sentence **TP** = **R**, taking particular care to explain the precise meaning of the relation ' = ' in this context.

135

Describe the transformations **J** and **K** if the images **J**(*F*) and **KJ**(*F*) are as shown in Figure 30. Describe **L** if **KJ** = **L**. What do you find about **JK**(*F*)?

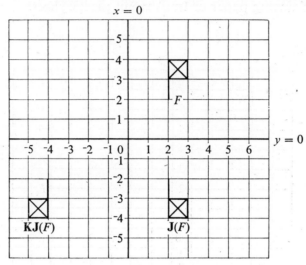

Fig. 30

Summary

We use capital letters in bold type to denote transformations. If **A** and **B** are transformations and *P* is a figure or an object, then:

(*a*) **A**(*P*) denotes the image of *P* under the transformation **A**;

(*b*) **BA**(*P*) denotes the image of **A**(*P*) under the transformation **B**.

The object may be omitted. **BA** = **C** means that transformation **B** applied *after* transformation **A** is equivalent to the single transformation **C**.

Exercise E

Questions 1–3 refer to Figure 31. Use tracing paper to help find the answers. △1 refers to triangle numbered 1 and so on.

1. (*a*) Let **A** denote a half-turn about (5, 3), **B** denote a half-turn about (5, ⁻3) and **C** denote the translation $\begin{pmatrix} 10 \\ 6 \end{pmatrix}$.

State which of the following are true, which false:

    (i) **A**(△1) = △2;   (ii) **A**(△2) = △1;

    (iii) **C**(△5) = △1;   (iv) **C**(△4) = **A**(△8).

(*b*) With the same data state which of the following are true, which false:

    (i) **AC**(△6) = △2;   (ii) **BC**(△6) = △2;   (iii) **CB**(△7) = **A**(△3);

    (iv) **AB**(△6) can be mapped onto △1 by a single translation.

(*c*) Let **X** denote reflection in the line *x* = 0, **Y** denote reflection in the line *y* = 0, and let **A**, **B** and **C** have their previous meanings. Copy and complete the following:

    (i) **X**(△4) =   ;   (ii) **Y**( ) = △2;

    (iii) **XB**( ) = △5;   (iv) (△2) = △8.

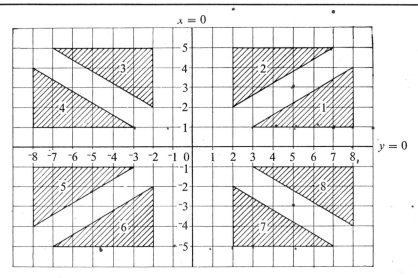

Fig. 31

2. (*a*) *Powers of a transformation.* If **C** still has the same meaning, what does **CC** mean? What will be the coordinates of the right-angled vertex of the image of $\triangle 1$ under the transformation **CCC**? Suggest a shorter notation for **CC** and **CCC**. What do you think **C⁴** should mean?

(*b*) With the above notation state which of the following are true, which false:

(i) $A^2(\triangle 2) = \triangle 8$;   (ii) $X^2(\triangle 4) = Y^2(\triangle 4)$;   (iii) $C^2(\triangle 6) = C(\triangle 1)$.

3. *The identity transformation.* This is the 'transformation' which moves nothing. Some call it the 'stay-put' transformation.

(*Note*: 'identical' means 'exactly the same'.)

We denote this transformation by the letter **I**. Which of the following describe the identity transformation for the pattern in Figure 31?

(*a*) The translation $\begin{pmatrix} 0 \\ 0 \end{pmatrix}$.

(*b*) Rotation through 180° about the pattern's centre of symmetry.

(*c*) Rotation through 360° about the pattern's centre of symmetry.

(*d*) Rotation through 360° about a vertex of one of the triangles.

(*e*) Reflection in one of the two lines of symmetry.

4. Let **X** denote reflection in $x = 0$, **Y** denote reflection in $y = 0$ and **H** denote a half-turn about O. Simplify the following:

(*a*) H^2;   (*b*) XY;   (*c*) HX;   (*d*) H^4;   (*e*) $X^2Y^2H^2$.

5. If **R** is *any* operation what can you say about **RI**?

Simplify: (*a*) IR;   (*b*) RI^2;   (*c*) I^4R^2.

6. Let **L** denote the translation $\begin{pmatrix} 4 \\ 5 \end{pmatrix}$ and **M** the translation $\begin{pmatrix} -3 \\ 6 \end{pmatrix}$. Describe translations **E** and **F** such that:

(*a*) $LM = E$;   (*b*) $LF = M$.

Find **G** given that $EFG = I$.

7. Let M_1 denote reflection in $x = 1$, M_2 denote reflection in $x = 2$, and so on. Let P be the point $(2, 2)$. Mark on a diagram the points P, $M_1(P)$, $M_2(P)$, $M_3(P)$. Discuss whether $M_2 = I$. If k is a number other than 1, is it possible to find k such that $M_1 = M_k$?

8. Let Q denote a rotation through a quarter-turn, H denote a rotation through a half-turn and T denote a rotation through a three-quarter-turn, the centre of rotation in all three cases being $(0, 0)$. If I is the identity transformation, complete the following operation table for the set $\{Q, H, T, I\}$ under the operation 'following'. Is the set closed under the operation?

'Following'	(First transformation) Q	H	T	I
(Second transformation) Q	H			
H				
T			T	
I				

9. An equilateral triangle can be mapped onto itself by rotations of 120° and 240° about its centre. Suppose:

I denotes the identity transformation,
P denotes a rotation of 120° about the centre of the triangle,
Q denotes a rotation of 240° about the centre of the triangle.

Copy and complete the table for the operation 'following'. Is the set $\{I, P, Q\}$ closed under this operation?

'Following'	(First transformation) I	P	Q
(Second transformation) I			
P		Q	
Q			

4. INVERSE TRANSFORMATIONS

If an object F is mapped by a one-to-one transformation A onto the image F', then there will always be a second transformation which will map F' onto F. This is called the *inverse* of A. It is written A^{-1}. Since transforming and then returning to the starting position is equivalent to 'staying put', we see that

$$A^{-1}A = I.$$

What can you say about AA^{-1}?

Example 1

Let R be the translation described by the vector $\binom{2}{3}$. Find:

(a) R^3;   (b) R^{-1};   (c) R^{-3}.

(a) $R^3 = RRR$, that is

$\binom{2}{3}$ following $\binom{2}{3}$ following $\binom{2}{3}$. This is clearly $\binom{6}{9}$.

138

(b) $\mathbf{R}^{-1}$ brings you back to your starting point after $\binom{2}{3}$, it is therefore $\binom{-2}{-3}$.

(c) $\mathbf{R}^{-3}$ has not been defined, but a sensible meaning would be 'the transformation which combined with $\mathbf{R}^3$ brings you back to your starting point'. It is therefore $\binom{-6}{-9}$.

Note that it is also $(\mathbf{R}^{-1})^3$, that is, $\binom{-2}{-3}$ following $\binom{-2}{-3}$ following $\binom{-2}{-3}$.

Summary

The identity or stay-put transformation leaves every point of a figure unchanged. It is denoted by $\mathbf{I}$.

Given a transformation $\mathbf{A}$, the transformation which undoes the effect of $\mathbf{A}$ is called the *inverse* of $\mathbf{A}$. It is written $\mathbf{A}^{-1}$.

$$\mathbf{A}\mathbf{A}^{-1} = \mathbf{A}^{-1}\mathbf{A} = \mathbf{I}.$$

Exercise F

1. If $\mathbf{T}$ denotes 'translate across the page from left to right through 2 units', describe:

(a) $\mathbf{T}^2$;   (b) $\mathbf{T}^3$;   (c) $\mathbf{T}^1$;   (d) $\mathbf{T}^{-1}$;   (e) $\mathbf{T}^{-3}$.

What do you think $\mathbf{T}^0$ should mean?

2. Let $\mathbf{S}$ denote the translation $\binom{2}{3}$ and $\mathbf{T}$ the translation $\binom{-1}{1}$.

Write down the coordinates of the points onto which $(0, 0)$ is mapped under the transformations:

(a) $\mathbf{ST}$;   (b) $\mathbf{TS}$;   (c) $\mathbf{T}^2$;   (d) $\mathbf{S}^{-1}$;
(e) $\mathbf{T}^{-2}$;   (f) $\mathbf{TST}$;   (g) $\mathbf{STS}^{-1}$.

3. Let $\mathbf{H}$ denote a half-turn about $(0, 0)$; let $\mathbf{M}$ denote reflection in the line $y = x$.
(a) What is the transformation $\mathbf{H}^{-1}$? Is there only one possibility?
(b) What is the transformation $\mathbf{M}^{-1}$? Is there only one possibility?
(c) Give the coordinates of the images of the point $(2, 3)$ under the transformations:
(i) $\mathbf{H}^{-1}$;   (ii) $\mathbf{H}^{-1}\mathbf{H}$;   (iii) $\mathbf{HH}^{-1}$;   (iv) $\mathbf{H}^{-3}$;   (v) $\mathbf{M}^{-2}$;
(vi) $\mathbf{M}^{-3}$;   (vii) $\mathbf{M}^{-1}\mathbf{H}^{-1}$;   (viii) $\mathbf{H}^{-3}\mathbf{M}^{-1}$.

4. Let $\mathbf{X}$ denote the translation $\binom{0}{1}$ and $\mathbf{Y}$ the translation $\binom{1}{0}$.

Write down the coordinates of the image of the point $(1, 1)$ under the transformations:

(a) $\mathbf{X}^2$;   (b) $\mathbf{Y}^3$;   (c) $\mathbf{X}^{-2}$;   (d) $\mathbf{XYXY}$;   (e) $\mathbf{X}^2\mathbf{Y}^2$;
(f) $\mathbf{X}^{-3}\mathbf{Y}^3$;   (g) $\mathbf{X}^4\mathbf{Y}$;   (h) $\mathbf{X}^{-4}\mathbf{Y}^{-5}$;   (i) $\mathbf{X}^a\mathbf{Y}^b$.

139

5. Let A represent a rotation of $45°$ about $(0, 0)$. Say which of the following statements are true and which are false:

(a) A^{-1} could be a negative $\frac{1}{8}$ turn about $(0, 0)$;
(b) A^{-1} could be a rotation of $315°$ about $(0, 0)$;
(c) $A^{-1} = A^{-9}$;
(d) $A^{-1} = A^7$;
(e) $A^{-9} = A^{-17}$;
(f) $A^{-1} A^9 = I$.

6. Draw a letter L by joining the points $(2, 6)$, $(1, 2)$ and $(5, 1)$.

(a) Write down the single translation that will map it onto the letter L joining $(-3, 3)$, $(-4, -1)$ and $(0, -2)$.
(b) Write down a pair of translations that will effect the same translation as that in (a).
(c) Write down a triple of translations that will also effect the same transformation.

(d) Write down the coordinates of the image of the first L under the translation $\binom{5}{5}$.

(e) If the first letter L is translated so that $(1, 2)$ maps onto $(3, -2)$, find the images of the other vertices.

5. FURTHER REFLECTIONS

Both the following sections are investigations leading to important results about reflections.

5.1   Reflections in parallel lines

In Figures 32 the flag has been reflected in parallel lines m_1 and m_2. The flag P and three of its images have been numbered 1, 2, 3, 4. We shall write M_1 to denote reflection in the line m_1 and M_2 to denote reflection in the line m_2.

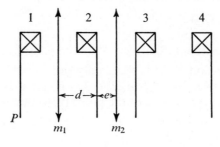

Fig. 32

(a) Copy Figure 32 fairly closely. Which distances have to be equal to those marked d and e? Take care over the position of 4. Label the appropriate images $M_1(P)$, $M_2(P)$, $M_1M_2(P)$, $M_2M_1(P)$. You will have to draw one further image. Which one? Label it 5.

140

(*b*) The images of objects under reflections are, of course, congruent to each other. The original *P* has the flag on the right of the pole, so does one of the images in Figure 32 (which?). *P* and this image are directly congruent. The other images have the flag on the left of the pole. These are oppositely congruent to *P*.

What single transformation will map 3 onto oppositely congruent 4? Will the same transformation map 4 onto 3? Give it an appropriate letter and suffix. What happens if you perform this transformation twice on 3? What is the inverse of this transformation?

(*c*) What single transformation will map 1 onto its directly congruent image 3? Is there any connection between the distance needed to describe this transformation and the distances *d* and *e*?

Specify exactly the single transformation that will map 1 onto its directly congruent image 5. What is the distance between the images involved?

Are there more images of *P* than those you have drawn? How many?

(*d*) Repeated reflection in two parallel lines produces a pattern that can be used for a frieze as shown in Figure 33. Copy this diagram, or better, make up a frieze of this type for yourself starting with your own basic pattern (or *motif*). How many parallel mirror-lines need there be? Can you put them where you like? Mark them on your diagram. It is obvious that this pattern can also be produced by repeated translations. What is the connection between the distance between the mirrors and the distance through which the motif has to be translated?

Fig. 33

(*e*) The point (−2, 0) is mapped onto (5, 0) after being reflected in a line and then reflected again in a parallel line. Find possible equations of the lines. How many pairs are there? How will they lie with respect to the points (that is, are they both between the points, or one between them and one outside, etc.)?

(*f*) Make a general statement about the combination of two reflections in parallel lines.

5.2 Reflections in intersecting lines

(*a*) Copy Figure 34. Let **P** denote reflection in *p* and **Q** denote reflection in *q*. *S* denotes the sword shown. (Note that the important part of it is the straight line representing the blade and handle, the hilt and guard are optional extras and can be added afterwards, freehand!)

Construct **P**(*S*) and **QP**(*S*). What single transformation will map *S* onto **QP**(*S*)? To specify it precisely you will need to measure an angle. What relation does this angle appear to have to the angle between *p* and *q*?

Construct **Q**(*S*) and **PQ**(*S*) as well. What single transformation maps *S* onto **PQ**(*S*)? Does the angle involved appear to have any relation to the angle you measured previously? Do **PQ** and **QP** map *S* onto the same image?

What is the connection between all this and Figure 30 (p. 136)? In that example you noted that **JK** = **KJ**. Can you see why **J** and **K** commute whereas **P** and **Q** do not?

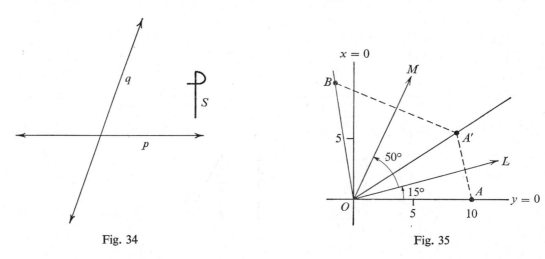

Fig. 34   Fig. 35

(*b*) In Figure 35 *OL* makes an angle of 15° with *y* = 0.

If *A′* is the reflected image of *A* (10, 0) in *OL* what is the angle *LOA′*?
OM is another mirror line such that ∠ *LOM* = 50°. If *B* is the reflected image of *A′* in *OM* what is ∠*MOB*? What is ∠*AOB*? How is its value related to the angle between the two mirror lines?

(*c*) Copy Figure 35. Make it large enough to have plenty of room to label angles. With the data of (*b*) above, take the point *A*(10, 0) and reflect it in *OM*, and then reflect its image in *OL* to give *D*. Find ∠*AOD*. What connection has this with ∠*AOB*?

(*d*) In Figure 36 the figure *P* has been reflected in *two* straight lines *v* and *w* respectively and its image is *P′*. Do you think the lines *v* and *w* are parallel? Copy Figure 36 and construct the mirror-line *v* of the reflection which maps *A* onto *A′*. Construct also the image of *P* under this reflection and discuss how you can now reflect it onto the final position P'. Hence construct the line w. Are these the only possible positions for lines v and w? Use tracing paper to construct the fixed point O of the rotation which maps P onto P'. What connection has O with v and w?

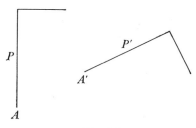

Fig. 36

(*e*) If M_1, M_2 are reflections in two intersecting lines, describe the transformations $M_1M_2M_2M_1$ and $(M_1M_2)^{-1}$. If $M_1M_2 = M_2M_1$, what can you say about the lines?

(*f*) Discuss the following *theorem*.

The product of reflections in two intersecting lines m_1 and m_2, in that order, is a rotation about their point of intersection through twice the angle from m_1 to m_2.

Discuss also how it can be proved and what special cases arise.

Summary

The following tables sum up what has been discovered so far about the way in which pairs of transformations combine to be equivalent to single transformations.

TABLE I. TRANSLATIONS

		Second transformation	
		Translation through t_2	Translation through $^-t_1$
First transformation	Translation through t_1	Translation through t_1+t_2	Identity

TABLE 2. REFLECTIONS

		Second transformation			
		Reflection in same line m	Reflection in line parallel to m	Reflection in line perpendicular to m	Reflection in line making an angle $\theta°$ with m
First transformation	Reflection in line m	Identity	Translation through twice the distance between lines	Half-turn about point of intersection of lines	Rotation through $2\theta°$ about point of intersection of lines

7

MATRICES

Man is a tool-using animal—without tools he is nothing, with tools he is all.

CARLYLE, Sartor Resartus

1. INFORMATION MATRICES

1.1 Relation matrices

The new boys in form $3S$ have joined various school clubs, and the relation 'is a member of' is illustrated in Figure 1.

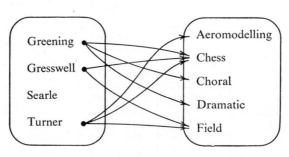

Fig. 1

This type of diagram can become confused if there is a large number of arrows, and an alternative way of displaying the information is in a table as follows:

	Aeromodelling	Chess	Choral	Dramatic	Field
Greening	0	1	1	1	0
Gresswell	.	.	.	.	.
Searle	.	.	.	.	.
Turner	.	.	.	.	.

This table of numbers arranged in a rectangular pattern is an example of a *matrix*. (The brackets can be thought of as 'parcelling up' the individual numbers— or *elements*—into 'one' matrix.)

(a) Copy and complete the above matrix from the information given in Figure 1.

(b) Which boy belongs to the most clubs?

(c) Which club has the most new boys from form 3S in it?

(d) Does the matrix have any symmetrical features?

(e) Can the information be represented graphically or by a Venn diagram? Which method (arrow diagram, matrix, graph or Venn diagram) do you consider the most effective?

The matrix that you have completed in (a) illustrates the relation 'is a member of' between the set of boys (the domain), which is listed at the side of the matrix, and the set of clubs (the range), which is listed along the top of the matrix.

1.2 Data matrices

On many occasions a large quantity of numerical data has to be stored for use, and often this data can most conveniently be arranged in matrix form.

A nurseryman's catalogue offers the following collections of spring bulbs. The 'complete' collection P contains 50 crocus, 25 daffodils, 10 hyacinths, 25 narcissi and 15 tulips. The 'shade' collection Q contains 40 crocus, 25 daffodils and 25 narcissi. The 'Green Fingers' collection R contains 30 crocus, 10 daffodils, 15 hyacinths and 20 tulips. This information can be displayed in matrix form as below, where the contents of an R collection have been entered.

	Crocus	Daffodils	Hyacinths	Narcissi	Tulips
P collection	.	.	.	.	.
Q collection	.	.	.	.	.
R collection	30	10	15	0	20

(a) What is the significance of the 0 in the bottom row?

(b) Is collection R 'related' to narcissi bulbs? State, in words, the relation between the collections, and the types of bulb.

(c) Copy and complete the matrix.

In this example we are interested not so much in whether or not a collection is related to a certain type of bulb, but in *how many* daffodils, for example, are contained in a Q collection. The idea of the relation is still present, but not in an obvious way.

A matrix that tabulates (in any way) the information about a relation between two sets is called an *Information Matrix*. In particular, a matrix (such as the one showing membership of school clubs) that answers the question 'Is x related to y?' by either 'yes' (1) or 'no' (0) is often called, simply, a *Relation Matrix*.

1.3 Route matrices

The mainline termini in London are connected by the network of Underground lines shown in Figure 2. (Intermediate stations have been left out of the diagram, and for the purpose of this example Euston and Euston Square have been considered as one station.)

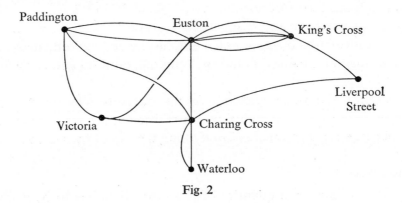

Fig. 2

(*a*) Can you go direct from Paddington to Liverpool Street without passing through any of the other stations shown in the diagram?

(*b*) Can you go direct from Paddington to Euston in more than one way? If so, in how many ways?

(*c*) In how many ways can you travel direct from Euston to Paddington?

(*d*) Copy and complete the following matrix, which lists the answers to such questions as these.

	to						
from	CC	Euston	KC	Liv. St	Paddington	Victoria	Waterloo
CC							
Euston							
KC							
Liv. St							
Paddington	1	2	0	0	0	1	0
Victoria							
Waterloo							

(*e*) Does this matrix have any symmetrical features?

(*f*) What important differences are there between this matrix and those in Sections 1.1 and 1.2?

An information matrix such as this is also called a *route matrix*. It lists the information for the relation 'is directly connected to' on a set of stations. (Notice that the set is related to itself.) Such matrices and their networks will be considered more fully in Chapter 15.

Exercise A

1. Construct a matrix similar to that in Section 1.1 for a domain consisting of yourself and three or four of your friends and with a range that is a set of interests such as soccer, pop music, scouts, James Bond novels, chess, model-making, etc.

2. Five shops hold the following stock of pop records. Shop *A* has 60 LPs, 87 EPs and 112 singles; shop *B* has 103 LPs, 41 EPs and 58 singles; shop *C* stocks only LPs and has 72 of them; shop *D* is selling out, and has 23 EPs and 12 singles left; and shop *E* specializes in EPs and singles and has 275 records in stock, of which 157 are singles. List the stocks in matrix form.

3. A firm that manufactures radio parts sells a number of 'do-it-yourself' kits for the amateur constructor. One kit, called the 'Beginner's Bijou', has 3 valves, 2 coils, 1 speaker, 7 resistors and 5 capacitors; another, the 'Straight Eight', has 8 valves, 6 coils, 2 speakers, 25 resistors and 24 capacitors; while the 'Super Sistor' has 6 transistors, 8 coils, 1 speaker, 23 resistors, and 16 capacitors. Tabulate this information in matrix form.

4. Susan, Bridget and Jennifer decide to make their own dresses for a party. Susan's pattern needs 4 m of velvet, $3\frac{1}{2}$ m of binding, a 35 cm zip-fastener and 3 buttons. Bridget's has a separate jacket; her dress and jacket need $1\frac{1}{2}$ m of velvet, $4\frac{1}{2}$ m of nylon, 2 m of binding, a 20 cm zip-fastener and no buttons. Jennifer favours a shift, and she requires 2 m of velvet, $1\frac{1}{2}$ m of binding and no zip-fasteners or buttons. State their requirements in matrix form.

5. Compile the route matrix for the section of the London Underground shown in Figure 3.

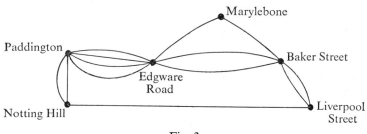

Fig. 3

147

2. MATRIX ADDITION

The marks obtained by two boys in the first and second halves of a term respectively are displayed in the following matrices.

	English	Maths	Science
John Black	52	47	55
Bill White	48	58	61

	English	Maths	Science
Black	54	62	59
White	60	53	57

It is convenient to have a convention which enables us to describe the shape of a matrix. Each of the above matrices has two *rows* and three *columns*; and we say that the *order* of these matrices is 2×3 (read as 'two by three').

(*a*) Write down a 2×3 matrix listing the total marks for the term (in each subject) for the two boys. To do this you will have to add together the numbers in the corresponding positions of the matrices for the first and second halves of the term. This combination of two matrices of the same order (by adding corresponding elements) to form a third matrix is known as *matrix addition*. What is the order of the *sum* of these two matrices?

(*b*) What is the order of **M**, and what is **M**, if

$$\begin{pmatrix} 3 & 2 & 5 & 0 \\ 6 & ^-1 & 4 & 2 \end{pmatrix} + \mathbf{M} = \begin{pmatrix} 6 & 3 & 5 & 4 \\ 7 & 2 & 3 & ^-6 \end{pmatrix}?$$

We use a capital letter in bold type to denote a matrix, and use the sign $+$ to denote matrix addition.

(*c*) Write down the orders of the following matrices. Which, if any, can be added together?

$$\mathbf{A} = \begin{pmatrix} 1 & 3 & 2 \\ 5 & 1 & 6 \end{pmatrix}; \quad \mathbf{B} = \begin{pmatrix} 2 & 1 & 4 \\ 0 & ^-1 & 0 \\ 1 & 3 & ^-5 \end{pmatrix}; \quad \mathbf{C} = \begin{pmatrix} 6 & 0 \\ 1 & ^-2 \\ 3 & 5 \end{pmatrix}; \quad \mathbf{D} = \begin{pmatrix} 5 \\ 7 \\ 0 \end{pmatrix};$$

$$\mathbf{E} = \begin{pmatrix} ^-4 & 6 \\ 0 & 1 \end{pmatrix}; \quad \mathbf{F} = (3 \ 8 \ 1 \ 4); \quad \mathbf{G} = (7); \quad \mathbf{H} = \begin{pmatrix} 0 & 1 \\ 1 & 2 \\ ^-3 & ^-4 \end{pmatrix}.$$

(*d*) Which of the matrices in (*c*) would you call *square*?

(*e*) A matrix of the form

$$\begin{pmatrix} 5 \\ 7 \end{pmatrix}$$

has already been met in Chapter 5. What name was given to it in that chapter?

Similarly, a vector could be a matrix containing just one row. Is it reasonable to consider vectors as special cases of matrices?

The addition of two vectors is again an application of matrix addition, since the law of combination is the same. For example

$$\begin{pmatrix} 3 \\ 2 \end{pmatrix} + \begin{pmatrix} -5 \\ 6 \end{pmatrix} = \begin{pmatrix} -2 \\ 8 \end{pmatrix}.$$

Exercise B

1. The four houses of a co-educational school are named after the planets Mars, Pluto, Saturn and Jupiter. The way in which the first-form boys and girls were allocated to these houses is described in the following matrices:

	Form 1A			Form 1B			Form 1C	
	M P S J			M P S J			M P S J	

Boys $\begin{pmatrix} 3 & 2 & 5 & 5 \\ 4 & 6 & 2 & 4 \end{pmatrix}$; Boys $\begin{pmatrix} 5 & 4 & 2 & 1 \\ 4 & 3 & 5 & 5 \end{pmatrix}$; Boys $\begin{pmatrix} 3 & 6 & 3 & 4 \\ 3 & 2 & 6 & 5 \end{pmatrix}$.
Girls   Girls   Girls

(a) Calculate from these matrices a 2×4 matrix describing the allocation of boys and girls by houses in the first form as a whole.

(b) Which house has the largest number of (i) boys; (ii) girls; (iii) first-formers?

2. A milkman sells three kinds of milk; red top, silver top and gold top.

As he makes his way down Topsham Road he notes his deliveries to the first four houses as follows:

	No. 1	No. 2	No. 3	No. 4
Gold	$\begin{pmatrix} 0 \\ 2 \\ 1 \end{pmatrix}$	$\begin{pmatrix} 2 \\ 0 \\ 0 \end{pmatrix}$	$\begin{pmatrix} 1 \\ 1 \\ 1 \end{pmatrix}$	$\begin{pmatrix} 2 \\ 1 \\ 2 \end{pmatrix}$
Silver				
Red				

(a) What single matrix represents his total deliveries to these four houses?

(b) What matrix represents the amount of milk bought weekly (including Saturdays and Sundays) by the housewife at No. 1 if her daily order remains the same throughout the week?

3. Let $\qquad A = \begin{pmatrix} 2 & 5 \\ 0 & 1 \end{pmatrix}, \quad B = \begin{pmatrix} 3 & 1 \\ 2 & 5 \end{pmatrix}$ and $C = \begin{pmatrix} 1 & -1 \\ 2 & 0 \end{pmatrix}$

(a) Calculate:

(i) $A+B$;   (ii) $B+C$;   (iii) $C+A$.

(b) Calculate:

(i) $2A$ (i.e. $A+A$);   (ii) $A+B+C$;   (iii) $3C$;   (iv) $2A+B$;   (v) $B-C$.

(c) Does (i) $A+B = B+A$;   (ii) $A+(B+C) = (A+B)+C$?

Will these results hold for any three matrices A, B and C, provided they are each of the same order?

4. Let $\qquad P = \begin{pmatrix} 3 & 2 \\ 5 & -1 \\ 4 & 0 \end{pmatrix}, \quad Q = \begin{pmatrix} 5 & 7 \\ -6 & 4 \\ 2 & -9 \end{pmatrix}$ and $O = \begin{pmatrix} 0 & 0 \\ 0 & 0 \\ 0 & 0 \end{pmatrix}.$

Find a matrix R such that $P+Q+R = O$. (O is described as a *zero matrix*.)

5. If

$$\begin{pmatrix} 3 & 2 & 5 \\ 1 & 4 & 6 \end{pmatrix} + \begin{pmatrix} a & b & c \\ d & e & f \end{pmatrix} = \begin{pmatrix} 4 & a & b \\ c & d & e \end{pmatrix},$$

find the values of a, b, c, d, e and f.

3. MATRIX MULTIPLICATION

In Section 1.2 we had the matrix listing the contents of three collections of bulbs

	Crocus	Daffodils	Hyacinths	Narcissi	Tulips
Collection P	50	20	10	25	15
Collection Q	40	25	0	25	0
Collection R	30	10	15	0	20

Mr Jones orders $1P$ collection, $2Q$ collections and $3R$ collections; and Mr Smith $3P$'s and $1Q$. These requirements can be displayed in the matrix

	P	Q	R
Mr Jones	1	2	3
Mr Smith	3	1	0

Note that this matrix relates customers to collections, and the first matrix relates collections to types of bulbs.

(*a*) How many crocus bulbs will Mr Jones have altogether? To answer this question we shall need a part of each of the above matrices. Which parts?

From the requirements matrix we take the row representing Mr Jones's order (1   2   3) which tells us how many of each collection he has bought.

This is combined with the column

$$\begin{pmatrix} 50 \\ 40 \\ 30 \end{pmatrix}$$

from the contents matrix, which tells us how many crocus bulbs there are in each P, Q and R collection.

From one P collection he gets	50 bulbs (1×50)
from two Q collections	80 bulbs (2×40),
and from three R collections	90 bulbs (3×30),
in all, a total of	220 bulbs $(50+80+90)$.

(*b*) This method of combining a row and a column is called *multiplication* and is written

$$(1 \quad 2 \quad 3) \begin{pmatrix} 50 \\ 40 \\ 30 \end{pmatrix} = (220)$$

Note that (usually) *no* sign is put between the two matrices to be multiplied, and that the *product* is written as a matrix.

If a row containing five elements is to be multiplied by a column, how many elements must there be in the column? What will be the order of the product matrix in this case?

(c) In the same way, the number of crocus bulbs obtained by Mr Smith is given by the multiplication

$$(3 \quad 1 \quad 0) \begin{pmatrix} 50 \\ 40 \\ 30 \end{pmatrix} = (190).$$

(d) These two calculations have something in common. What is it?

If we wanted to do both calculations together we can set out the work as follows:

$$\begin{matrix} (1 & 2 & 3) \\ (3 & 1 & 0) \end{matrix} \begin{pmatrix} 50 \\ 40 \\ 30 \end{pmatrix} = \begin{matrix} (220) \\ (190). \end{matrix}$$

But the two rows on the left make up the original requirements matrix, and it is usual to set out the work as

$$\begin{pmatrix} 1 & 2 & 3 \\ 3 & 1 & 0 \end{pmatrix} \begin{pmatrix} 50 \\ 40 \\ 30 \end{pmatrix} = \begin{pmatrix} 220 \\ 190 \end{pmatrix}.$$

We now have the multiplication of a 2×3 matrix and a 3×1 matrix. What is the order of the result?

(e) Similarly, the numbers of daffodils obtained by each is given by the multiplication

$$\begin{pmatrix} 1 & 2 & 3 \\ 3 & 1 & 0 \end{pmatrix} \begin{pmatrix} 20 \\ 25 \\ 10 \end{pmatrix} = \begin{pmatrix} 100 \\ 85 \end{pmatrix}.$$

If we wish to know the numbers of crocus *and* daffodils obtained by each, this calculation can be linked with that in (d) to give

$$\begin{pmatrix} 1 & 2 & 3 \\ 3 & 1 & 0 \end{pmatrix} \begin{pmatrix} 50 & 20 \\ 40 & 25 \\ 30 & 10 \end{pmatrix} = \begin{pmatrix} 220 & 100 \\ 190 & 85 \end{pmatrix}.$$

Can you still see where the number 85 has come from? What information does it tell us?

(f) The rest of the multiplication of the two original matrices follows in the same manner, resulting in a matrix which lists for both customers how many of each type of bulb he has. Copy and complete the calculation started below. (As a help, the row and column enclosed in the dots are those which are multiplied to give the number of narcissi bought by Mr Smith.)

$$
\begin{array}{cc}
& P\ Q\ R \\
\text{Mr Jones} & \begin{pmatrix} 1 & 2 & 3 \\ 3 & 1 & 0 \end{pmatrix} \\
\text{Mr Smith} &
\end{array}
\begin{array}{c}
P \\ Q \\ R
\end{array}
\begin{array}{ccccc}
\text{crocus} & \text{daffodils} & \text{hyacinths} & \text{narcissi} & \text{tulips} \\
\begin{pmatrix} 50 \\ 40 \\ 30 \end{pmatrix} & \begin{matrix} 20 \\ 25 \\ 10 \end{matrix} & \begin{matrix} 10 \\ 0 \\ 15 \end{matrix} & \begin{matrix} 25 \\ 25 \\ 0 \end{matrix} & \begin{matrix} 15 \\ 0 \\ 20 \end{pmatrix} \end{matrix}
\end{array}
$$

$$
=
\begin{array}{c}
\text{Mr Jones} \\ \text{Mr Smith}
\end{array}
\begin{array}{ccccc}
\text{crocus} & \text{daffodils} & \text{hyacinths} & \text{narcissi} & \text{tulips} \\
\begin{pmatrix} 220 \\ 190 \end{pmatrix} & \begin{matrix} 100 \\ 85 \end{matrix} & \begin{matrix} \cdot \\ \cdot \end{matrix} & \begin{matrix} \cdot \\ \cdot \end{matrix} & \begin{matrix} \cdot \\ \cdot \end{pmatrix} \end{matrix}
\end{array}
$$

This method of combining two matrices is known as *matrix multiplication*, the final matrix being the *product* of the first two.

What connection is there between the orders of the three matrices?

What connection is there between the domains and ranges of the three relations involved in these three matrices?

If we denote the first matrix by **A** and the second by **B**, the product is denoted by **AB**, and we say that we have multiplied **B** on the left by **A** (or that we have premultiplied **B** by **A**).

3.1 Compatability

(a) If **R** is a 5×3 matrix and **S** is a 3×2 matrix, what is the order of **RS**?

(b) If the matrix $\mathbf{A} = \begin{pmatrix} 3 & 5 & 2 \\ 1 & -1 & 4 \end{pmatrix}$ can be multiplied on the left by a matrix **B**, why must **B** have only two columns? Is there any restriction on the number of rows in **B**?

(c) If the matrix **A** above can be multiplied on the right by a matrix **C** (that is, if a matrix **C** can be multiplied on the left by **A**), what can you say about the number of rows and columns of **C**?

$$
\begin{bmatrix} \text{Customers} \rightarrow \\ \quad\quad \text{collections} \end{bmatrix}
\begin{bmatrix} \text{Collections} \rightarrow \\ \quad\quad \text{types of bulb} \end{bmatrix}
\longrightarrow
\begin{bmatrix} \text{Customers} \rightarrow \\ \quad\quad \text{types of bulb} \end{bmatrix}
$$

$$
[2 \times 3]\ [3 \times 5] \qquad\qquad \longrightarrow \qquad\qquad [2 \times 5]
$$

Fig. 4

You will have seen from answering these questions that two matrices can be multiplied only if the matrix on the left has as many columns as the matrix on the right has rows. When this is the case, the matrices are said to be *compatible* for multiplication. If the orders of the matrices involved are written out as in Figure 4, then the numbers encircled must be the same for compatibility, and the outer figures give the order of the product. (This is sometimes known as the 'domino rule'.) Note also that the same pattern exists between the domains and ranges of the relations that are involved in the matrices.

(d) Let $\quad \mathbf{P} = \begin{pmatrix} 2 & 5 \\ 0 & 1 \end{pmatrix}, \quad \mathbf{Q} = \begin{pmatrix} 3 & 1 \\ 2 & 4 \\ 0 & 5 \end{pmatrix} \quad \text{and} \quad \mathbf{R} = \begin{pmatrix} 1 & 3 & 5 \\ -1 & 2 & 4 \end{pmatrix}.$

(i) Which of **PQ** and **QP** are possible products? Work out the one that is.
(ii) Repeat for **PR** and **RP**.
(iii) Are both **QR** and **RQ** possible products?
　　　Are they the same?

In general, even if two matrices **A** and **B** can be multiplied 'both ways round' the resulting products are not the same.

Can you find two 2×2 matrices **J** and **K** such that **JK** = **KJ**?

Exercise C

1. Calculate: (a) $(1 \quad 3 \quad -2) \begin{pmatrix} 4 \\ 6 \\ 5 \end{pmatrix}$; (b) $(3 \quad -1) \begin{pmatrix} -2 \\ -10 \end{pmatrix}$; (c) $(5 \quad 0) \begin{pmatrix} 0 \\ -6 \end{pmatrix}$;

(d) $(5 \quad 2 \quad 3 \quad 4) \begin{pmatrix} 1 \\ 0 \\ -1 \\ 6 \end{pmatrix}$; (e) $(3 \quad 0 \quad 1 \quad 4) \begin{pmatrix} 1 \\ 1 \\ 1 \\ 1 \end{pmatrix}$.

2. In an athletics match between Town School and the County School each school had two competitors in every event, and the individual results were as follows:

	1sts	2nds	3rds	4ths
Town School	(8	3	6	7)
County School	(4	9	6	5)

(a) Add these two rows together and explain the result.
(b) Why is the sum of the elements in each row 24?
(c) The home team, Town School, arranged 5 points for a 1st, 3 points for a 2nd, 1 point for a 3rd and nothing for a 4th.
Express this points scheme as a column matrix and calculate each team's score. Which team won?
(d) Afterwards the captain of the County School team felt that they had been unlucky to lose, so he worked out the match score on the basis of 4 points for a 1st, 3 for a 2nd, 2 for a 3rd and 1 for a 4th. What did he discover?

3. Let $\mathbf{U} = (2 \quad 3)$, $\mathbf{V} = (-1 \quad 4)$ and $\mathbf{W} = \begin{pmatrix} 3 \\ -1 \end{pmatrix}$.

(a) Calculate **UW**, **VW** and **U**+**V**, and show that **UW**+**VW** = (**U**+**V**)**W**.
(b) Show similarly that (3**U**+5**V**)**W** = 3**UW**+5**VW**.

4. Find the value of x if:

(a) $(3 \quad x) \begin{pmatrix} 5 \\ 1 \end{pmatrix} = (18)$;   (b) $(x \quad 7) \begin{pmatrix} 4 \\ x \end{pmatrix} = (22)$;

(c) $(^-2 \quad x \quad 4) \begin{pmatrix} x \\ 3 \\ 5 \end{pmatrix} = (15)$.

5. Calculate the following products:

(a) $(2 \quad 3) \begin{pmatrix} 5 & 0 & 1 \\ 1 & 4 & 2 \end{pmatrix}$;   (b) $\begin{pmatrix} 5 & 0 & 1 \\ 1 & 4 & 2 \end{pmatrix} \begin{pmatrix} 1 \\ 2 \\ 1 \end{pmatrix}$;   (c) $(2 \quad 3) \begin{pmatrix} 5 & 0 & 1 \\ 1 & 4 & 2 \end{pmatrix} \begin{pmatrix} 1 \\ 2 \\ 1 \end{pmatrix}$;

(d) $(1 \quad 3 \quad 4) \begin{pmatrix} 2 & 5 \\ 0 & 3 \\ 1 & 2 \end{pmatrix}$;   (e) $\begin{pmatrix} 3 & 0 & 1 \\ 5 & 7 & 0 \\ -1 & 0 & 2 \end{pmatrix} \begin{pmatrix} 5 \\ 1 \\ 2 \end{pmatrix}$;   (f) $\begin{pmatrix} 3 & 1 & 0 & 1 \\ 2 & 0 & 5 & -2 \end{pmatrix} \begin{pmatrix} 3 \\ 2 \\ 2 \\ 5 \end{pmatrix}$.

6. A factory produces three types of portable radio sets called Audio 1, Audio 2 and Audio 3. Audio 1 contains 1 transistor, 10 resistors and 5 capacitors; Audio 2 has 2 transistors, 18 resistors and 7 capacitors; and Audio 3 has 3 transistors, 24 resistors and 10 capacitors. Arrange this information in matrix form and find the factory's weekly consumption of transistors, resistors and capacitors if its weekly output of sets is 100 Audio 1's, 250 Audio 2's and 80 Audio 3's.

7. 'Bildit' is a constructional toy with standard parts called flats (F), pillars (P), blocks (B), rods (R) and caps (C). It is boxed in sets, numbered 1 to 4. Set 1 has $1F$, $4P$, $8B$, $14R$ and $2C$; set 2 has $2F$, $10P$, $12B$, $30R$ and $4C$; set 3 has $4F$, $24P$, $30B$, $60R$ and $10C$; and set 4 has $10F$, $40P$, $72B$, $100R$, and $24C$.

(a) Tabulate this information in a 4×5 matrix.

(b) The manufacturer gets an order for 20 sets of 1, 25 of set 2, 10 of set 3 and 6 of set 4. Write this order as a matrix in an appropriate form, and hence, by arranging a suitable multiplication, find the total numbers that will be required of each part.

(c) If flats cost 6p each, pillars 4p, blocks 1p, rods 2p and caps 3p, write down a price matrix, and find the cost (in new pence) of each type of set.

(d) Hence find the total cost of the order that the manufacturer received. Express this cost as a product of three matrices.

8. Let $A = \begin{pmatrix} 0 & 1 & 0 \\ 0 & 0 & 1 \\ 1 & 0 & 0 \end{pmatrix}$, $B = \begin{pmatrix} 0 & 0 & 1 \\ 1 & 0 & 0 \\ 0 & 1 & 0 \end{pmatrix}$, $C = \begin{pmatrix} 1 & 0 & 0 \\ 0 & 1 & 0 \\ 0 & 0 & 1 \end{pmatrix}$, $X = \begin{pmatrix} 2 \\ 4 \\ 6 \end{pmatrix}$

and $Y = (1 \quad 3 \quad 5)$.

(a) Find the products AX, BX, CX, YA, YB and YC. What do you notice about the elements of the products?

(b) Find a matrix D for which $D \begin{pmatrix} 2 \\ 4 \\ 6 \end{pmatrix} = \begin{pmatrix} 2 \\ 6 \\ 4 \end{pmatrix}$.

Work out the product YD.

(c) Find a matrix I such that

$$(1 \quad 3 \quad 5) I = (1 \quad 3 \quad 5).$$

Work out the product IX.

154

9. Calculate the following products:

(a) $\begin{pmatrix} 2 & 5 \\ 3 & 2 \end{pmatrix} \begin{pmatrix} 1 & 0 \\ 2 & -1 \end{pmatrix}$;
(b) $\begin{pmatrix} 4 & 3 \\ 0 & -1 \end{pmatrix} \begin{pmatrix} 1 & 2 & 0 \\ 0 & 3 & 4 \end{pmatrix}$;
(c) $\begin{pmatrix} 1 & 2 & 0 \\ 0 & 3 & 4 \end{pmatrix} \begin{pmatrix} 7 & 2 & 0 \\ 0 & 3 & 1 \\ 1 & 4 & 1 \end{pmatrix}$;

(d) $\begin{pmatrix} 4 & 3 \\ 0 & -1 \end{pmatrix} \begin{pmatrix} 1 & 2 & 0 \\ 0 & 3 & 4 \end{pmatrix} \begin{pmatrix} 7 & 2 & 0 \\ 0 & 3 & 1 \\ 1 & 4 & 1 \end{pmatrix}$;
(e) $(3 \quad 1 \quad 5) \begin{pmatrix} 7 & 2 \\ 1 & 4 \\ 0 & 1 \end{pmatrix} \begin{pmatrix} 3 & 0 \\ 0 & 2 \end{pmatrix}$.

10. In a triangular athletics match between the schools A, B and C points were awarded as follows: 6 for a 1st; 4 for a 2nd; 3 for a 3rd; 2 for a 4th and 1 for a 5th. The individual results are shown in the following matrix:

	1sts	2nds	3rds	4ths	5ths
School A	4	2	3	2	6
School B	2	5	4	4	3
School C	4	3	3	4	1

An alternative points scheme of 8 for a 1st; 5 for a 2nd; 3 for a 3rd; 2 for a 4th; and 1 for a 5th was proposed. Express the two points schemes in a 5×2 matrix and find the positions in which the schools finished under each scheme.

11. (a) What is the order of matrix $\mathbf{B}$ if

$$(3 \quad 4 \quad 2)\,\mathbf{B} = (2 \quad 1 \quad 0 \quad 3 \quad 6)?$$

(b) If $\mathbf{P}$ is a 3×4 matrix and $\mathbf{Q}$ is a matrix such that both products $\mathbf{PQ}$ and $\mathbf{QP}$ are possible, what is the order of $\mathbf{Q}$?

(c) $\mathbf{X}$ is a matrix of order 3×2, $\mathbf{Y}$ is of order 2×2 and $\mathbf{Z}$ is of order 2×3. Write down two products which will give matrices of order 2×2. Make up your own matrices $\mathbf{X}$, $\mathbf{Y}$ and $\mathbf{Z}$ to see whether these two products are the same.

12. $\mathbf{A} = \begin{pmatrix} 3 & 0 \\ 0 & 3 \end{pmatrix}$, $\mathbf{B} = \begin{pmatrix} 0 & 2 \\ -1 & 1 \end{pmatrix}$ and $\mathbf{C} = \begin{pmatrix} 1 & 2 \\ 3 & 4 \end{pmatrix}$.

(a) Work out $\mathbf{BC}$ and $\mathbf{A(BC)}$; and $\mathbf{AB}$ and $\mathbf{(AB)C}$. Does $\mathbf{A(BC)} = \mathbf{(AB)C}$?
(b) Does $\mathbf{AB} = \mathbf{BA}$? Does $\mathbf{BC} = \mathbf{CB}$?

13. Matrices of the form $\begin{pmatrix} a & 0 \\ 0 & b \end{pmatrix}$, where a and b can be any numbers, are called P matrices,

and matrices of the form $\begin{pmatrix} 0 & y \\ z & 0 \end{pmatrix}$, again where y and z can be any number, are called Q

matrices. What can you say about the products of:
(a) a P matrix and a P matrix;
(b) a P matrix and a Q matrix (does the order matter?);
(c) a Q matrix and a Q matrix?

Summary

1. A *matrix* is a rectangular array of numbers or *elements*, and is denoted in brief by a bold capital letter.

$$\text{For example,} \quad \mathbf{A} = \begin{pmatrix} 3 & 0 & ^-6 & 3 \\ 2 & 2 & 10 & 1 \\ 5 & 1 & ^-7 & 4 \end{pmatrix}.$$

2. An *information matrix* contains information about the relation between the elements of a domain (listed at the side of the matrix) and the elements of the range (listed at the top). If the matrix is stating only whether there is a relation or not, by answering 'yes' (1), or 'no' (0), then it is specifically known as a *relation matrix*.

3. The matrix above has 3 *rows* (the domain of the relation that it illustrates has 3 elements), and 4 *columns* (the range of the relation has 4 elements), and the matrix is said to have *order* 3×4. A square matrix has the same number of rows as columns.

4. Two matrices may be *added* only if they have the same order (and the matrix *sum* will also have this order). The combination is carried out by adding together corresponding elements. For example, if

$$\mathbf{B} = \begin{pmatrix} 3 & 2 & 0 \\ 0 & 1 & ^-5 \end{pmatrix} \quad \text{and} \quad \mathbf{C} = \begin{pmatrix} ^-1 & 4 & 6 \\ 2 & 7 & 1 \end{pmatrix}, \quad \text{then the sum is} \quad \mathbf{B+C} = \begin{pmatrix} 2 & 6 & 6 \\ 2 & 8 & ^-4 \end{pmatrix}.$$

$$\text{Also, for example,} \quad 3\mathbf{B} = \begin{pmatrix} 9 & 6 & 0 \\ 0 & 3 & ^-15 \end{pmatrix}.$$

5. Two matrices may be *multiplied* only if the first has as many columns as the second has rows. When this is so, the matrices are said to be *compatible* for multiplication. This is shown in the 'domino rule', which also shows the order of the *product*. For example, for the matrices $\mathbf{A}$ and $\mathbf{B}$ above, $\mathbf{B}$ and $\mathbf{A}$ are compatible for multiplication (in this order) and we have

$$\mathbf{B\,A} \quad = \quad \mathbf{D} \quad \text{say,}$$
$$[2 \times 3].[3 \times 4] \rightarrow [2 \times 4]$$

(Notice that $\mathbf{A}$ and $\mathbf{B}$ are not compatible in the order $\mathbf{A\,B}$).

6. In a matrix product, each element is the combination of a row from the left-hand matrix and a column from the right-hand matrix. For example, the element in the second row and third column of $\mathbf{D}$ is the result of combining the second row of $\mathbf{B}$ with the third column of $\mathbf{A}$.

$$\begin{pmatrix} \cdot & \cdot & \cdot \\ 0 & 1 & ^-5 \end{pmatrix} \begin{pmatrix} \cdot & \cdot & ^-6 & \cdot \\ \cdot & \cdot & 10 & \cdot \\ \cdot & \cdot & ^-7 & \cdot \end{pmatrix} = \begin{pmatrix} \cdot & \cdot & \cdot & \cdot \\ \cdot & \cdot & 45 & \cdot \end{pmatrix},$$

since $\quad (0 \times ^-6) + (1 \times 10) + (^-5 \times ^-7) = 45.$

8

GEOMETRICAL SETS

So many paths that wind and wind.

ELLA WHEELER WILCOX, *The World's Need*

1. LOCUS

(*a*) Three boys lit a firework and immediately ran off at the same speed in different directions. Figure 1 shows their positions when the firework exploded.

Puzzle: Where did the firework explode?

To solve this we must find a point that is the same distance from each of *A*, *B* and *C*.

First of all, consider *A* and *B*. Copy Figure 1 and use your compasses to mark several points which are the same distance from *A* as from *B*. If you could mark all such points, what would the set of points form? Suppose we call this set of points *p*.

B

C

A

Fig. 1

If *q* is a set of points that are the same distance from *B* as from *C*, what can you say about *p* ∩ *q*?

Construct *p* and *q* and hence solve the puzzle.

The latin word *locus* (strictly meaning 'place' or 'position') is often used instead of the phrase 'set of points'. In the puzzle above we would say 'the locus of a point that is the same distance from A as from B is the *mediator* of AB', and so on.

It is important to realize that a locus need not necessarily be a line or a point; it can be a curve, or a region, or a solid object. Describe a circle in a sentence beginning, 'A circle is a set of points ...'. Look carefully at your description. Could it also apply to the surface of a sphere? We have to be very careful to explain what the universal set, $\mathscr{E}$, is in each case.

(*b*) If the three boys lighting a firework had been Martian boys, equipped with wings, lighting a firework (or something stronger) somewhere on a TV transmission tower, could we have found out where the firework was?

$\mathscr{E}$ here is the whole of space. Consider first A and B. The locus $\{P : AP = BP\}$ is the infinite plane π_1 shown in Figure 2. Describe the locus π_2 in set language. What is $\pi_1 \cap \pi_2$? Does this fix the position of the firework? What further information do we need? If the TV tower is the set of points T, describe the position of the firework in set language.

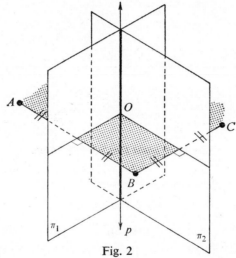

Fig. 2

(*c*) Sometimes loci in two dimensions have to be drawn by finding a number of positions of the point under consideration. If appropriate, these may then be joined together by a straight line or a smooth curve; otherwise, if they can be seen to occupy a region, the boundaries are identified and the region shaded in. We then try to find the name of the curve or region involved by comparing it with known curves or regions.

The final step in investigating a locus is to try to ensure that your impression of its nature is correct. A demonstration by experiment can only suggest the truth of something and can never be said to prove it beyond all doubt. We shall leave such proofs until later books.

Exercise A

In this exercise, *A*, *B*, *C*, ... are always fixed points, while *P*, *Q*, *R*, ... are variable ones.

1. In Figure 3, *p* represents part of a line, *q* represents part of a half-line, and *r* represents a line segment *AB*. Copy each part of the figure separately and sketch the locus of a point in the plane of the page which is always 1 cm from (*a*) the line *p*; (*b*) the half-line *q*; (*c*) the line segment *r*.

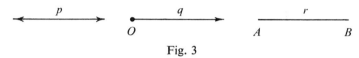

Fig. 3

Describe also these three loci if the point is now able to be in 3-D space.

2. (*a*) Figure 4 shows two half-lines, *p* and *q*, and a point *Z* that is equidistant from them. Copy the figure and mark on it by eye four or five further possible positions for the point *Z*. What does the locus of *Z* appear to be?

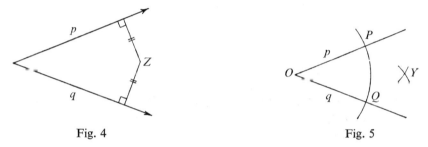

Fig. 4   Fig. 5

(*b*) In Figure 5 the point *Y* has been constructed as follows. The point of a pair of compasses was put at *O* and an arc drawn cutting *p* and *q* at *P* and *Q*. Then with centres *P* and *Q*, two equal arcs were drawn cutting at *Y*. What can you say about the symmetry of the figure as a whole? If *OY* is joined, what can you say about any point on *OY* in relation to *p* and *q*? Describe *OY* as a locus. Why is it called the bisector of ∠*POQ*?

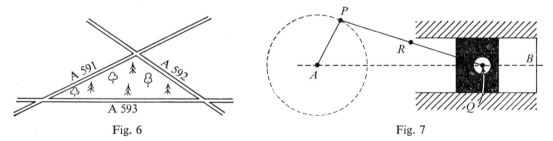

Fig. 6   Fig. 7

(*c*) Figure 6 shows three busy main roads surrounding a common.
A caravanner wishes to put his caravan on the common but as far as possible from each road. Use your answers to (*a*) and (*b*) to help him find the best site.

3. Figure 7 represents the working of a piston engine. *AP* rotates about *A* and *Q* slides along *AB*. Draw a diagram with *AP* = 3 cm and *PQ* = 9 cm. By finding about 20 different positions of *R*, the mid-point of *PQ*, draw its locus. Can you identify it?

4. Figure 8 represents the label from a cocoa tin. Sketch the path, in 3-D, of Fred the fly if he crawls from A to C when the label is on the tin. Name some familiar objects that possess this shape.

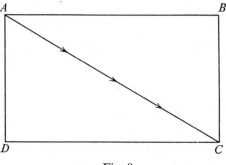

Fig. 8

5. (*a*) Mark two points A and B about 5 cm apart and towards the centre of a piece of paper. Stick a pin or drawing pin through each point. Cut out of card a triangle PQR which has an acute angle at P. (It is best to avoid angles that are nearly 0° or 90°.) Make each side of the triangle at least 5 cm. Place the triangle so that PQ rests against A and PR rests against B as shown in Figure 9. Mark a dot on the paper at the point occupied by the vertex P. Move the triangle to a new position, still keeping PQ against A and PR against B and mark the new position of P. Repeat many times keeping P on the same side of AB and trace the locus of P. What do you find? Do you think the locus is part of a circle? Test it, stating your method (using your eye is not good enough!) Describe the locus if P can lie on either side of AB. What relation has AB to the new locus?

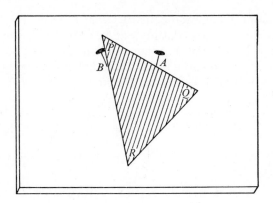

Fig. 9

(*b*) Repeat part (*a*) for a triangle PQR that has an obtuse angle at P.
(*c*) Repeat part (*a*) for a triangle PQR that has a right angle at P.

6. *The Parabola*

Figure 10 shows a set of circles, all with centre at (2, 0). Copy the figure on squared paper and draw a few more circles of the set. Mark the points where the circle of radius 2 cuts the line $x = 2$, where the circle of radius 3 cuts the line $x = 3$ and so on (i.e. mark points

160

equidistant from (2, 0) and the line $x = 0$). What is special about the intersection of the circle of radius 1 with the line $x = 1$? Construct also the points where the circle of radius $1\frac{1}{2}$ cuts the line $x = 1\frac{1}{2}$. Join all the points together by a smooth curve.

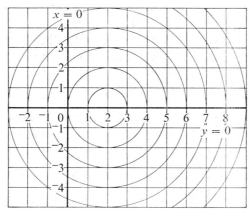

Fig. 10

7. *The Ellipse*

(*a*) In Figure 11, A and B are fixed 10 cm apart and the perimeter of the triangle is 28 cm. Draw the locus of P. This can be easily done by making a loop of string 28 cm long and looping it around two drawing pins at A and B. A pencil, as shown, keeps the loop taut and draws the locus of P.

The ellipse is the curve which you see if you shut one eye and look at a circle whose plane is not at right-angles to your line of sight. Describe its symmetries. Describe the locus of P in set language. What would be the locus of P in 3-D space?

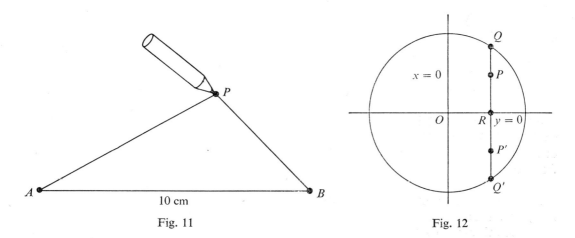

Fig. 11   Fig. 12

(*b*) Draw a circle with radius 5 cm and centre (0, 0), as in Figure 12, and draw a number of lines, such as QQ', parallel to $x = 0$. On QQ' mark points P and P' such that $PR = \frac{3}{5}QR$ and $P'R = \frac{3}{5}Q'R$. Sketch in the complete locus of P.

(*c*) On the edge of a piece of paper mark P, Q and R such that $PQ = 3$ cm and $QR = 2$ cm. Keeping Q on $y = 0$ and R on $x = 0$, as in Figure 13, mark sufficient positions of P to draw its locus. Show that the coordinates of P are $(5 \cos \theta°, 3 \sin \theta°)$, where $\angle OQR = \theta°$.

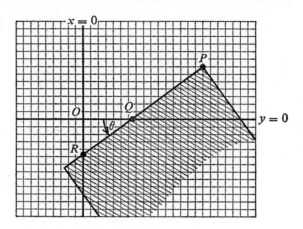

Fig. 13

8. *The searchlight game*

Interesting examples of loci may be found by considering the beams from 'double-ended' searchlights. A double-ended searchlight is shown in Figure 14. The beam shines out horizontally in two directions 180° apart.

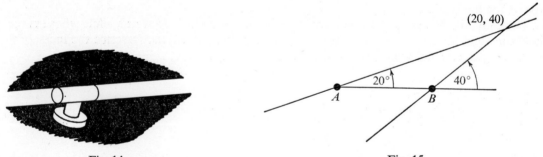

Fig. 14   Fig. 15

(*a*) Two such searchlights are set up, one at A and the other at B on a horizontal plane, as shown in Figure 15. The figure also shows two beams; the beam from A makes an angle of 20° and that from B an angle of 40°, measured anticlockwise, with the half-line drawn from A to pass through B.

The point where the two beams cross can conveniently be called the point (20, 40). This is an ordered pair of numbers (why?) and 20 and 40 are the first and second coordinates of the point. This is a new system of coordinates. What systems have you met before? We shall find this system very handy for denoting points on a locus although there are drawbacks to the system which you will discover as we go along.

(*b*) In Figure 16 there are four beams and their angles with AB have been marked. The point marked R is on the 100° beam from A and the 18° beam from B and so is marked (100, 18). Write down the coordinates of P, Q and S. What are the coordinates of A and B?

What can you say about a point (θ, θ)? Discuss the coordinates of a point on AB between A and B. What about points on AB produced.? Why is this coordinate system less useful for general purposes than the Cartesian system?

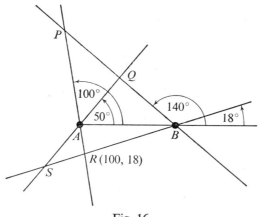

Fig. 16

(c) The operators of the two searchlights agree that they will rotate their beams according to pre-arranged plans. While A rotates at a steady speed, B will vary his rotation in relation to A's, for example, rotate at twice the speed, or rotate at the same speed, but in the opposite direction. In fact they will agree on a function mapping A's angle onto B's angle.

Plan number 1. They decide that A will start in the $0°$ direction and B in the $60°$ direction and that they will rotate their beams at the same speed. After a few moments if A is in a direction $\theta°$, in what direction is B shining? The function is $\theta \rightarrow \theta + 60$. Two of the points where their beams cross have coordinates (20, 80), (40, 100). Fill in the other coordinate of the point (160,   ). Is this a clock arithmetic? Write down the coordinates of three more points where their beams cross. Use your protractor to plot a number of points. What does the locus appear to be?

Plan number 2. The beams are rotated so that the locus passes through (40, 160), (60, 140), (80, 120), etc. Write down two further points on the locus. Write down the relation between the coordinates for any point on the locus (remember mod 180!). Give the functional relation between the coordinates. Plot a number of points, including the point (5,   ). What do you find about the point (10,   )? Try to find the name given to this locus.

Make up plan number 3 for yourself. You will have to give three facts about the plan:
(i) the starting directions;
(ii) the speed of rotation of B in relation to that of A;
(iii) whether the rotations are in the same direction or in opposite directions.

2. ENVELOPES

In the first section we considered sets of points. A corresponding idea is to consider sets of lines or curves. If you have done some curve stitching with coloured thread or wool, you will already be familiar with this idea. A pinboard, with pins arranged either in a square lattice or in a circle, also helps one to produce attractive results.

(a) Figure 17 shows a circle with 24 points equally spaced around it. Lines are drawn joining 0 to 6, 1 to 7, 2 to 8, ... What can you say about the lengths of these lines? They appear to form another smaller circle inside. Discuss the relation between the lines and this new circle. We say that the lines *envelop* the circle or that the circle is the *envelope* of the lines.

What would happen if lines were drawn joining

$$0 \text{ to } 9, \quad 1 \text{ to } 10, \quad 2 \text{ to } 11, \dots?$$

What would happen if lines were drawn joining

$$0 \text{ to } 12, \quad 1 \text{ to } 13, \quad 2 \text{ to } 14, \dots?$$

Can a set of lines envelop a point?

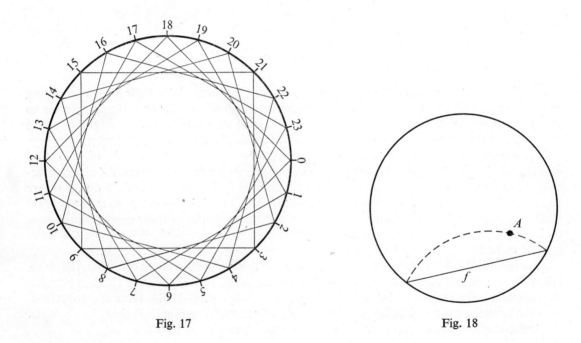

Fig. 17   Fig. 18

Notice that we are using arithmetic mod 24 on the set $\{0, 1, 2, \dots, 23\}$. Figure 17 has been constructed using the function $x \to x+6$, e.g. 2 is joined to $2+6 = 8$, and 21 is joined to $21+6 = 3$.

(b) Paper folding enables us to obtain an envelope quickly. The folds do not show up very well and it is best to run over them in pencil. Cut a circle out of tracing paper or use a circular filter paper. Mark a point A inside the circle and fold the paper over so that the circle edge passes over A, as shown in Figure 18. This gives the fold f. Repeat this about 15 times around the circle. What shape do the folds envelop?

Exercise B

1. In Figure 19, AP rotates about A and $AP = 8$ cm; and l is a line through P such that the angle between AP and the line l is always $30°$. Draw the envelope of l. Explain why it is a circle and find its centre and radius.

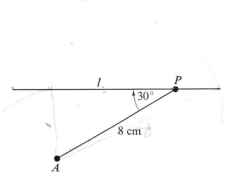

Fig. 19

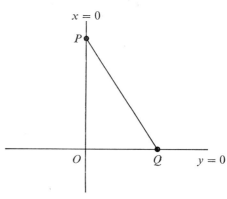

Fig. 20

2. (a) In Figure 20, $PQ = 4$ cm, P is always on $x = 0$ and Q is always on $y = 0$. (Imagine a slipping ladder.) Draw the envelope of PQ in the first quadrant, that is when P and Q are on the *positive* parts of $x = 0$ and $y = 0$.

(b) Extend the envelope, if P and Q are now allowed to move on the *negative* parts of $x = 0$ and $y = 0$. (The complete curve has four branches and is called an *astroid*.)

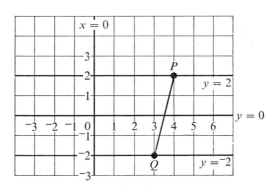

Fig. 21

3. In Figure 21, P is the point $(h, 2)$ and is always on $y = 2$; Q is the point $(k, {}^-2)$ and is always on $y = {}^-2$. Draw the envelope of PQ if $hk = 12$, by taking, say, 14 suitable values of h between ${}^-12$ and 12 inclusive.

(In the figure, P is at $(4, 2)$ and Q at $(3, {}^-2)$; thus $hk = 4 \times 3 = 12$.)

4. Draw the envelope of PQ in Figure 21 if $h + k = 12$. What do you find?

5. (a) A very attractive envelope can be drawn using a set of circles. Draw a circle of radius 2·5 cm so that the centre is about 3 cm to the right of the centre of the page. Mark a

point *A* on the extreme right-hand edge of the circle. Mark any other point *P* on the circle, and draw a circle with centre *P* and radius *PA*. Do this with a number of different positions of *P* on the base circle. The final result looks more artistic if these positions are equally spaced. This envelope is called a *cardioid*, which means heart-shaped.

(*b*) Draw a similar envelope but, instead of taking point *A* on the base circle, mark it at a distance of 1 cm outside the base circle. Do you find that this envelope is also a cardioid?

6. Mark 24 points at equal distance around a circle and label them 0, 1, 2, ..., 23. Join 1 to 2, 2 to 4, 3 to 6, etc., until you come to 11 to 22, then join 12 to 0, 13 to 2, etc., until every point has been joined. (Notice that you have been considering $x \rightarrow 2x$ (mod 24).) Compare the result with the cardioid referred to above.

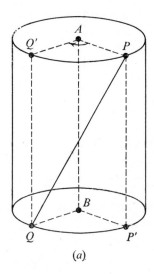
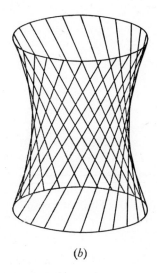

(*a*)   (*b*)

Fig. 22

7. *Project.* Figure 22(*a*) represents a circular cylinder. Points *P* and *Q* move on the top and bottom edges of the cylinder respectively, in such a way that *PAQ'* is a constant angle. The envelope of the line segment *PQ* may be inferred from Figure 22(*b*). A model to illustrate this may easily be made by constructing two congruent circular discs of hardboard and drilling a number of holes equally spaced round their edges. Shirring elastic is then threaded through the holes as shown. The discs may be mounted on a wooden axle, made from dowelling; alternatively Meccano wheels and rods may be used. A surface of this nature is called a *ruled surface*. Other examples of ruled surfaces are the 'hyperbolic paraboloid' roofs of some modern buildings.

8. Construct a paper cube of side 5 cm. Pierce it along an axis of rotational symmetry with a long needle and spin it. Repeat with other axes of symmetry. Try to predict the envelopes you notice and to account for them when you have made your observations.

3. ORDERINGS AND SOLUTION SETS

3.1   Orderings

(*a*) If we take an (unknown) number x and compare it with a known number such as 2 the numbers x and 2 may be related in various ways, as follows:

(i) x is less than 2. This is written as $x < 2$, and the solution set $\{x : x < 2\}$ of the relation is represented on the number line. (See Figure 23(*a*).)

(ii) x is equal to 2. (See Figure 23(*b*).)

(iii) x is greater than 2. (See Figure 23(*c*).)

(iv)   The first two may be combined to give x less than or equal to 2, (i.e. x is not greater than 2. (See Figure 23(*d*).)

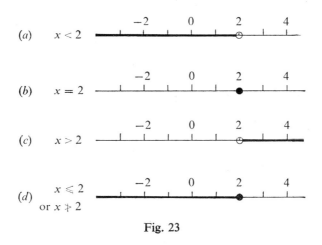

Fig. 23

(v)   What are the corresponding statements when (ii) and (iii) are combined?

Relations of the type (i), (iii), (iv) and (v) are known as *orderings*. (The relation in (ii) is, of course, an *equation*).

(*b*) Similar results exist in two (and more) dimensions. In Chapter 2 (Exercise B, Question 8) you were asked to plot some football results as ordered pairs on an x, y graph.

What is the relation connecting the home team's score (x) and the away team's score (y) when the result is a draw?

The set of points that represent home wins will all be in one colour and will satisfy the ordering $x > y$. If we call this the set H, then $H = \{(x, y) : x > y\}$. This will be read as 'H is the set of points with coordinates (x, y) such that x is greater than y'.

Write down in the same way the definition for the set A of points that represent away wins.

3.2   Solution sets

John went shopping with 30 new pence to buy fireworks. Bangers cost 2p and rockets cost 3p. His parents insist that the number of rockets that he buys must be more than half the number of bangers. What is the locus of possible combinations of bangers and rockets?

We can illustrate this graphically on a grid, or lattice. Across the page we show the number of bangers b, and up the page we show the number of rockets r. Of course

$$b \geqslant 0,$$
$$r \geqslant 0.$$

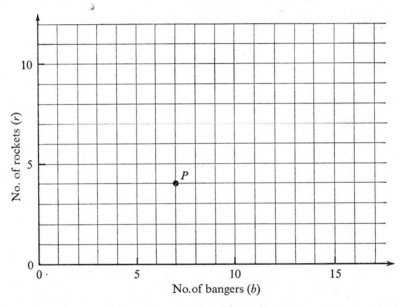

Fig. 24

(*a*) The intersections of the lines in Figure 24 show the possible combinations of fireworks. For example, P represents the combination of 7 bangers and 4 rockets.

He will not have enough money to buy many of these combinations.

Could he buy 7 bangers and 4 rockets?

How many rockets can he buy if he gets 7 bangers?

Copy Figure 24 and put a ring round those combinations that he can afford. (Remember that he does not have to spend all his money.) Does the set of ringed points form any pattern? This set is the locus of combinations that John can pay for.

(*b*) If he buys b bangers at 2p, how much does he spend? If he buys r rockets at 3p, how much does he spend? If he buys b bangers and r rockets, what is the total cost? Since he has only 30p, we know that

$$2b+3r \leqslant 30.$$

This *ordering* together with $b \geqslant 0$ and $r \geqslant 0$ describes the same locus as you have drawn.

(c) But he must also make sure that the number of rockets which he buys is more than half the number of bangers. On your copy of Figure 24 mark with a cross all those combinations that satisfy this condition.

(d) What is the relation between r and b here? Since r must be more than one-half of b,

$$r > \frac{b}{2}.$$

(e) Some of the points in Figure 24 will have both a cross and a circle. These points (the intersection of the two sets) will be the locus of combinations that John is able to buy. Thus if

$$P = \{(b, r): 2b+3r \leqslant 30\},$$

$$Q = \left\{(b, r): r > \frac{b}{2}\right\},$$

then John can buy members of the set $P \cap Q$.

(f) If he wants to buy as many bangers as possible, how many rockets will he buy; and how much change will he have?

Exercise C

1. The ways in which two dice (they are assumed to be green and yellow for easy identification) can land may be conveniently represented as lattice points on a graph as in Figure 25. Make a copy of this figure.

(a) Draw a circle around each member of the set

$$P = \{(g, y): g < y\}.$$

(b) Draw a small square around each member of the set.

$$Q = \{(g, y): g+y > 7\}.$$

(c) Describe in your own words the sets P, Q and $P \cap Q$.

(d) How many members has $P \cap Q$?

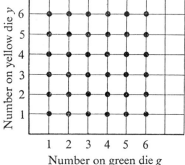

Fig. 25

2. A box contains 4 blue beads and 7 red beads. Each blue bead counts 2 points while each red bead counts $^{-}1$ point. A game is played in which each player in turn takes some beads from the box without looking, works out his points total, and then returns the beads to the box. The player scoring the highest total wins.

(a) Plot points (b, r) on a graph to represent the possible ways in which a person can take some beads from the box (b is the number of blue beads taken and r the number of red beads taken.)

(b) Draw a ring around each dot on your graph representing a turn that produces a positive total of points. (For example (3 blue, 4 red), which gives a total of 2 points.)

(c) Draw a square around each dot representing a turn in which 7 beads are removed from the box.

What fraction of these gives a positive total?

(d) Draw a triangle around each dot representing a turn in which 2 beads are taken. What fraction of these gives a positive total.

(e) When playing this game would it be better to remove 2 or 7 beads from the box?

3. A father with a family of 6 children decided to give each of them either a book or a record for Christmas.

(a) If he buys r records and b books, plot on a graph the points (r, b) representing the ways in which the presents could be bought.

(b) What is the equation connecting r and b?

(c) The father's choice is limited by the fact that he wishes to spend at most £5 while the books and records in which he is interested cost £1 and £0·62 each respectively. Put a circle around those points you have already plotted which represent ways fulfilling this further condition.

(d) If he also decides to buy at least 2 books, what alternatives are now left open to him?

4. A car ferry has room for up to 12 cars. A bus (or lorry) takes up the space of 3 cars. If c is the number of cars and b the number of buses taken on one crossing, then

$$c \geqslant 0, \quad b \geqslant 0 \quad \text{and} \quad c + 3b \leqslant 12. \quad \text{(Why?)}$$

(a) Graph the points (c, b) satisfying these orderings.

(b) The tolls for a car and a bus are 10p and 15p respectively, while the running expenses of the ferry are 75p a crossing. On your graph draw in the line $2c + 3b = 15$ and hence put a small circle around each point on your graph which represents a profitable crossing.

(c) Assuming the ferry makes a profit, what is (i) the largest, (ii) the smallest number of vehicles it can take on a crossing?

(d) What is the greatest profit that can be made when (i) one bus and some cars, (ii) two buses and some cars are carried?

4. GRAPHING EQUATIONS AND ORDERINGS

If the number of possibilities becomes large, the method we have used in Section 3 becomes very tiring. It is possible to simplify the work. There are four orderings which John must satisfy when buying his fireworks. These are illustrated by the sketches in Figure 26.

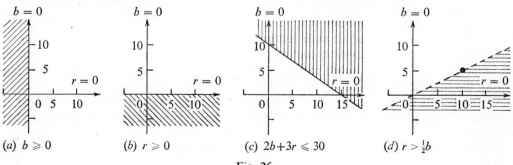

(a) $b \geqslant 0$ (b) $r \geqslant 0$ (c) $2b + 3r \leqslant 30$ (d) $r > \frac{1}{2}b$

Fig. 26

170

Notice that it is the unshaded area in which we are interested. In each case we have a line dividing the plane into two regions called *half-planes*. In (*a*), (*b*) and (*c*) the line is also included in the unshaded region, and is drawn as a full line. In (*d*) it is not included, so is drawn as a broken line. What is different about the ordering in (*d*)? What are the equations of the lines dividing the plane?

If we place these four sketches one on top of the other (this could be done by tracing each part of Figure 26) then the area still left unshaded will be the intersection of the four sets, and will contain the locus of points that we need. The locus will be points with whole number coordinates only (the meeting points of the grid lines), since John cannot buy fractions of a firework. This is illustrated in Figure 27.

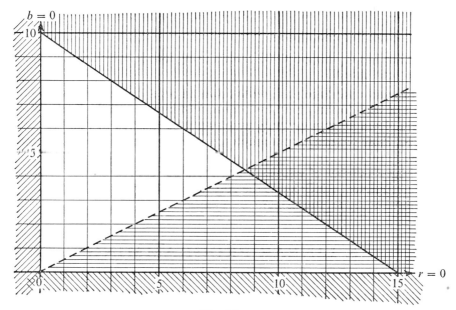

Fig. 27

Exercise D

1. On a single sheet of graph paper and using the same scales draw the lines representing the following equations. Indicate clearly which is which. (Show values of x from $^-4$ to 5 and of y from $^-4$ to 4.)

(*a*) $x = 2$; (*b*) $y = ^-2$; (*c*) $y = 2x$;
(*d*) $x+y = 3$; (*e*) $3x-y = ^-7$; (*f*) $5x-4y = 6$.

Refer to the graph you have drawn in Question 1 when answering Questions 2–4.

2. Draw separate graphs on a small scale, to show the regions representing the following orderings. (Shade the parts of the plane containing points whose coordinates do not satisfy the orderings.)

(*a*) $x > 2$; (*b*) $y < ^-2$; (*c*) $y > 2x$;
(*d*) $x+y > 3$; (*e*) $3x-y > ^-7$; (*f*) $5x-4y < 6$.

171

3. On small graphs show (by shading as in Question 2) the regions:

(a) $\{(x, y): x > 2, \quad y > {}^{-}2\}$; (b) $\{(x, y): x+y < 3, \quad x > 2\}$;

(c) $\{(x, y): 5x-4y < 6, \quad x+y < 3, \quad x > 0\}$;

(d) $\{(x, y): x+y > 3, \quad y < 2x, \quad x < 2\}$.

4. In the graph you drew to answer Question 1 there should be a triangle with vertices at the points $({}^{-}3, {}^{-}2)$, $({}^{-}1, 4)$ and $(5, {}^{-}2)$. Write down the three orderings whose solution set is the triangle and the points of the region bounded by the triangle.

5. (a) Draw a graph to show the region of points whose coordinates satisfy all five of the following orderings:

$$x > 0; \quad y > 0; \quad 2x+y > 4; \quad x+3y > 9; \quad x+y < 6.$$

(b) Put a dot on your graph for each member of the set

$$\{(x, y): x > 0, \quad y > 0, \quad 2x+y > 4, \quad x+3y > 9, \quad x+y< 6\},$$

which has whole number coordinates.

6. A post office has to transport 900 parcels using lorries, which can take 150 at a time, and vans which can take 60.

(a) If l lorries and v vans are used, write down an ordering which must be satisfied.

(b) The cost of each journey is £5 by lorry and £4 by van and the total cost must be less than £44. Write down another ordering which must be satisfied by l and v.

(c) Represent these orderings on a graph and dot in the members of the solution set.

(d) What is:

(i) the largest number of vehicles which could be used;

(ii) the arrangement which keeps the cost to a minimum;

(iii) the most costly arrangement?

5. LINEAR PROGRAMMING

We are now in a position to look at some problems, which, although artificial, give some idea of the kind of situations where the techniques of linear programming (linear because all the conditions involve straight lines) are applicable.

Linear programming deals with problems in which a large number of simple conditions have to be satisfied at the same time. By drawing graphs we first find a polygonal region containing the points that represent possible solutions. When the solution set is found it is the manager's or production engineer's task to decide which solution to take. This is usually done by trying to make the profit as large as possible, or the time taken for the process as short as possible.

The following example illustrates the method.

Example

A haulage contractor has 7 six-tonne lorries, 4 ten-tonne lorries and 9 drivers available. He has contracted to move a minimum of 360 tonnes of coal from a pit-head to a power station daily. The six-tonne lorries can make 8 journeys a day and the ten-tonne lorries can make 6 journeys a day.

How should the contractor organize the use of his lorries to run the lorries at minimum cost if a six-tonne lorry costs £5 a day and a ten-tonne lorry costs £8 a day? Suppose the contractor uses

$$x \text{ six-tonne lorries and } y \text{ ten-tonne lorries.}$$

Then, because of the limitation on the number of lorries and drivers available:

$$x \leqslant 7,$$
$$y \leqslant 4,$$
$$x + y \leqslant 9.$$

Working on a daily basis, a six-tonne lorry can carry as much as 48 tonnes a day, while a ten-tonne lorry can carry as much as 60 tonnes a day. Thus, to fulfil the contract of a minimum of 360 tonnes a day

$$48x + 60y \geqslant 360.$$

(This condition can be simplified by dividing by 12 to give $4x + 5y \geqslant 30$.)

Now that we have the conditions expressed as algebraic orderings they can be graphed to find the solution set. We remember that $x \geqslant 0$, $y \geqslant 0$ and that only points with whole number coordinates represent possible solutions (see Figure 28).

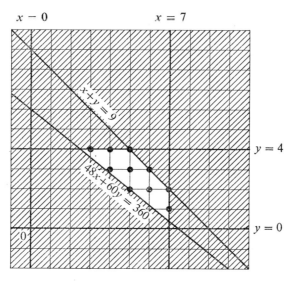

Fig. 28

From the graph it is clear that the points
(3, 4), (4, 4), (4, 3), (5, 4), (5, 3), (5, 2), (6, 3), (6, 2), (7, 2), (7, 1)
all represent possible solutions.

It now remains to see which of these solutions best suits the contractor's policy.

The cost of using (x, y) lorries is £$(5x+8y)$. By substituting in turn each of the ordered pairs above it is seen that this is smallest when

$$(x, y) = (5, 2)$$

giving a cost of £41.

In practice you will be able to find the best solution without having to consider each ordered pair separately. Later in *Book 5*, we shall investigate ways of finding the best solution algebraically.

Exercise E

In this exercise the numbers will all be positive or zero.

1. A factory makes cricket bats and tennis rackets. A cricket bat takes 1 h of machine time and 3 h of craftsman's time while a tennis racket takes 2 h of machine time and 1 h of craftsmen's time. 28 h of machine time and 24 h of craftsman's time are available each day.

(*a*) If the factory makes x bats and y rackets on a particular day, write down two orderings satisfied by x and y based on: (i) machine time, (ii) craftsman's time.

(*b*) Represent these orderings graphically taking values of x from 0 to 28 and of y from 0 to 24.

(*c*) What is the largest number of: (i) bats, (ii) rackets, which could be made in a day?

(*d*) What numbers of bats and rackets must be made if the factory is to work at full capacity?

(*e*) The profits on a bat and on a racket are £1 and 50p respectively. Find the maximum profit to the factory on a day when it produces: (i) only bats, (ii) only rackets, and (iii) works at full capacity.

2. In an airlift it is required to transport 600 people and 45 tonnes of baggage. Two kinds of aircraft are available: the Albatross, which can carry 50 passengers and 6 tonnes of baggage; the Buzzard, which can carry 80 passengers and 3 tonnes of baggage.

(*a*) If a Albatrosses and b Buzzards are used, explain why

$$5a+8b \geqslant 60 \quad \text{and} \quad 2a+b \geqslant 15.$$

(*b*) Only 8 Albatrosses and 7 Buzzards are available. Represent, on a graph, the possible arrangements of aircraft which can supply the necessary transport. Dot in the members of the solution set.

(*c*) What is the smallest number of aircraft that can be used?

3. A factory manager has to decide how many of each of two types of machine to install. The facts about them are as follows:

	Factory floor space (m²)	Labour needed per machine (men)	Output per week (units)
Machine X	500	9	300
Machine Y	600	4	200.

The factory has 4500 m² of floor space, but only 54 skilled workers are available to work the machines.

(*a*) If he buys x of machine X and y of machine Y, write down two orderings satisfied by x and y based on (i) floor space and (ii) labour.

174

(b) Represent the possible solutions open to the manager graphically. Dot in each member of the solution set clearly.

(c) How many machines of each type will he buy to: (i) achieve maximum output, (ii) give work to all the skilled workers?

(d) If he satisfies (c) (ii) by how much will the factory's output be below the maximum possible?

4. A market gardener intends to split an 18-hectare field between lettuces and potatoes. The relevant details are as follows:

	Lettuces	Potatoes
Cost per hectare including labour (pounds)	5	3
Labour per hectare (man days)	3	I

Only £60 and 30 man days are available

(a) If he plants l hectares of lettuces and p hectares of potatoes, then

$$l + p \leqslant 18.$$

Write down two other orderings that must be satisfied by l and p.

(b) Represent these three orderings graphically.

(c) Calculate his profit, on the basis of £12 per hectare for lettuces and £8 per hectare for potatoes, for the solutions represented by each vertex (other than $(0, 0)$) of the polygon bounding the solution set. Which solution gives the greatest profit?

5. A manager of a theatre, which holds 600 people, sells the seats at two prices: 15p and 25p. To cover his expenses he has to take at least £60 at the box office for each house.

(a) If he sells x seats at 15p and y seats at 25p, write down orderings based on: (i) the capacity of the theatre and (ii) the seats that have to be sold to make a profit.

(b) It is the manager's policy to have at least 200 seats available at the cheaper price. Express this as an ordering involving y, and represent all the orderings you have found graphically.

(c) (i) What profit would be made if the theatre were filled by holders of 15p tickets?
(ii) What is the smallest number of seats that can be sold without sustaining a loss?
(iii) What is the maximum profit that can be made on one house?

Summary

1. A set of points that satisfy a given condition, or combination of conditions, is called a *locus*. The locus may take the form of a line, curve, or region.

2. Sometimes a curve emerges when a set of lines is drawn under some given condition. In these circumstances the curve is called an *envelope*, as the lines envelop the curve.

3. Some particular curves, which often arise in locus problems, are shown in Figure 29.

4. *Linear Programming* involves finding the solution set of simultaneous orderings. Simple problems can be solved graphically by shading out the regions that do

not satisfy the given conditions; the points in the region left (if any) being the solution set. (*Linear* because the boundaries of these regions are usually straight lines.)

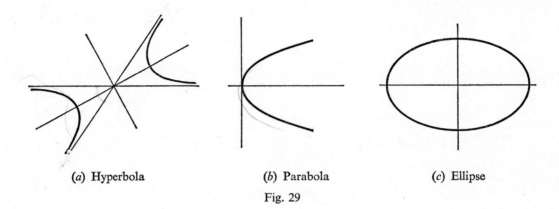

(*a*) Hyperbola   (*b*) Parabola   (*c*) Ellipse

Fig. 29

REVISION EXERCISES

SLIDE RULE SESSION NO. 3

Give all answers to 3 s.f.

1. 23×67.8.
2. 1.91×9.8.
3. $2.12 \div 1.56$.
4. $230 \div 16.2$.

5. $\sqrt{39.5}$.
6. $\sqrt{395}$.
7. 1.67^2.
8. 16.7^2.

9. $\dfrac{1.23 \times 2.31}{3.12}$.
10. $\dfrac{34.5}{45.3 \times 5.34}$.

SLIDE RULE SESSION NO. 4

Give all answers to 3 s.f.

1. 19×37.
2. 0.045×77.
3. $19 \div 37$.
4. $77 \div 0.045$.

5. $19 \times \sqrt{61.5}$.
6. $\sqrt{(220.5 \times 8)}$.
7. 0.0468^2.
8. $\pi \times 8.4^2$.

9. $1.23 \times 2.34 \times 345$.
10. $\dfrac{541 \times 7.67}{23}$.

G

1. A ship sets out to reach a small island which lies 400 km away on a bearing of 065°. The navigator makes a mistake and the ship sails the first 300 km on a bearing of 056°. Find, by scale drawing, how far the ship then is from the island, and how the course should be changed so that the ship will still reach its destination.

2. Copy Figure 1 and construct on your diagram the displacements:

(a) $\mathbf{m} + \mathbf{n}$; (b) $^-\mathbf{n}$; (c) $\mathbf{m} + {}^-\mathbf{n}$;

(d) $\mathbf{m} + 2\mathbf{n}$; (e) $2\mathbf{m} + \mathbf{n}$.

Label these displacements clearly.

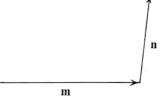

Fig. 1

3. A certain reflection maps (1, 1) onto (1, 1) and (5, 3) onto (3, 5). What will be the image of the following points under the same reflection?

(a) (2, 6); (b) (7, 7); (c) (2, $^-$3);

(d) ($^-$5, $^-$4); (e) (p, q).

4. Points $A(2, 1)$, $B(5, 1)$ and $C(5, 4)$ form one triangle, and points $A'(6, 5)$, $B'(3, 5)$ and $C'(3, 2)$ form another. Triangle ABC can be mapped onto triangle $A'B'C'$ by a rotation. Give the centre of rotation and also the angle. Describe the symmetry of the figure made up of both triangles.

177

5. In Figure 2, O is the centre of the circle whose radius is 2 cm. Calculate

(a) $\angle BOD$, (b) area of $\triangle BOD$, (c) area of sector BOD, (d) area of the minor segment whose chord is BD.

6. A, B, C, D are four points in space. A, B, C are in a plane π; B, C, D are in a different plane π'. What is $\pi \cap \pi'$?

7. Calculate the following matrix products where possible:

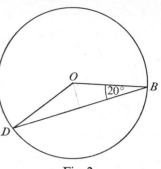

Fig. 2

(a) $(1 \quad 2)\begin{pmatrix} 1 \\ -2 \end{pmatrix}$;

(b) $(1 \quad 2)\begin{pmatrix} 1 & 3 \\ -2 & 4 \end{pmatrix}$;

(c) $(1 \quad 2 \quad 0)\begin{pmatrix} 1 & 3 \\ -2 & 4 \end{pmatrix}$;

(d) $\begin{pmatrix} 1 & 3 & 0 \\ 2 & -4 & 6 \end{pmatrix}\begin{pmatrix} -5 \\ 1 \\ 2 \end{pmatrix}$;

(e) $\begin{pmatrix} 0 & 1 \\ 0 & 6 \end{pmatrix}\begin{pmatrix} 1 & -3 \\ 0 & 0 \end{pmatrix}$;

(f) $\begin{pmatrix} 0 & 1 \\ 0 & 6 \end{pmatrix}(1 \quad 2 \quad 3)$.

8. An equilateral triangle ABC of edge 4 cm, made of thin cardboard, rests with AB on a horizontal table. The plane of the triangle is inclined at 20° to the table. How high is C above the table?

9. At Mahiti International Airport a building 20 m high is 200 m from the end of the main runway. Assuming that a plane is to climb along a straight path once it has taken off, calculate: (a) the length of the flight path, assuming the aircraft takes off at the end of the runway.

(b) the minimum safe angle of ascent.

10. A market gardener intends to split a 20 hectare field between lettuces (L) and potatoes (P). The relevant details are:

	L	P
Cost per hectare including labour	£5	£3
Labour per hectare (in man days)	2	1
Estimated profit per hectare	£9	£6

If £75 and 30 man days are available, how should he allocate the land for maximum profit? (He need not use the whole of the 20 hectares.)

H

1. A circle has a radius of 4 cm and a centre O. Two parallel chords are 4 cm apart. What is the total length of arc between these chords when:

(a) one chord passes through the centre;
(b) the chords are symmetrical about the centre?

2. A triangle with vertices at (2, 3), (6, 6), (2, 8) is reflected in the line $y = 0$. Write down the coordinates of the vertices of the image triangle. Are the two triangles congruent? Could the second triangle be the image of the first: (a) under a translation; (b) under a rotation? Explain briefly.

3. *P*, *Q*, and *R* are three points with coordinates (0, 2), (1, 0) and (4, 3), respectively.

(*a*) Give the vector representing **QR**.

(*b*) If **PS**=**QR**, give the coordinates of *S*.

(*c*) Is it true that **QP** = **RS**?

(*d*) Is it true that **PR** = **QS**?

(*e*) What kind of quadrilateral is *PQRS*?

4. Using Figure 3,

(*a*) Express **AC** in terms of **a, b** and **c** in two different ways.

(*b*) Using your answer for (*a*), write down **b** in terms of **a** and **c**.

(*c*) If **MC** = **AD**, where must *M* be?

5. The Yugoslavian town Split is 180 km south and 140 km east of Krk. By means of an accurate scale drawing, find (*a*) the direct distance, and (*b*) the bearing, of Split from Krk.

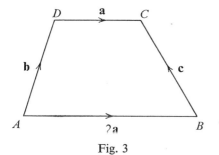

Fig. 3

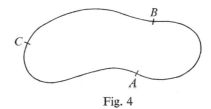

Fig. 4

6. A rocket rises 30 km vertically, 40 km at 30° from the vertical, and then 100 km at 35° from the vertical. Calculate to the nearest kilometre how high it then is above the launching pad. (Ignore the curvature of the earth.)

7. *A*, *B* and *C* are three cats sitting on the edge of a goldfish pond (see Figure 4). Copy the figure and shade in the part of the pond in which a goldfish is nearer to *A* than to either *B* or *C*.

8. Packets of Wosh are sold in giant, large and standard sizes; these packets contain 1 kg, 0·7 kg, 0·5 kg respectively, and their costs to the shopkeeper are 13p, 10p, 7p respectively.

(*a*) Express these facts as column matrices, one for mass and one for cost.

(*b*) If a shopkeeper ordered 20 giant, 30 large and 10 standard packets, write down the mass of Wosh ordered as the product of two matrices, then work it out.

(*c*) In a similar way, find the total cost of the order in £.

9. **R** denotes a rotation through 80° about (0, 0).

(*a*) Describe the transformations $\mathbf{R}^2$, $\mathbf{R}^{-1}$, $\mathbf{R}^4$.

(*b*) If **U** denotes a rotation through 40° about (0, 0), find two integers *x* such that $\mathbf{R}^x = \mathbf{U}$.

10. A piece of paper is to be cut into the form of a rectangle *x* cm wide and *y* cm long, such that:

(*a*) the length is to be more than double the width;

(*b*) the area is to exceed 48 cm²;

(*c*) the length of each diagonal is to be less than 15 cm.

Write down three orderings involving *x* and *y*, and give one pair of whole numbers satisfying them all.

I

1. 'Pirates Cove to Cutlass Creek is $\begin{pmatrix} -1 \\ 2 \end{pmatrix}$. For the gold, start at Pirates Cove. then

$$\begin{pmatrix} 3 \\ 0 \end{pmatrix}, \ \begin{pmatrix} 1 \\ 1 \end{pmatrix}, \ \begin{pmatrix} -2 \\ 1 \end{pmatrix}, \ \begin{pmatrix} 2 \\ 2 \end{pmatrix}, \ \begin{pmatrix} -2 \\ 1 \end{pmatrix}, \ \begin{pmatrix} -1 \\ 1 \end{pmatrix}.'$$

Where, if you came across this note, would you look for the treasure on the island in Figure 5?

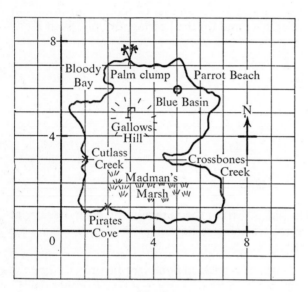

Fig. 5

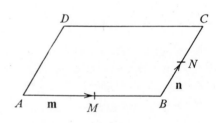

Fig. 6

2. $ABCD$ is a parallelogram, and M, N are the mid-points of AB, BC, respectively (see Figure 6). $AM = \mathbf{m}$ and $BN = \mathbf{n}$. Give, in terms of $\mathbf{m}$ and $\mathbf{n}$:

(a) AB; (b) MB; (c) MN; (d) AC; (e) DC.

3. The coordinates of the vertices of triangle T_1 are (1, 0), (3, 0) and (3, 1). Rotation through 180° about (1, 1) maps T_1 onto T_2, and similarly rotation through 180° about (2, 2) maps T_2 onto T_3, whilst rotation through 180° about (3, 3) maps T_3 onto T_4.

(a) Show on a diagram the positions of T_1, T_2, T_3 and T_4.
(b) Describe the single transformation which would map (i) T_1 onto T_3, (ii) T_1 onto T_4.
(c) Can any one of the triangles be mapped onto one of the others by means of a reflection? Justify your answers.

4. The vertices of a triangle are $O(0, 0)$, $A(4, 0)$, $B(0, 4)$. P is a point whose images in the three sides of the triangle OAB are Q, R and S. Find the position of P if triangle QRS is the image of triangle OAB under a reflection in the line $x+y = 2$.

180

5. (i) Give examples of:

(a) a row matrix; (b) a column matrix; (c) a square matrix;
(d) a 2 × 3 matrix; (e) a 3 × 2 matrix.

(ii) **A** is a 3 × 2 matrix, **B** is a 3 × 5 matrix, **C** is a 2 × 3 matrix. Why is it impossible to work out the product **ABC**? Which product of these three matrices *can* be worked out? State the order of each product matrix which results.

6. A triangle has vertices at the points $P(2, 3)$, $Q(5, 3)$ and $R(4, 6)$. Write down the images P', Q' and R' of these points in the mirror line $x = 1$. Can triangle PQR be mapped onto triangle $P'Q'R'$ by a rotation?

7. A straight line segment connects $O(0, 0)$ and $P(3, 0)$. It is rotated about O through an angle of $\theta°$. What are the new coordinates of P? Make a table of the coordinates of P for the values $\theta = 0, 30, 45, 60, 90, 120, 135, 150, 180$.

8. Mark the points $A(4, 5)$, $B(2, 1)$ and $C(4, 0)$ on graph paper.

(a) Show on your diagram the result of applying the translation $\binom{-2}{1}$ to ABC. Label this $A'B'C'$.

(b) $A'B'C'$ maps onto $A''B''C''$ when rotated 60° anticlockwise about O. Show $A''B''C''$ on the diagram.

(c) Find by construction the centre of the rotation that maps ABC straight onto $A''B''C''$.

9. Figure 7 shows a regular octahedron. Say which of the following are true and which false.

(a) The solid has exactly eight faces, six vertices and eight edges.

(b) The angle $ACE = 60°$

(c) $EC = AF$.

(d) There are exactly four distinct routes from A to B along the edges of the solid and passing through just two other vertices.

10. It is required to transport 500 troops and 42 tonnes of equipment. Two types of aircraft are available in sufficient numbers; an Albatross can carry 50 men and 5 tonnes of equipment, while a Kestrel can carry 40 men and 3 tonnes of equipment. For the journey, the operating cost for each Albatross is £1050 and for each Kestrel is £900. How many of each are required if the total cost is to be a minimum?

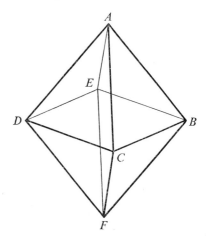

Fig. 7

J

1. For each triangle illustrated in Figure 8, write down an expression for the length marked with the letter. DO NOT WORK IT OUT. For example, the answer to the first would be given in the form $x = 2/\cos 32°$ and left at that. Now do the rest.

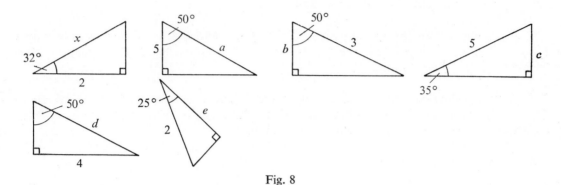

Fig. 8

2. An 8 m ladder is sliding down a wall (see Figure 9). What angle does it make with the ground when its base is:

(a) 2 m;   (b) 3 m;   (c) 6 m;   (d) 8 m,

from the bottom of the wall?

3. Carry out the matrix multiplication

$$\begin{pmatrix} 1 & 2 \\ -3 & 0 \end{pmatrix} \begin{pmatrix} 4 & 5 \\ 6 & 7 \end{pmatrix}.$$

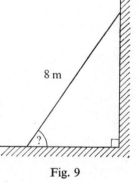

Fig. 9

4. Draw the triangle $A(0, 1)$, $B(2, {}^-1)$, $C({}^-2, {}^-1)$ accurately on squared paper, taking a scale of 1 cm to 1 unit. O is the origin. Find the coordinates of:

(a) the point O_1 which is the reflection of O in AB;
(b) the point O_2 which is the reflection of O_1 in BC;
(c) the point O_3 which is the reflection of O_2 in CA.

Would you arrive at the same point in the end if the reflections had been done first in CA, then in BC, then in AB?

5. In Figure 10, ABC and $A'B'C'$ are congruent triangles, and $BCC'B'$ is a straight line.

(a) Describe a single transformation that would map A, B, C onto A', B', C' respectively.
(b) What can you state about the relation between the lines AA' and CC'.

6. C_1 is a circle of radius 1 unit with its centre at the origin. The translation $\begin{pmatrix} 2 \\ 0 \end{pmatrix}$ maps C_1 onto C_2; a rotation through 120° about the origin maps C_2 onto C_3; and reflection in the line $y = 0$ maps C_3 onto C_4. The figure consisting of the four circles so far obtained is now reflected in the line $x = 0$. With a scale of 1 cm to 1 unit draw a figure showing the original circle and all the images. Describe its symmetry.

7. Copy Figure 11.

(a) Draw on your diagram the locus of points which are within the triangle and are equidistant from *AB* and *BC*.

(b) Shade the set of all points which are nearer to *BC* than to *AB* and which are less than 2·0 cm from the point *A*.

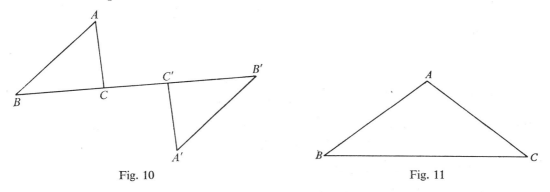

Fig. 10   Fig. 11

8. A helicopter flies due north at 180 km/h for 1 h. It then changes course to due east and continues for 20 min. On what course should it fly to return directly to the starting point and how long will this take if the same speed is maintained?

9. (a) Let
$$\mathbf{M} = \begin{pmatrix} 1 & -2 \\ 3 & 1 \end{pmatrix} \quad \text{and} \quad \mathbf{N} = \begin{pmatrix} 2 & 1 \\ 0 & 1 \end{pmatrix}.$$

Find **MN** and **NM**. Are they equal?

(b) *ABCD* is a parallelogram. If the coordinates of *A*, *B*, *C* are (0, 0), (2, 1) and (3, 3) respectively, what are the coordinates of *D*?

10. Find the points with whole-number coordinates which satisfy all three of the orderings $x < 4$, $3y - x \leqslant 6$, $3y + 2x > 6$.

K

1. Find (to 3 s.f.) the numbers represented by the letters in Figure 12.

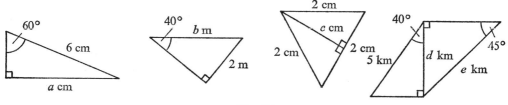

Fig. 12

2. A rocket rises 16 km vertically, 80 km at 30° from the vertical and 64 km at 70° from the vertical. Calculate how high it is then above the launching pad. (Ignore the curvature of the earth.)

3. Draw the triangle whose vertices are (0, 0), (3, 0) and (3, 2). Draw the same triangle after an anticlockwise rotation of 90° about the origin. How could the same result be obtained by doing two reflections instead? Can you find more than one way of doing this?

4. $O(0, 0)$, $A(3, 0)$ and $B(0, 2)$ are the vertices of a triangle. This triangle is translated to a new position $O'A'B'$, the displacement vector being $\binom{2}{2}$.

(a) Give the coordinates of O', A' and B'.
(b) Name a vector equivalent to $\mathbf{OO'}$.
(c) What kind of quadrilateral is $OAA'O'$?

5. The directed-route matrix

$$\begin{array}{c} \\ \text{from} \end{array} \begin{array}{c} \\ A \\ B \\ C \end{array} \overset{\begin{array}{ccc} A & B & C \end{array}}{\begin{pmatrix} 0 & 2 & 1 \\ 1 & 0 & 1 \\ 0 & 1 & 0 \end{pmatrix}}$$

represents a network of roads. Without drawing the network say if it is possible, in any way at all, to reach A from C. If so, describe the route.

6. Give the result of squaring the matrix in Question 5.

In Questions 7 and 8, say which statements are true and which are false.

7. The distance between two parallel planes is 6 m. The locus of a point equidistant from these planes and 3 m away from a line in one of these planes is:

(a) a single point; (b) a single line; (c) two lines; (d) none of these.

8. A triangle ABC in which no two sides are equal is given a clockwise turn equal to the angle BAC about the vertex B. Then:

(a) the angle between the old and new directions of AC is equal to $\angle BAC$;
(b) the angle between the old and new directions of BC is equal to $\angle ABC$;
(c) B is an element of the mediator of AA', where A' is the new position of A;
(d) no side of the new triangle is parallel to a side of the original triangle.

9. The height of the tide (h m) on a certain day is given by $h = 15 + 8 \cos x°$, where $x = 30n$, n being the number of hours after midnight. Make out a table of values for h for the period midnight to noon the same day, and graph the function $n \to h$.
At what times on that day will the height of the tide be 13 m?

10. State in each case whether the description is of a line, a half-line or a line segment:

(a) $\{(x, y): x = 0, y \geqslant 0\}$;
(b) $\{(x, y): y = 4\}$;
(c) the longest side of the triangle whose vertices are $(2, 1)$, $(3, 7)$, $(5, 4)$.
Give an example of an equation for a half-plane.

L

1. AB and DC are parallel sides of a quadrilateral $ABCD$ in which $\angle A = 52°$, $\angle B = 46°$, $AB = 12$ cm and $AD = 3$ cm. Calculate:

(a) the length BC; (b) the length DC.

2. Find all the angles of a triangle whose sides are 3, 4, and 5 cm long.
What are the angles of a triangle whose sides are 9, 12 and 15 cm long?

3. Find four different single transformations which will map the equilateral triangle *ABC* onto the other equilateral triangle in Figure 13. (*ACE* is a straight line.)

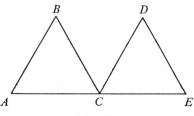

4. Describe the sets of points (x, y) which are mapped onto themselves by the following transformations:

(*a*) rotation through 90° about the origin;

(*b*) reflection in the line $x = y$;

(*c*) the translation given by the vector $\begin{pmatrix} 2 \\ 1 \end{pmatrix}$;

(*d*) reflection in the line $x = 1$ followed by reflection in the line $y = 2$.

Fig. 13

5. In a previous question, we called a matrix of the form $\begin{pmatrix} a & 0 \\ 0 & b \end{pmatrix}$ a *P* matrix and a matrix of the form $\begin{pmatrix} 0 & c \\ d & 0 \end{pmatrix}$, a *Q* matrix.

Form the operation table for the *P*-matrices and *Q*-matrices under matrix multiplication. State the identity element (if any).

6. The point *P* moves on the line *AB* (see Figure 14). A transformation maps *P* onto the point *P'* such that *P* is the mid-point of *OP'*. Copy the figure and draw on it as much as you can of the locus of *P'*. Mark the transforms *C'*, *D'* of *C* and *D* and state their coordinates.

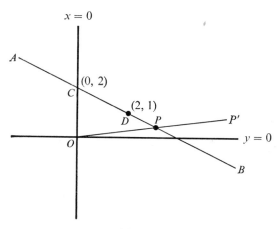

Fig. 14

7. On the same graph draw the lines $y = x$ and $y = 2x$, using a scale of 2 cm to 1 unit. With a protractor measure the acute angle between the lines. If *A* is the region for which $y > x$ and *B* is the region for which $y < 2x$, shade the region $A \cap B$.

To which of the regions *A*, *B* or ($A \cap B$) do each of the following points belong:

(*a*) (0, 1);   (*b*) (2, 3);   (*c*) (5, 6);   (*d*) (2, 5);   (*e*) (2, 1).

8. Show that the matrices **M** and **I**, where $\mathbf{M} = \begin{pmatrix} -1 & 2 \\ 0 & 3 \end{pmatrix}$ and $\mathbf{I} = \begin{pmatrix} 1 & 0 \\ 0 & 1 \end{pmatrix}$, satisfy the equation $\mathbf{M}^2 - 2\mathbf{M} - 3\mathbf{I} = 0$.

9. A football moves from A to E by passes described by the following displacements:

$$\mathbf{AB} = \begin{pmatrix} 4 \\ 3 \end{pmatrix}, \quad \mathbf{BC} = \begin{pmatrix} -3 \\ 4 \end{pmatrix}, \quad \mathbf{CD} = \begin{pmatrix} 2 \\ 0 \end{pmatrix}, \quad \mathbf{DE} = \begin{pmatrix} 5 \\ -1 \end{pmatrix}.$$

Calculate **AE**, and indicate the ball's motion on a diagram. Calculate the length of **AE**.

10. A man has a straight hedge 80 m long, and he buys an electrical cutter with which to trim it. The nearest electricity point is 30 m away from the hedge, half-way along. Unfortunately the cable on the cutter is only 40 m long. What length of the hedge can the man reach? How much should he extend the cable so that he can trim the whole hedge?

9

TRANSFORMATIONS II

Change is not made without inconvenience.
JOHNSON, *Preface to the 'English Dictionary' as from Hooker*

1. ENLARGEMENT

You are all familiar with films projected onto a screen, whether at a cinema or at school, and the connection between the film and the projected picture is essentially a difference in size. Many other instruments have been developed to enlarge objects, from a simple magnifying glass which perhaps doubles the dimensions of an object, to an electron microscope which magnifies an object thousands of times.

(*a*) What other instruments for enlarging objects do you know?

People often find it necessary to make a copy of a drawing such as a map, a house plan, or a dress pattern on a different scale, and an instrument, the pantograph, has been designed to do just this. However, not everyone has a pantograph available and it is useful to know how to copy a diagram on a different scale without it.

(*b*) What does a pantograph look like?

Find out and make one if you have the time.

Figure 1 shows a drawing of a teacher (i), and a '2 times' enlargement, (ii). If (ii) is regarded as the original, then (i) is said to be a '½ times' enlargement. (The number of 'times' is called the *scale factor* of the transformation.)

Fig. 1

(*c*) Draw a straight line on tracing paper and hold the paper so that the line passes through a pair of corresponding points in (i) and (ii), for example, the tip of the teacher's nose. What point will the line always pass through no matter which pair of corresponding points you pick?

We call this point the *centre of enlargement*.

(*d*) Take any point P in (i) and measure the distance OP. Now measure OP' where P' is the corresponding point in (ii). Calculate the value of OP'/OP to 1 decimal place for several pairs of points. What do you notice?

(*e*) How does the enlargement in Figure 2 differ from that in Figure 1?

If (i) is taken as the original then (ii) is said to be a '$^-\frac{1}{2}$ times' enlargement. Another way of saying this is that the scale factor is $^-\frac{1}{2}$. Alternatively, if (ii) is taken as the original, then the factor of the enlargement is $^-2$.

This may seem a strange method of enlargement but it is very much like the action of a camera or our eyes.

(*f*) Can you suggest why the scale factors of the enlargements of Figure 2 are negative?

(*g*) When is an enlargement equivalent to a rotation?

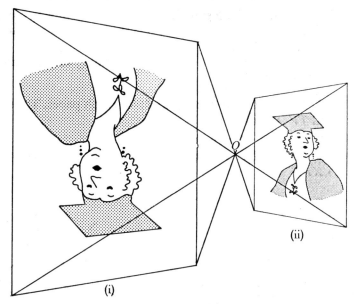

Fig. 2

Exercise A

1.

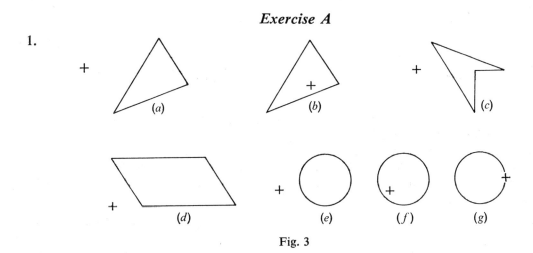

Fig. 3

Make copies of the shapes in Figure 3 on plain paper. Using the given centres, enlarge them by factors of: (i) 3; (ii) ⁻2; (iii) ½.

2. Copy the following table and write down the coordinates of *A, B, C* and *D* after enlargements with centre *O* (0, 0) and with the given scale factor.

Scale factor	A(4, 0)	B(0, 3)	C(4, 3)	D(3, ⁻2)
3				
⁻1				
¼				

3. (a) Draw two different sized triangles with equal angles and having corresponding sides parallel. Draw lines joining corresponding vertices. Do these lines pass through the same point? Would they, if you extended them?

(b) Repeat (a) but arrange the triangles so that corresponding sides are no longer parallel. To what conclusion do you come?

4. Copy the points L, M, N, T and V (see Figure 4) on graph paper. The exact coordinates do not matter provided TV is parallel to LN.

(a) Find W, assuming that TWV is an enlargement of LMN.

(b) Find X, assuming that VXT is an enlargement of LMN. What can you say about the triangles TWV and VXT?

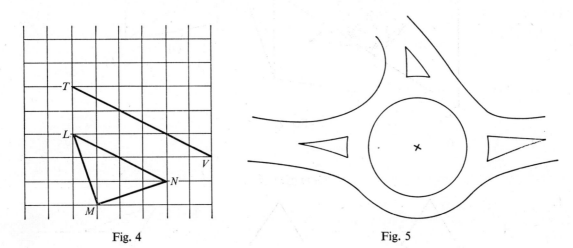

Fig. 4   Fig. 5

5. At the National Road Research Laboratory, Crowthorne, the road engineers try out road and junction layouts by marking them temporarily on a large tarmac area and observing the traffic flow through them. Figure 5 shows the design of an experimental roundabout drawn on a scale of 1 cm to 5 m. How many times larger than the drawing would the roundabout be? Explain carefully how you could mark this roundabout out on your school playground so that it is 100 times as large as the figure.

6. (a) Draw, preferably on graph paper, two squares with a pair of sides equal and parallel. How many different centres of enlargement can you find?

(b) If the squares in (a) are of different size, how many centres of enlargement can you find?

7. (a) Draw the square whose vertices are the points (1, 1), (2, 1), (2, 2) and (1, 2). Enlarge it, with scale factor 2 and with (0, 0) as centre of enlargement.

(b) Enlarge the resulting figure, with scale factor 3 and the same centre of enlargement.

(c) What single enlargement would have mapped the original square onto the final one?

8. (a) Draw a square with vertices at (4, 0), (10, 4), (6, 10) and (0, 6). Enlarge this square by a scale factor of -0.5, with centre of enlargement (0, 0).

(*b*) Now enlarge the resulting figure by a scale factor of ⁻3, using the same centre of enlargement.

(*c*) What is the scale factor of the enlargement which would map the original figure onto the final one?

(*d*) If an enlargement of scale factor *h* is followed by an enlargement of scale factor *k*, what is the scale factor of the combined transformation?

1.1   Similarity

We have said that enlargement of a figure produces one which is the same shape but a different size. We shall investigate this relationship between figures in more detail.

Whenever you look at a house plan, use a dress pattern or make a scale model aeroplane, you are making use of similar shapes. We shall investigate more ways in which we can make use of this idea. Unfortunately, in everyday use the word 'similar' often means 'roughly the same', whereas in mathematics 'similar' has this much more exact meaning.

(*a*) How can you recognize the kind of tree from which the leaves in Figure 6 were copied?

The leaves from an oak tree are recognizable because they always have roughly the same shape, although it may be difficult to find two leaves which have the same shape and size or even have exactly the same shape. You all know a sheep when you see one, but although you would find it difficult to distinguish between them, a shepherd could, for no two sheep are exactly the same.

(*b*) Does the size of a television screen make any difference to the shape of the people you see pictured on it?

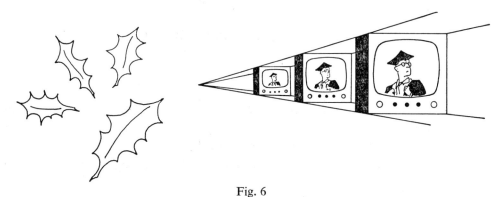

Fig. 6

Pictures on different sized television screens, and photographs of different sizes printed from the same negative, have *exactly* the same shape and (unlike two leaves or two sheep) are similar in the mathematical sense.

(c) Which of the following are *necessarily* similar in the mathematical sense:

    (i)   two squares of different size;

    (ii) any two rectangles;

    (iii) a 'Matchbox' car and the original from which it was copied;

    (iv) two electric light bulbs;

    (v) two footballs;

    (vi) a violin and a double-bass;

    (vii) two regular tetrahedra;

    (viii) a rectangle of length 6 cm and width 4 cm and a rectangle of length 18 cm and width 12 cm;

    (ix) a triangle in which two angles are 85° and 72°, and a triangle in which two angles are 23° and 85°?

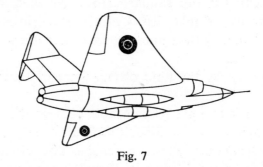

Fig. 7

(d) Figure 7 is a drawing of a scale model of a jet aircraft—the Gloster Javelin. The model has been carefully constructed so that externally it has exactly the same shape as the original aircraft. The scale used was $\frac{1}{50}$.

What does this mean?

(e) (i) The Javelin has a length of approximately 20 m. What is the length of the model?

    (ii) The wing span of the model is 25 cm. What is the wing span of the original?

    (iii) What height, approximately, would a model of a man be if it were made to the same scale?

    (iv) The angle between the leading edge and the trailing edge of each wing on the model is 50 degrees. What can you say about the corresponding angle of the original?

    (v) Will the surface area of the wings of the model be $\frac{1}{50}$ of the surface area of the original plane?

(f) Two space capsules have the dimensions shown in Figure 8. (*Note*: these diagrams are not drawn to scale.)

Are they similar?

Work out the scale factors for the corresponding lengths given. What do you find?

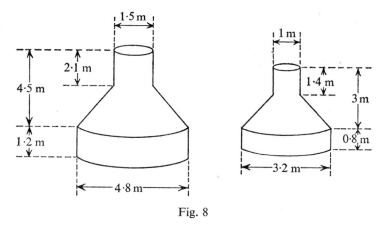

Fig. 8

Exercise B

1. A toy car is 3 cm long and constructed to a scale of $\frac{1}{120}$th of the original. What is the length of the original car?

2. A scale model of a double-decker bus is 75 mm high. What is the scale if the original bus is 5 m high?

3. A photograph is 8 cm long and 6 cm wide. What will be the dimensions of a smaller print of the same photograph if it is known that one side is to be 2 cm long?

4. The vertices of three right-angled triangles are given by the following sets of Cartesian coordinates:

$$A = \{(^-3, \, ^-3), (3, 5), (3, \, ^-3)\};$$
$$B = \{(^-1, 3), (1, 5), (2, 4)\};$$
$$C = \{(^-2, 3), (1, \, ^-1), (^-2, \, ^-1)\}.$$

Which two triangles are similar?

5. A toy shop had two models of a popular family car in stock. The first model was 12 cm long, 6 cm wide and 4 cm high, while the second model was 4·5 cm long, 2·1 cm wide and 1·5 cm high. Both models were advertised as 'accurate scale models'. Could this be true?

6. A photograph which is 6 cm wide and 9 cm long is mounted on a piece of white card so as to leave a border 1 cm wide all the way around. Is the card similar in shape to the photograph? Give reasons for your answer.

7. The Cunard liner 'Queen Elizabeth II' has the following specifications:

Length overall	290 m
Beam	32 m
Draught	9 m
Passengers	2 000
Crew	1 000
Gross mass in tonnes	59 000
Service speed	$28\frac{1}{2}$ knots

If a model is built of the ship on a scale of $\frac{1}{100}$th, what will be its specifications?

8. Not all football pitches have the same dimensions. Do they differ in shape as well as size? Is this true of hockey pitches?

9. Figure 9 shows an outline map of Australia. The map has been covered by a grid of lines spaced 1 cm apart. Make use of this grid to draw a similar map whose dimensions are double that of the figure.

How could you improve the accuracy of your enlargement using this method?

How could you draw a map of Australia to $\frac{7}{10}$ of the scale of the figure?

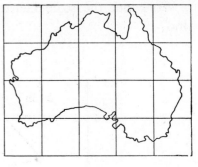

Fig. 9

10. Use what you know about similar figures to find the lengths represented by letters in Figure 10. The figures are not drawn to scale.

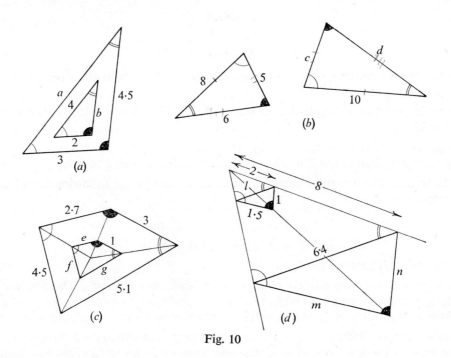

Fig. 10

1.2   Areas of similar figures

(a) Figure 11 shows three prints taken from the same negative. The small one is the same size as the negative, while the larger ones are '2 times' and '$2\frac{1}{2}$ times' enlargements. What are the areas of the three prints?

(b) An art master buys sheets of drawing paper which are 1 m long by 0·6 m wide. How many pieces whose dimensions are (i) $\frac{1}{2}$, (ii) $\frac{1}{5}$, that of the original size can he cut from a single sheet?

194

Fig. 11

Exercise C

1. (a) Make a copy of the triangular tessellation in Figure 12 and mark on it 3 triangles similar to the basic triangle of the pattern but of different size.

(b) What is the scale factor in each case?

(c) How many small triangles are needed to make up each of the different similar triangles?

(d) If the basic triangle of the pattern is taken as the unit of area, what are the areas of the similar triangles with dimensions, 2, 3, 4, 5, 6 and 7 times that of the basic triangle?

(e) If one triangle is a k times enlargement of another, how many times as big is its area?

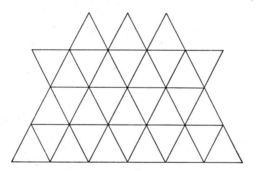

Fig. 12

2. Draw two rectangles which are not similar. Are equiangular triangles necessarily similar? Are Rhombuses? Are Circles?

3. A handy man calculated that he needed 72 lino tiles, each 30 cm square, to tile his kitchen floor. When he went to buy them he saw some tiles each 15 cm square which he liked better, so he bought 144 of them instead. Why was he surprised when he laid them?

4. Two triangles are similar. A pair of corresponding sides are 2 cm and 3 cm long. If the first triangle has an area of 16 cm², what is the area of the second?

195

5. On a map whose scale is 2 cm to the kilometre, Hangman's Wood is represented by a green patch of area 8 cm². What would be the corresponding area on a map whose scale is 5 cm to the kilometre?

6. Two photographs from the same negative are such that the area of one is 49 times the area of the other. If the smaller photograph is $1\frac{1}{2}$ cm × 2 cm, what are the dimensions of the larger photograph?

7. A balloon is approximately spherical. How many times does its surface area increase when it is blown up from a diameter of 5 cm to one of 25 cm?

8. A model village was made on a scale of $\frac{1}{10}$th that of the real village. Copy and complete the following table of dimensions.

Object	Model village	Real village
Length of cottage	1 m	—
Height of cottage	—	5 m
Floor area of cottage	0·7 m²	—
Length of football pitch	11 m	—
Width of football pitch	7 m	—
Area of football pitch	—	7700 m²
Area of village green	15 m²	—
Area of duck pond	—	60 m²
Area of fire station doors	0·5 m²	—
Cost of painting doors	2p	—

9. Figure 13 shows a tessellation of congruent triangles on which are marked out some similar triangles.

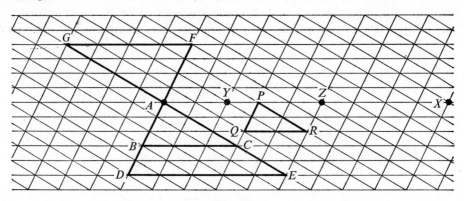

Fig. 13

(a) Give the centre and the scale factor of the enlargement which transforms:

(i) *ABC* into *ADE*;   (ii) *PQR* into *ABC*;

(iii) *PQR* into *AFG*;   (iv) *ADE* into *PQR*;

(v) *AFG* into *ABC*;   (vi) *ADE* into *AFG*.

(b) Take one of the basic triangles of the tessellation as the unit of area, and find the number of units of area of each of the triangles *PQR*, *ABC*, *ADE* and *AFG*.

To what set do these numbers belong? Give a reason why this is so.

196

1.3 Volumes of similar solids

Here is an extract from *Gulliver's Travels*, by Swift. 'I then made a sign that I wanted Drink. They found by my eating that a small quantity would not suffice me; and being a most ingenious people, they slung up with great Dexterity one of their largest Hogsheads, then rolled it toward my hand and beat out the Top: I drank it off at a draught, which I well might do, for it hardly held half a Pint, and tasted like a small Wine of Burgundy, but much more delicious.'

We are also told that 'a typical Lilliputian was a human creature not six inches high'.

How many Lilliputian pints did the hogs-head contain?

If a Lilliputian is a little less than 6-in high and a human being is a little less than 72-in high, then we have a scale factor of $\frac{1}{12}$. Figure 14 shows two similar hogsheads the larger one being a 12 times enlargement of the smaller one. What features of the larger are 12 times the corresponding ones of the smaller? Quite clearly from the drawing the larger contains much more than 12 times the smaller.

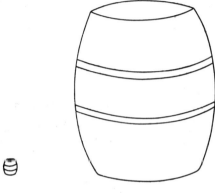

Fig. 14

For the larger one the area of the base is $12 \times 12 = 144$ times that of the smaller, and the height is 12 times as much.

By what factor will the volume have increased?

The smaller one, we are told, holds $\frac{1}{2}$ pint. Using this information can you find how many pints the large one holds? The small one will contain an equal number of Lilliputian pints.

Example

If a solid lead sphere of diameter 9 cm has a mass of 4 kg, what would be the mass of a solid lead sphere of diameter 10 cm weigh?

For the larger sphere, the diameter has been increased by a scale factor $\frac{10}{9}$ or 1·11 (to 3 s.f.).

Mass is related to volume, hence the mass of the larger (similar) sphere will be

$$(1 \cdot 11)^3 \times 4 = 5 \cdot 48 \text{ kg (by slide rule).}$$

We measure areas in *square* metres, *square* centimetres etc.; this should remind us that if the lengths of a figure are multiplied by a scale factor k, then the area is multiplied by k^2 or 'k squared'.

Similarly volumes are measured in *cubic* metres, *cubic* centimetres, etc; thus the volume is multiplied by $$k^3 \quad \text{or} \quad \text{'}k \text{ cubed'.}$$

Exercise D

1. A housewife mixes enough ingredients to make a cake for her family but discovers that some relatives are coming to tea. Because of this, she uses a cake tin which is the same shape but twice the dimensions of the tin she normally uses. What additional quantity of ingredients will she need to mix in order to make the larger cake?

2. If two similar jugs hold 1 litre and 8 litres, and the height of the smaller jug is 15 cm, find the height of the larger one.

3. A small tube of toothpaste 6 cm long costs 8p. What should be the price of a 'family size' tube of the same shape and 9 cm long? In fact it would probably be slightly less than the price you have worked out. Why?

4. A sphere of radius r cm has volume 5 cm³. What is the volume of a sphere of radius $2r$ cm?

5. The diameter of a sphere is 3 times that of another. How many times greater is:

(*a*) the surface area of the larger than the smaller;

(*b*) the volume of the larger than the smaller sphere?

6. A solid bronze model of a statue is 50 cm high and weighs 30 kg. How much would the full size statue of height 4 m weigh if it were made of solid bronze?

7. A 1 m long scale model of a swimming bath is made and found to hold 10 litres of water. If the real bath is 25 m long, how much water is required to fill it? The bath is 3 m deep at one end. How deep should it be in the model?

8. A boy designed and made a car using Meccano and a small diesel engine. The car was 20 cm long, had a mass of 1 kg, the engine had a capacity of 1 cm³, and the car travelled at 50 km/h. The boy argued that, if he had the materials, he could make a similar car 10 times as large which would be 2 m long, have a mass of 10 kg, have an engine of capacity 10 cm³ and travel at 500 km/h. Why has no one been able to build a car like this?

Summary

An enlargement with scale factor k and centre O can be constructed by drawing lines from O through points A, B, C, ... of the figure and marking off A', B', C', ... where $OA' = k.OA$, $OB' = k.OB$, $OC' = k.OC$, ...

When a figure is enlarged in this way the enlarged figure is similar to the original and corresponding lines are parallel. (See Figure 15.)

When two figures are similar:

(*a*) the lengths of one figure are always a constant factor (the scale factor) times the corresponding lengths of the other;

(*b*) corresponding angles are equal.

If two similar figures are such that lengths in the first are k times the corresponding lengths in the second, then the *area* of the first is k^2 times the area of the second, and the *volume* of the first is k^3 times the volume of the second.

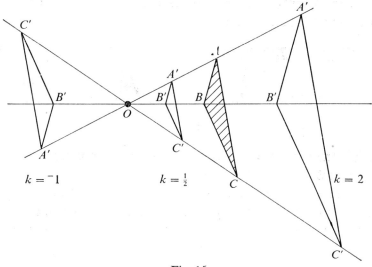

Fig. 15

2. THE SHEARING TRANSFORMATION

In our study of geometry so far we have dealt with reflections, rotations, translations and enlargements. Now we shall consider another transformation, shearing, which, amongst other things, has important applications to areas and volumes.

(a) In Figure 16(a) we see a side view of a pile of 30 exercise books. In Figure 16(b) we see the same pile but this time pushed sideways.

Approximately what shape is Figure 16(b)?

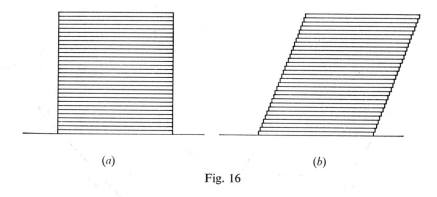

(a)   (b)

Fig. 16

If you measured the heights of the two piles, what would you expect to find? Are any parts of the books visible in one pile but not in the other? Are the areas of the parts of the two piles you can see the same? What other features of the pile of books have remained unaltered? Such features are said to be *invariant*.

199

(b) Figure 17 shows another example of the same idea, this time with a pile of thin cards such as a pack of playing cards.

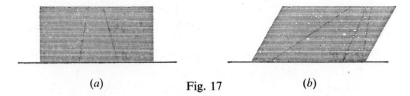

(a)   Fig. 17   (b)

In Figure 17(b) the jagged edges of Figure 16(b) have virtually disappeared and what we see is almost exactly a parallelogram.

The idea of shearing is illustrated by Figures 16 and 17 and one of the most important features of this transformation is that area is preserved, even though shape is not. (See also the picture at the beginning of the chapter.) This point will be considered in greater detail later in the chapter.

First, we shall examine other aspects of the shearing transformation.

2.1   Describing a shear

Make a copy of Figure 19, and, in order to help you answer the questions of this section, shade the various parallelograms and triangles mentioned using different colours.

(a) Suppose that in the tessellation in Figure 19, we represent the 'pile of books' by the parallelogram BFKE. Keeping EK fixed, that is, like the bottom book in the pile, we can shear parallelogram BFKE onto parallelogram FLKE. Similarly, keeping RQ fixed we can shear parallelogram RQVW onto parallelogram RQUV. Find two other parallelograms onto which RQVW can be sheared and state the side which is kept fixed in each case.

We can see that under a shear one line remains fixed in position, as, for example, the bottom book of the pile does in Figure 16. This line is called *the invariant line*. So far it has always coincided with one side of the object which has been sheared. However, this need not be the case. In Figure 18, *l* is the invariant line of the shear which maps ABCD onto A'B'C'D'.

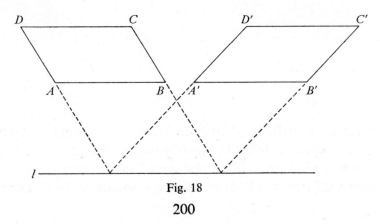

Fig. 18

200

(*b*) Consider how the parallelogram *LSRK* in Figure 19 could be mapped onto parallelogram *JPNH*.

It could, for example, be done by combining two shears of the type discussed in (*a*) above.

(i) *LSRK* → *LPNK* with the line *LK* invariant;

(ii) *LPNK* → *JPNH* with the line *PN* invariant;

the intermediate position, *LPNK*, being shown dotted. However, by considering where *SL* meets *PJ* and *RK* meets *NH*, can you see how the mapping can be accomplished using only one shear?

What would be the invariant line of this shear?

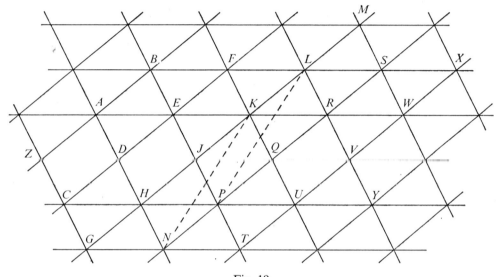

Fig. 19

(*c*) Where is the invariant line of the single shear which maps triangle *EFJ* onto triangle *DEJ*?

(*d*) Where is the invariant line of the single shear which maps parallelogram *BFUP* onto parallelogram *LSPH*?

What can you say about points on opposite sides of the invariant line in a shearing transformation?

(*e*) If *FE* is the invariant line of a shear, find the images of:

(i) parallelogram *PQUT* if *P* → *R*; (ii) parallelogram *AFKD* if *A* → *B*.

This shows us that a shear is defined if we are given its invariant line and the image of one point not on that line.

Exercise E

1. If the top book of 30 in Figure 16(*b*) has been pushed 6 cm to the right, approximately how far have the following books been moved:

(*a*) tenth from the bottom; (*b*) twentieth from the bottom?

2. When a certain figure is sheared, a point 2 cm from the invariant line moves 1·5 cm to the right. How far and in which direction do points move which are:

(a) 4 cm from the invariant line, and on the same side of it as the first point;
(b) 6 cm from the invariant line, and on the same side;
(c) 6 cm from the invariant line and on the *other* side of it?

Questions 3–10 *refer to Figure* 19 *and require a copy of this figure without the dotted lines. Extra lines may be required in some questions.*

3. Show, by marking the intermediate position in red, how two successive shears could be used to map parallelogram *ABFE* onto parallelogram *AEJD*.

Show also, using different colours, how it could be done by combining (a) three shears, (b) four shears.

4. Construct a sequence of shears (using as few as possible) which would:

(a) map triangle *GHC* onto triangle *NHG*;
(b) map triangle *LRW* onto triangle *RQU*.

Indicate intermediate images on your diagram.

5. Parallelogram *HPTN* can be mapped onto parallelogram *JQUP* by a translation. How can it be done by combining two shears?

6. Show on your diagram how *LSRK* can be mapped onto *JQUP* by:

(a) a shear followed by a translation;   (b) a shear followed by a rotation.

Give at least two methods in each case.

In (a) state the invariant line of the shear and a displacement defining the translation. In (b) state the invariant line of the shear and the centre and angle of rotation.

7. Describe the invariant lines of the shears which map:

(a) *FLRK* onto *DKRJ*;   (b) *FDQ* onto *FGL*.

8. Find the image under a shear of:

(a) *BFKE*, if *FE* is the invariant line and $B \rightarrow A$;
(b) *KHU*, if *JQ* is the invariant line and $K \rightarrow E$.

9. Describe four quadrilaterals onto which parallelogram *EKQJ* can be mapped by a single shear.

Are they all parallelograms?

Do you think you could shear a parallelogram onto a quadrilateral which is not a parallelogram?

What geometrical property does this suggest to be invariant under shearing?

10. We saw in Section 2.1 (b) how the combination of the two shears

$$LSRK \rightarrow LPNK, \text{ with the line } LK \text{ invariant, and}$$

$$LPNK \rightarrow JPNH, \text{ with the line } PN \text{ invariant,}$$

is equivalent to the single shear

$$LSRK \rightarrow JPNH \text{ with the line } AB \text{ invariant.}$$

Give another example in which the combination of two shears is equivalent to a single shear. What conditions must the invariant lines satisfy in each case?

11. Shear the 'man' in Figure 20 keeping $y = 1$ invariant and mapping A onto (7, 6).

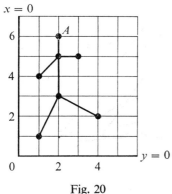

Fig. 20

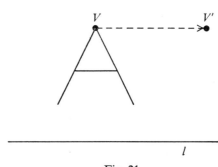

Fig. 21

12. Copy Figure 21 and, using only a straight-edge and set-square, draw the image of the letter A under the shear which has invariant line l and maps V onto V'.

13. In a shear the line $y = 0$ is invariant and the point (0, 1) is mapped onto (2, 1). What is the image of the line $x = 0$? What is the image of (0, 3)?

14. In a shear the line $x = 0$ is invariant and the line $y = 0$ is mapped onto the line $y = x$. Find the images of (0, 0), (0, 1), (1, 0) and (1, 1). What is the image of the square formed by these four points? Find also the images of the points (2, 2), (3, 1), (0, 4), (⁻1, 2) and (⁻1, ⁻1).

15. A shear has $y = 0$ as its invariant line. What can you say about the images of lines parallel to $y = 0$? Can you say that such lines are also in some way invariant? If so, how does this way differ from the invariance of *the* invariant line?

16. Draw the square whose vertices have the coordinates (0, 0), (1, 0), (1, 1), (0, 1). Transform each vertex into the point with the same x-coordinate as the original, but double the y-coordinate (for example (0, 1) → (0, 2)). What kind of figure do you obtain? What is its area?

This transformation is called a *one-way stretch*. The scale factor for this particular stretch was 2, and $y = 0$ was the invariant line.

What figure is obtained if we stretch the original square by a scale factor of 2, keeping $x = 0$ invariant?

What transformation is equivalent to the combination of the two stretches we have used?

2.2 Area of the parallelogram and triangle

If the base of a parallelogram is of length b cm and the perpendicular height is of length h cm, then:
$$\text{area} = b \times h \text{ cm}^2.$$

This can be found in two different ways. Any one of the sides can be regarded as the base. Figure 22 (*a*) shows the two rectangles onto which the parallelogram $ABCD$ can be sheared. One base and height are labelled. Notice that the directions of the shears are not across the page.

In Figure 22(b) triangle *PQR* can be sheared onto triangle *PMR*. Its area is therefore one-half of that of the rectangle *PRML*.

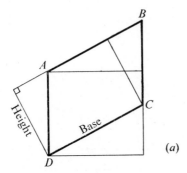

Fig. 22

If the base of a triangle is of length *b* cm and the perpendicular height is of length *h* cm, then its

$$\text{area} = \frac{b \times h}{2} \text{ cm}^2.$$

This can be found in three different ways. One case is shown in Figure 22(b).

Exercise F

1. Figure 23 shows a side view of a ream of 500 sheets of paper. The edges of the paper seen are 20 cm long and each sheet is 0·005 cm thick. What is the area of the shaded region? How does the transformation from the original rectangular shape differ from what we have called a shear?

Fig. 23

2. Draw a parallelogram with sides 3 cm and 4 cm and one angle 64°. Measure the distance between each pair of opposite sides and calculate the area of the parallelogram in two ways.

3. Draw a triangle with sides 6 cm, 7 cm and 10 cm. Draw three 'heights', measure them and calculate the area of the triangle in three different ways. How would you estimate the actual area from these answers?

4. *ABCD* is a rectangle in which *AB* = 7·8 cm and *AD* = 10·4 cm. *P* and *Q* lie on *AB* and *AD* respectively, with *AP* = 4·3 cm and *AQ* = 3·9 cm. Calculate the area of triangle *CPQ*. (This can be done without drawing or measurement.)

5. Draw accurately a quadrilateral *ABCD* in which *AB* = 6 cm, *BC* = 3·2 cm, *CD* = 5·6 cm, ∠*B* = 104° and ∠*C* = 136°. Calculate the area of *ABCD*.

6. A trapezium has sides of length 3 cm, 10 cm, 24 cm and 17 cm, with the shortest and longest sides parallel and 8 cm apart. Calculate its area.

3. SHEARING IN THREE DIMENSIONS

In Section 2 we considered the side view of a pile of books. We now apply the same idea to the pile of books as a whole.

Figure 24 shows how a rectangular cuboid is sheared parallel to one of its faces. Has the volume altered? Is the height still the same? Which faces are still rectangles? What are the shapes of the other faces? We can also shear the cuboid parallel to

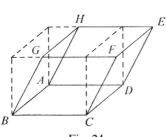

Fig. 24

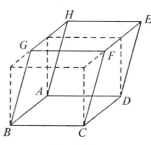

Fig. 25

another face (see Figure 25). The cuboid could be sheared obliquely (see Figure 26). Which faces are now still rectangles? How would you have to saw through this sheared cuboid to obtain a rectangular cross-section?

These cuboids are all *prisms*, that is, they have a constant cross-section parallel to a pair of end faces. They all have the same base area, the same perpendicular height and the same volume.

In particular:

$$\text{volume} = (\text{area of base}) \times (\text{perpendicular height}).$$

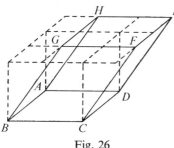

Fig. 26

Fig. 27

3.1 Cylinders

In a similar manner we can shear a cylinder. In Figure 27 the sheared pile of pennies on the right has the same volume as the pile on the left. The volume of both cylinders is the same, namely,

$$(\text{area of base}) \times (\text{perpendicular height}),$$

or $\pi r^2 h$ where the radius is r units and the height h units.

205

The results of this section—prisms on the same base and with the same height are equal in volume—is a special case of Cavalieri's Principle. This states that if two solids have equal areas of cross-section at the same distances from the base then they have equal volumes. Even if the sides of the pile of pennies are not straight the volume is still the same.

3.2   Pyramids

(*a*) In Figure 28 we see three congruent square-based pyramids fitted together to form a cube.

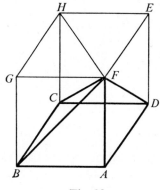

Fig. 28

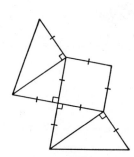

Fig. 29

The net for one of these pyramids is shown on a small scale in Figure 29. The lengths marked are all equal to the length of edge of the cube. (Tabs need adding. The net is easily marked out on graph paper and with some adjustment of the triangular faces three pyramids can be made together as a single net.)

Make three such pyramids to form a 5 cm cube. Since the volume of the cube is

$$(\text{area of } ABCD) \times (\text{height } AF),$$

the volume of each pyramid must be

$$\tfrac{1}{3}(\text{area of base}) \times (\text{perpendicular height}).$$

(*b*) Is this true for all pyramids? Before answering this question we note the following two facts.

(i) By shearing, we can see that pyramids with the same base *ABC* and equal heights have equal volume (see Figure 30).

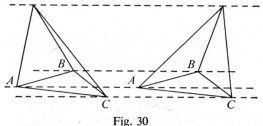

Fig. 30

206

(ii) Any pyramid can be dissected into tetrahedra as in Figure 31.

Consider, therefore, the tetrahedron *ABCL* drawn in thick outline in Figure 32. This can be regarded as part of the triangular prism *ABCNML* where *ABC* and *LMN* are congruent triangles in parallel planes. What shape are the faces *ABML*, *BCNM* and *ACNL*?

Why are the triangles *ABL* and *MBL* congruent?

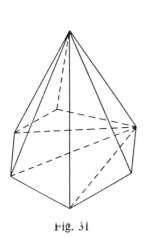

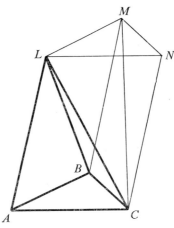

Fig. 31   Fig. 32

The prism is divided into the three tetrahedra *ABCL*, *LBMC* and *LMNC*. The tetrahedra *ABCL* and *LBMC* have congruent bases *ALB* and *LBM*. They have the same vertex *C*, and hence, the same height.

This implies that they have the same volume.

Why do the tetrahedra *ABCL* and *LMNC* have equal volume?

If the three tetrahedra are equal in volume and together they make the prism, then the volume of each tetrahedron is

$$\tfrac{1}{3}(\text{area of base}) \times (\text{perpendicular height}).$$

This, using facts (i) and (ii) above, implies that the same is true for all pyramids.

3.3   Cones

Just as a circle is the limiting form of a polygon with many sides, in the same way a cone is the limiting form of a pyramid with many faces. Its volume can therefore be found in a similar manner to that of the pyramids which approximate to it. Figure 33 shows a right circular cone and its volume is

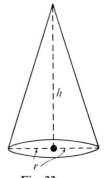

$$\tfrac{1}{3}(\text{area of base}) \times (\text{perpendicular height}) = \tfrac{1}{3}\pi r^2 h.$$

Fig. 33

207

Exercise G

1. Take a cuboid of cake in which the length, breadth and height differ (see Figure 34). Slice it into two wedges *ABCDEF* and *CDHEFG*. Divide the top wedge into three *differently* shaped tetrahedra by cuts along the planes *CAE* and *CAF*. Repeat for the bottom wedge. Why will the tetrahedra have equal volumes?

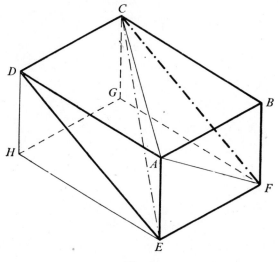

Fig. 34

2. How many $\frac{1}{4}$ kg packets of butter 9 cm × 5 cm × 3 cm can fit into a box 40 cm × 36 cm × 24 cm? Explain which way they would have to fit for your answer to be correct.

3. A water tank has a base 1 m square, and contains water to a depth of 60 cm. How far will the water level rise if a rock of volume 0·1 m³ is put in the tank and is completely submerged?

4. A drainpipe has external diameter 10 cm and internal diameter 9 cm. Find the volume of metal in a 6 m length of pipe.

5. An open wooden box has external measurements 60 cm × 50 cm × 25 cm high. The bottom and sides are 1 cm thick. Find the capacity of the box and the volume of wood used to make it.

6. The slanting edges of an 8 cm square-based pyramid are 9 cm long. Find:

(*a*) the area of each slanting face; (*b*) the volume of the pyramid.

7. Calculate the volume of a tetrahedron whose height is 6 cm and whose base is an equilateral triangle of side 4 cm.

8. Calculate the volume of a cone of radius 1·7 cm and height 3·4 cm.

9. Calculate the height of a cone of radius 3·2 cm and volume 39 cm³.

10. A cone has volume 400 cm³ and height 12·1 cm. Find:

(*a*) the area of its base; (*b*) the radius of its base.

Summary

A shear is a transformation in which:

(*a*) the points of a certain line, for example the line *l* in Figure 35, are invariant;

(*b*) all other points move **parallel to** the invariant line—the lines *PP'*, *QQ'* and *RR'* are all parallel to *l*;

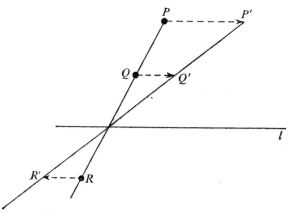

Fig. 35

(*c*) the distance moved by points is proportional to their distance from the invariant line;

(*d*) straight lines map onto straight lines, for example, the line *PQ* onto the line *P'Q'*. A line and its image will meet on the invariant line unless parallel to it.

A shear is defined once we know:

(i) the invariant line; (ii) the image of a point not on this line.

(*e*) Area of parallelogram   = (base) × (perpendicular height).

     Area of triangle   = $\frac{1}{2}$(base) × (perpendicular height).

     Volume of prism   = (area of base) × (perpendicular height).

     Volume of cylinder   = $\pi r^2 h$.

     Volume of pyramid   = $\frac{1}{3}$(area of base) × (perpendicular height).

     Volume of circular cone = $\frac{1}{3}\pi r^2 h$.

10

COMPARISONS

There is no excellent beauty that has not some strangeness in the proportion.

FRANCIS BACON, *Of Beauty*

1. RATIO AND PROPORTION

1.1   Ratio in everyday language

You are using the concept of ratio whenever you say 'twice as high', 'half as much' or 'two-thirds of the way there'.

1.2   Ratios for comparison

Last year the school hockey team won 10 of the 15 matches played. This year they won 13 matches out of an extended fixture list of 18. Which was the better season? Why was it not satisfactory merely to answer that this year was better because more matches were won?

To answer the question we must compare the number of matches won with the number played.

Last year: 10 won *compared with* 15 played or (in short) 10:15.
This year: 13 won *compared with* 18 played or   13:18.

These are called the *ratios* of matches won to matches played. The ratio 10:15 is read 'ten to fifteen'. To compare these ratios we look at the fractions that go with them. With the ratio 10:15 goes the fraction $\frac{10}{15}$ or $\frac{2}{3}$, and with the ratio 13:18 goes $\frac{13}{18}$. Which is the larger fraction, $\frac{2}{3}$ or $\frac{13}{18}$? This is easily seen. Since $\frac{2}{3}$ is equivalent to $\frac{12}{18}$, the second of the two is the larger fraction and so we say that this year's record is slightly better than last year's.

A ratio expresses a comparison between quantities. If I say 'Your hair is twice as long as mine', I am expressing the fact that (in my opinion) the ratio of your hair length to mine is 2:1. What two things could I say if I thought the ratio was 1:2?

1.3   Working with ratios

When we want to work with ratios we usually have to consider the associated fractions—the fractions that go with them. We saw this in the previous section and the following examples will make the method clear. We may or may not use the slide rule, depending on the nature of the fractions we have to simplify.

Example 1

Which is the better bargain:
(a) 7p off a hand towel, normal price £0·49; or
(b) 24p off a bath towel, normal price £1·44?
We can express each as a ratio. (Notice that, unless the two quantities are in the same units, it is meaningless to talk of the ratio between them.) The ratios are
(a) 7p:49p with associated fraction $\frac{7}{49} = \frac{1}{7}$.
(b) 24p:144p with associated fraction $\frac{24}{144} = \frac{1}{6}$.
Offer (a) saves you 1p for each 7p spent; offer (b) saves 1p for every 6p spent, so (b) is on this basis the better buy. What other factors could influence your decision?

Example 2

A man driving along a motorway at 85 km/h is overtaken by a car travelling at 95 km/h. What is the ratio of the speeds of the two cars?
We are not asked which way round to express this ratio and so we may put either speed first, so long as we make it plain which we are doing.

$$\text{Speed of slower car: speed of faster car} = 85 \text{ km/h}: 95 \text{ km/h}$$
$$= 85:95 = 17:19.$$

We could now write the associated fraction

$$\frac{17}{19} \quad \text{or} \quad \frac{17/19}{1} \quad \text{or} \quad \frac{1}{19/17}, \quad \text{i.e.} \quad \frac{17}{19} = \frac{Y}{1} = \frac{1}{X}$$

where X, Y may be read from the slide rule setting (see Figure 1).

$$\frac{17}{19} = \frac{0·89}{1} = \frac{1}{1·12} \text{ to slide rule accuracy.}$$

The position of the decimal point is found from an approximate answer.

So we can say that the speed of the slower car is 0·89 times that of the faster, or that the speed of the faster car is 1·12 times that of the slower. When we express a ratio in the form $1:n$, then n is the associated *scale factor*.

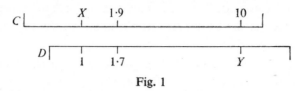

Fig. 1

Example 3

The ratio of the masses of two boys is 5:7. The lighter has a mass of 40 kg, what is the mass of the heavier?

$$\frac{5}{7} = \frac{1}{7/5} = \frac{1}{1\cdot4}; \quad \text{so the scale factor is } 1\cdot4.$$

Then the heavier boy has a mass of 1·4 × 40 kg, which is 56 kg.

Exercise A

1. Express the following ratios in their simplest terms and find the associated scale factors:

(a) 15p to 75p;   (b) 4 days to 6 weeks;
(c) £1·10 to 11p;   (d) 4·8 km to 12 m;
(e) 56 cm³ to 18 litres;   (f) 85000 people to 247 people.

2. A school has 500 children of which 40 are on the school committee. Express the ratio of committee members to children in the form $1:n$. Use the scale factor to find the appropriate number of committee members (under this arrangement) for schools of (a) 300 children, (b) 800 children.

3. A clothes line 10 m long is supposed to stretch 0·17 m. If the stretch ratio is the same, will a 25 m line be long enough to go between two posts 25 m apart? Assume the line is straight but that an extra 0·50 m is required for fixing.

4. Find the missing member of the ordered pairs in the following examples in which the ratios are given:

(a) 4:1, first member is 2 m;   (b) 1:3, first member is 5 km/h;
(c) 16:9, second member is 3 kg;   (d) 2:15, second member is £6·45.

5. In a certain school the ratio of boys to girls is 4:5. There are 240 boys. How many girls are there? In Form 1 A there are 32 children; can you say how many of them are boys?

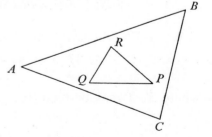

6. Measure the corresponding sides of the triangles in Figure 2. Work out in the form $1:n$

(a) longest side of ABC:longest side of PQR;
(b) middle side of ABC:middle side of PQR;
(c) shortest side of ABC:shortest side of PQR.
Are the triangles similar? If so, what is the scale factor?

Fig. 2

212

7. The triangles ABC and XYZ are similar (with corresponding vertices in the order stated). $AB = 4.63$ cm, $BC = 5.71$ cm. If $XY = 13.5$ cm, find the scale factor from ABC to XYZ and calculate YZ.

8. The fraction associated with the ratio of a given distance on a map to the corresponding distance on the actual ground is called the Representative Fraction (R.F.) of the map and is written in the form $1/n$. Is n a scale factor? If so, of what?

(*a*) What common map has an R.F. of $\dfrac{1}{63360}$?

(*b*) If a distance of 3·2 cm on a map represents a distance of 160000 cm on the ground, find the R.F. of the map.

(*c*) On a third map the R.F. is $\dfrac{1}{25000}$. What distance in kilometres does a distance of 3·45 cm on this map represent?

9. A boy was having his fourth birthday when his sister was born. Write down a table showing the age of the boy and his sister after 1, 2, 4, 8, 20, 80 years have passed. Work out the ratio of her age to his in the form $1:n$ in each case. Discuss whether she is 'catching him up'.

Summary

Ratio compares like quantities. With a ratio such as $3:5$ is associated the fraction $\frac{3}{5}$.

If a ratio is expressed in the form $1:n$, then n is the scale factor of the ratio.

1.4   Simple proportion

Consider the following table:

Number of slide rules bought (S)	1	2	3	4	5	6
Total cost in pounds (T)	1·80	3·60	5·40	7·20	9·00	10·80

What are the ratios of corresponding pairs?

They are in fact all equal $1:1.80 = 2:3.60 = 3:5.40 = \ldots$.

We say that the cost is *proportional* to the number of slide rules. This is an example of *direct* proportion.

We shall meet other types of proportionality later.

If S is the number of slide rules and T the total cost in pounds, write down the equation connecting T and S. What would the graph of the function $S \, \scriptstyle\rangle \, T$ look like?

Discuss which of the following pairs of quantities are normally connected by a proportion relation:

(*a*) number of kilometres travelled; time taken (the speed being constant).

(*b*) age of boy; size of cap worn.

(*c*) number of exercise books; height of stack.

1.5 Proportional sets of numbers from the slide rule

Make the setting on the C and D scales of your slide rule indicated diagrammatically in Figure 3.

What do you find under 2, 3, 4, 5 on the C scale? What would you expect under 6 on the C scale, and how would you confirm this?

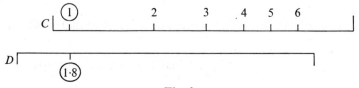

Fig. 3

The connection between the *ordered* set $A = \{1, 2, 3, 4, 5, 6\}$,

and the *ordered* set $B = \{1{\cdot}8, 3{\cdot}6, —, —, —, —\}$

can now be described in two ways. We can say:

(a) the ratio of each pair of corresponding elements of the two sets is the same;
or

(b) the elements of the two sets are proportional (more briefly, the two sets are proportional).

What is the function that maps the first set onto the second? How can the second set be obtained from the first by using a scale factor?

1.6 Writing down proportion

If x, y and z are three numbers such that $x:y = 1:2$ and $y:z = 2:5$, what is the ratio $x:z$?

We can write $x:y:z = 1:2:5$, a form which includes all three ratios considered.

If $p:q:r:s = 1:2:3:4$, pick out the ratios of

(a) $p:q$,   (b) $p:r$,   (c) $p:s$,   (d) $q:s$,   (e) $r:s$.

If $p = 3$, what are the values of q, r and s?

We can now express the proportionality of the sets in Section 1.5 either in the form

$$1:2:3:4:5:6 = 1{\cdot}8 : 3{\cdot}6 : 5{\cdot}4 : 7{\cdot}2 : 9{\cdot}0 : 10{\cdot}8$$

or in the form
$$\frac{1}{1{\cdot}8} = \frac{2}{3{\cdot}6} = \frac{3}{5{\cdot}4} = \frac{4}{7{\cdot}2} = \frac{5}{9{\cdot}0} = \frac{6}{10{\cdot}8}.$$

Example 4

Figure 4 shows a pair of similar figures.

Express the relation between the lengths of corresponding sides in two different ways.

214

(i) Now (*b*) is an enlargement of (*a*), consequently the length of any side of (*b*) is obtained from that of the corresponding side of (*a*) by multiplying by the scale factor *n*.

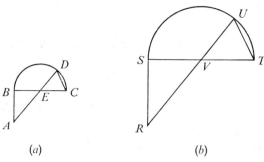

(*a*)   (*b*)

Fig. 4

That is
$$\frac{ST}{BC} = \frac{UT}{DC} = \frac{RV}{AE} = \ldots = n.$$

(ii) Also $$ST:UT:RV:\ldots = BC:DC:AE:\ldots.$$

Looking further into this question we may note that it is equally true that the arc lengths $\overset{\frown}{BD}$, $\overset{\frown}{SU}$, ... are related in the same way. (We write the length of an arc with a curved sign over the top to distinguish it from the length of the straight line segment.)

For example,
$$\frac{\overset{\frown}{US}}{\overset{\frown}{DB}} = \frac{\overset{\frown}{UT}}{\overset{\frown}{DC}} = \frac{ST}{BC} = n,$$

or $$\overset{\frown}{US}:\overset{\frown}{UT}:ST = \overset{\frown}{DB}:\overset{\frown}{DC}:BC.$$

Example 5

Divide 60 hens between three henhouses in the ratio 5:3:2.

The hens have to be shared out into $5+3+2 = 10$ shares. Each will consist of

$$\tfrac{1}{10} \times 60 = 6 \text{ hens.}$$

In the first henhouse we shall put $5 \times 6 = 30$ hens,

in the second $\qquad\qquad\qquad\qquad 3 \times 6 = 18$ hens,

in the third, $\qquad\qquad\qquad\qquad 2 \times 6 = 12$ hens.

We check that $30+18+12 = 60$, the total number. We can write out this problem like this:

$$5:3:2 = \tfrac{5}{10}:\tfrac{3}{10}:\tfrac{2}{10}$$

$$= \tfrac{5}{10} \text{ of 60 hens}:\tfrac{3}{10} \text{ of 60 hens}:\tfrac{2}{10} \text{ of 60 hens}$$

$$= 30 \text{ hens} : 18 \text{ hens} : 12 \text{ hens.}$$

Example 6

Three authors write an article and agree to share the fee of £80 in the ratio of the approximate number of words each writes. Ashworth writes 2200, Baker writes 1700 and Campbell writes 3650. How much does each get?

Since an approximate answer is required, we can use the slide rule. The total number of words is

$$2200 + 1700 + 3650 = 7550,$$

and this corresponds to £80. The setting is shown in Figure 5.

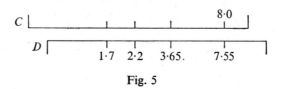

Fig. 5

Ashworth gets £23, Baker £18 and Campbell £39, to the nearest pound.

Exercise B

1. Are the following pairs of sets proportional to within slide rule accuracy?

(a) $\{5, 2, 4\}$, $\{2\frac{1}{2}, 1, 2\}$;   (b) $\{5, 6, 7\}$, $\{6, 7, 8\}$;

(c) $\{4, 5\cdot65, 9\cdot45\}$, $\{1\cdot195, 1\cdot69, 2\cdot82\}$.

If you think they are, describe the function that maps the first set onto the second.

2. Use your slide rule to find the values of the letters in the following pairs of proportional sets:

(a) $\{23, 7\}$, $\{12, x\}$;   (b) $\{3, 190\}$, $\{y, 62\}$;

(c) $\{2\frac{1}{2}, p, q\}$, $\{1, 3, 5\}$;   (d) $\{r, 0\cdot9, s\}$, $\{13\cdot7, 5\cdot2, 0\cdot86\}$.

3. Use your slide rule to write down the set onto which $\{4\cdot2, 7\cdot8, 10\cdot9\}$ is mapped under

$$x \rightarrow \tfrac{12}{23}x.$$

4. On a certain journey the set of distances in kilometres from A to B, B to C, C to D is $\{63\cdot5, 91, 55\}$. In the road map that the motorist consults, 1 cm represents 2·5 km. Find the distances apart in centimetres of A and B, B and C, C and D on the map.

5. Write down an ordered set of values of x containing 4 elements and the corresponding set of values of y if they are connected by the relation $3y = x$. Are the elements of the sets proportional?

6. $\frac{1}{4} = \frac{3}{12} = \frac{4}{16} = \frac{8}{32}$; write down at least 10 other relations, for example $3 = \frac{1}{4} \times 12$ and $\frac{8}{3} = \frac{32}{12}$.

7. Write the following in fraction form. Where possible find the scale factor.

(a) $3 : 1 : 4\frac{1}{2} = 6 : 2 : 9$;   (b) $LM : MO : NP = DE : EG : FH$;

(c) $x : 6 : y : z = 2 : 3 : 5 : 9$;   (d) $30° : 45° : 90° = a° : b° : 360°$.

216

8. In the triangle ABC, AB = 2 cm, BC = 3 cm and CA = 4 cm. Write down the lengths of the sides of a triangle similar to ABC:

(a) given that its smallest side is 10 cm long;

(b) given that its largest side is 6 m long.

9. $ABCD$ and $PQRS$ are two quadrilaterals in which

$$\frac{AB}{PQ} = \frac{BC}{QR} = \frac{CD}{RS} = \frac{DA}{SP} = \frac{CA}{RP}.$$

Make a sketch of them and say whether or not they must be similar.

10. Figure 6 shows two circles with the same centre O, and radii at 60°. Which of the following statements are true? If you think that any are false, say why.

(a) $\widehat{CB}:CB:CA = \widehat{RQ}:RQ:RP$;

(b) $CR:\widehat{RQ} = AP:\widehat{PQ}$;   (c) $\dfrac{\widehat{CB}}{RQ} = \dfrac{\widehat{CA}}{RP}.$

Fig. 6

11. Do you get more land as the junior partner in a 3:2 share out of 85 hectares, or as the senior partner in a 4:1 share out of 45 hectares?

12. Divide 360° into 4 angles in the ratio 1:2:3:4.

13. In a certain week a boy spent 5p on chocolate, 10p on fares, 15p on his stamp collection and 20p on a visit to the cinema. Write down and simplify the ratio of these sums of money. Use your answer to the previous question to help you draw a pie chart to illustrate his spending.

14. Divide £16·50 between three people A, B and C so that A gets £3 more than B who gets £3 more than C.

15. Four shareholders hold 380, 150, 515 and 640 shares respectively in a company. Divide £7500 between them to the nearest £ in the ratio of their holdings.

Summary

If $a:p = b:q = c:r = \ldots$, then the ordered sets $\{a, b, c, \ldots\}$ and $\{p, q, r, \ldots\}$ are proportional.

Also

$$\frac{a}{p} = \frac{b}{q} = \frac{c}{r} = \ldots$$

Adjacent pairs of numbers on corresponding scales of a slide rule are in the same ratio. The sets of numbers on the two scales are proportional. (See Figure 7.)

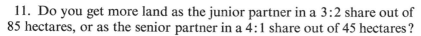

Fig. 7

Sharing in the ratio 5:3:2 involves a total of 10 shares, and so the corresponding fractions are

$$\frac{5}{10}, \frac{3}{10}, \frac{2}{10}.$$

To show quantities on a pie chart, for example the composition of a town council with 5 Conservatives, 6 Liberals and 4 Labour, we divide the 360° at the centre of a circle in the ratio 5:6:4 and divide the area inside the circle into appropriate sectors.

2. PERCENTAGES

2.1 A new way of comparing fractions

(*a*) Consider the fractions $\dfrac{6}{25}$ and $\dfrac{4}{17}$.

How can we decide which is the larger? One way is to replace them by a pair of equivalent fractions which have the same denominator (bottom number). What will this be? Find which is the larger using this method.

(*b*) Now express both the fractions as decimals to as many places as are needed. What decides the number of places? Does this check the answer you obtained by the previous method?

(*c*) We could express them both as fractions with a standard denominator. It is convenient to take this as 100. We can see that

$$\frac{6}{25} = \frac{24}{100} \text{ (easily done in the head);}$$

while

$$\frac{4}{17} = \frac{23 \cdot 5}{100} \text{ (to slide rule accuracy).}$$

It is clear that $\frac{6}{25}$ is the larger. Does this agree with your findings? When we take 100 as the denominator, we call the numerators of these standard fractions *percentage*s. We use the symbol % to mean 'divided by 100', often in the sense of 'per hundred' whence the term. You will note that while the fraction $\frac{6}{25}$ has an exact percentage corresponding to it, namely, 24, the percentage for $\frac{4}{17}$ can only be approximate, namely 23·5 or, to a higher degree of accuracy, 23·53.

Figure 8 shows a number line between 0 and 1 divided in three equivalent ways:

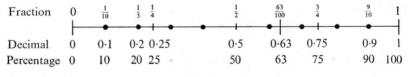

Fig. 8

What is the relation between any fraction *F* and the associated percentage *P*.

A few fractions have been marked on Figure 8. Read off some other fractions or decimals and convert into percentages and vice versa. Write some of your answers using the % sign, and some as fractions with denominator 100.

Example 7

A town of 55 000 men of working age has 1750 unemployed. Is there much cause for alarm?

Expressing the data in fraction form we get:

$$\text{the fraction of men unemployed is } \frac{1750}{55\,000} = \frac{7}{220}.$$

Another way is to express the original fraction with the standard denominator, 100:

$$\frac{1750}{55\,000} = \frac{7}{220} = \frac{3\cdot14}{100};$$

that is, $3\cdot14\,\%$ of the men of working age are unemployed. This makes it easy to compare it with, for instance, the national figure. Is $3\cdot14$ a low figure? How much could it be due to men changing their jobs and taking some time to find new ones? The percentage tells us that, for every 100 men of working age, $3\cdot14$ men are unemployed. Does the idea of $0\cdot14$ of a man seem silly? How many men are unemployed per 1000? How many per 10000? Discuss the use of the $0\cdot14$.

2.2 Percentages backwards

(a) Calculate $20\,\%$ of £10 and $20\,\%$ of £12. Why do your results differ?

(b) On Monday the price of a transistor radio was £10. On Tuesday the shop-keeper raised the price by $20\,\%$. On Wednesday he announced a SALE and reduced all prices by $20\,\%$. Explain clearly, using your answers to (a) above, why the price on Wednesday was not £10.

Remembering that a percentage represents a fraction with a denominator of 100, it is plain that the most important thing about a percentage is what it is a percentage *of*. While $95\,\%$ may sound a lot, $95\,\%$ of $0\cdot001$ is a small number, much smaller than $1\,\%$ of 100.

Example 8

A boy's pocket money was increased by $25\,\%$ to 25p. How much was it before? It is important to realise that the 25p is *not* the value on which the $25\,\%$ is calculated.

If the original pocket money were xp (i.e. $100\,\%$ of xp), then after a $25\,\%$ increase its value would be $\frac{125}{100}x$p (i.e. $125\,\%$ of xp).

So $\frac{125}{100}x = 25$.

The inverse of $\frac{125}{100}$ under multiplication is $\frac{100}{125}$. (See Chapter 4, page 76.)

So $x = \frac{100}{125} \times 25$.

That is, $x = 20$.

His pocket money was 20p.

Example 9

The total number of cars produced in October was 230000. This was a decrease of $8\,\%$ from the September total.

How many cars were produced in September?

Let the number of cars produced in September be x (100 % of x).

Then in October $\frac{92}{100}x$ (i.e. 92 % of x) will be produced.

So, $\frac{92}{100}x = 230000$.

$x = \frac{100}{92} \times 230000 = 250000$. ($\frac{100}{92}$ is the inverse of $\frac{92}{100}$ under multiplication.)

In September 250000 cars were produced.

Exercise C

1. Express as fractions: 50%, 75%, 140%, 66% and 400 %.

2. Express as percentages: $\frac{2}{5}$, $\frac{5}{8}$, 0·35, 0·3, $2\frac{1}{2}$.

3. Find the values of:
(a) $12\frac{1}{2}$ % of 88p;   (b) 200 % of £600;
(c) 17·4 % of 15 kg;   (d) $88\frac{1}{4}$ % of 350 million people.

4. An examination is marked out of 180 and 45% is required to pass. What is the highest mark obtainable by a candidate who fails?

5. If all prices are to be reduced by 15% find the new prices for items at present marked:
(a) £1·25;   (b) £0·45;   (c) £4·05.
Is a slide rule appropriate for this question?

6. The speedometer of a car overreads by 5 km/h when its true speed is 60 km/h. What is the percentage error? What will the speedometer read when the true speed is 50 km/h if the percentage error is the same?

7. Find the original value if:
(a) a 40% increase gives £180;   (b) a 200 % increase gives 450 kg;
(c) a 50 % decrease gives 45 cm;   (d) a $2\frac{1}{2}$ % decrease gives 198 volts.

8. If the foreman's wages were increased by 60 % and the manager's decreased by 40 %, they would both get £1800. What does each get now?

9. 'Fifteen per cent more powder' says the advertising blurb on a detergent packet. It now holds 350 g. How much did it hold before?

10. A car cost £770 when new. Its value decreased by 25 % in the first year and by about 12 % each year in the next three years. If this is calculated each year on the value at the beginning of that year, estimate the value of the car after four years to the nearest pound.

Summary

20 % means $\frac{20}{100}$.

20 % of 500 means $\frac{20}{100} \times 500$.

If when X is increased by 20 % the result is 500, then $\frac{120}{100}X = 500$,

$$\text{that is, } X = \frac{100}{120} \times 500.$$

(Multiplying by the multiplicative inverse of $\frac{120}{100}$.)

If when Y is decreased by 20 % the result is 500, then $\frac{80}{100}Y = 500$,

$$\text{that is, } Y = \frac{100}{80} \times 500.$$

3. GRADIENTS

3.1 Contours

You will have learned about contour lines from your geography lessons. Figure 9 shows a contour map of a small region. Describe the physical features of the region.

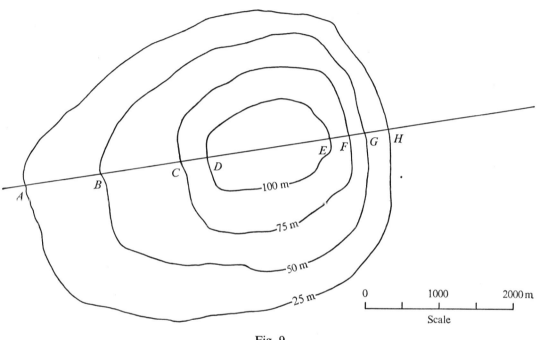

Fig. 9

If the line AH represents a straight road describe a journey along it. Taking the road as $y = 0$ and a vertical line (out of the plane of the page) through A as $x = 0$, write down the coordinates of A, B, C, D, E, F, G and H, and plot them on graph paper. Do you need the same scale on both lines? Should the points be joined by a straight line?

Where do you think the highest point of the road is? Which part of the road is steepest? Can you tell immediately from the contour map?

3.2 Gradient of a road

Assume for the present that the slope between C and D is steady. The road rises 25 m in about 350 m horizontally. We call this a *gradient* of 25 m in 350 m, or 1 m in 14 m. The gradient measures the steepness of the slope. Is this the same gradient as 1 km in 14 km?

221

The gradient is usually referred to as simply 1 in 14, which means 1 *up* for 14 *horizontally*. As the road will not usually have exactly the same slope between any two points, it is called an *average* gradient over a section of road. Later we shall discuss an actual gradient at a particular point.

In mathematics, it is more convenient to have a gradient expressed as a fraction or, if this is awkward, as a decimal. The gradient above would be expressed as a gradient of $\frac{1}{14}$ or 0·07, to 2 significant figures.

On the Continent, the gradient 1 in 14 above would be expressed as a percentage, $\frac{1}{14} \times 100 = 7 \cdot 1 \%$ to 2 significant figures.

Exercise D

1. (*a*) From the contour map in Figure 9, express the average gradient between the points *B* and *C*: (i) in the form 1 in *n*, (ii) as a Continental gradient, (iii) as a mathematical gradient.

(*b*) Write down the average mathematical gradient of the road between:

(i) *A* and *B*, (ii) *F* and *G*, (iii) *A* and *D*.

Discuss how we might distinguish uphill from downhill as we go from *A* to *H*.

(*c*) What is the average mathematical gradient between *A* and *H*. Discuss the meaning of this.

(*d*) If a car can only climb a hill whose slope is no greater than 1 in 9 can it travel from *H* to *A*?

2. Convert the following gradients into mathematical gradients, writing your answers in decimal form:

(*a*) 1 in 5;   (*b*) 1 in 4;   (*c*) 1 in 100;   (*d*) 1 in 1000.

Could the gradient of a horizontal road be written in the form 1 in *n*?

3. Write the following percentage gradients in the form 1 in *n*: (*a*) 5 %, (*b*) 25 %, (*c*) 8 %.

4. What gradient (mathematical) would describe a rise of 80 m in 0·5 km?

5. Draw an accurate contour map, giving a scale and labelling the contours, containing: (*a*) a road of constant gradient 15 % (but not necessarily straight) and (*b*) a road whose gradient varies from 0 to 15 %.

Summary

Road gradient 1 in *d*, that is 1 unit up for *d* units horizontally (see Figure 10).

Mathematical gradient $1/d$, or expressed as a decimal; it is a pure *number*.

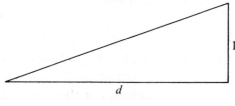

d

1

Fig. 10

4. RATES OF CHANGE

4.1   Average rate of change

(*a*) The graph in Figure 11 shows the sad story of a boy trying to catch a bus. It shows the graph of the function

'time since he left home → distance covered'.

Between what times was he moving (i) fastest, (ii) slowest? Describe what you think happened.

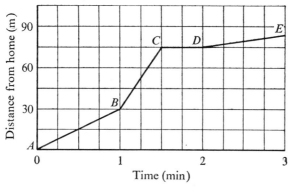

Fig. 11

(*b*) In the section *BC* of the graph the boy walks 45 m in $\frac{1}{2}$ min. We say that he walks at a *rate* of 45 m per $\frac{1}{2}$ min, that is his position is changing at a rate of 45 m per $\frac{1}{2}$ min.

45 m per $\frac{1}{2}$ min is therefore a *rate of change*.

Is this the same rate as 45 cm per $\frac{1}{2}$ sec? Or 45 km per $\frac{1}{2}$ h?

Clearly it is necessary to include the units since a rate of change compares two different quantities, in this case distance in metres with time in minutes.

A convenient way of writing this rate of change (or speed) is to write 45 m/$\frac{1}{2}$ min or, simplifying in the same way as with ratios, we can write this as 90 m/min. What is his speed in m/s? Write down the speed of the boy in m/s in the sections *AB*, *CD*, *DE* of the graph. Notice that the rate of change is the gradient of the line with the 'distances' measured from the scales and the units included.

Discuss the difference it would make to these speeds if the graph in Figure 11 were plotted on a different scale.

Exercise E

1. Figure 12 shows a graph of the volume of water in a bath. Describe what you think was happening, giving rates in the proper units where appropriate.

223

2. Figure 13 shows the temperature out of doors in England on a certain day. Write down the average rate of change of temperature between noon and 14.00 h, 14.00 h and 18.00 h, 18.00 h and 20.00 h, and 20.00 h and 22.00 h. What time of year do you think it was? Copy the graph and make up some suitable figures for the hours from 8.00 h to noon. Discuss how we could indicate whether the rate of change means that it is becoming hotter, or colder.

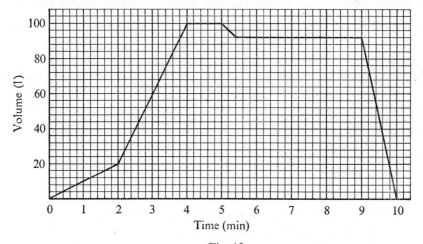

Fig. 12

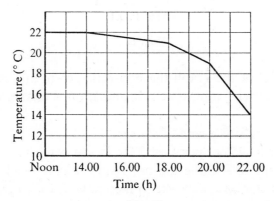

Fig. 13

3. On the same diagram draw the graphs which describe a 4 h journey made by a man called Arthur who walks at 7 km/h, and by a man called Benjamin who walks at 8 km/h but stops for 15 min rest after every hour of walking. Assuming that they leave town at the same time, who arrives first at the 8th, 16th and 24th kilometre posts? Describe the sort of walking race at which each excels.

4. Bacteria are multiplying in a culture tray. Originally there are about 2000000. The rate of increase is approximately constant for 10 h. Conditions are then varied and for the next 10 h the rate of increase is again constant and about twice the previous rate. At the end of 20 h there are about 11000000 bacteria. Sketch the graph of growth. Find the rates of increase in millions of bacteria per hour. Is this the way you would expect the number of bacteria to increase?

Summary

Average rate of change of, say, temperature over a certain interval of time is

$$\frac{\text{change in temperature (degC)}}{\text{change in time (min)}}.$$

When simplified it may be expressed in degC/min.

A rate of change must have its units stated.

A rate of change is the mathematical gradient of a graph, the distances across and up the page being measured in the units of their respective scales.

4.2 Rate of change at an instant

The speedometer of a car is a familiar sight. What does a reading of 50 km/h on it actually mean?

The speedometer in a car was covered up and the following experiment was carried out to estimate speed of the car at different intervals on a journey of 7 km.

The distance recorder was first set to zero, and then the distance travelled at intervals of 2 minutes was recorded. The results were:

Distance (km)	0	0·2	0·7	1·6	2·8	4·3	5·5	6·3	6·8	7·0
Time (min)	0	2	4	6	8	10	12	14	16	18

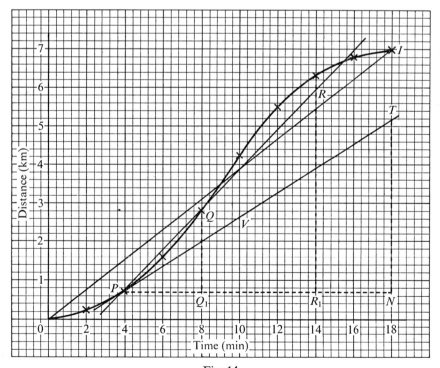

Fig. 14

225

Assuming that the distance was continuously changing during each 2 minute interval, the points in the graph may be joined by a smooth curve, as in Figure 14.

(*a*) What can we deduce from this graph?

The average speed for the whole journey (7 km in 18 min) is represented by the gradient of the line *OI*. This is

$$\frac{7}{18} \text{ km/min} \quad \text{or} \quad \frac{7}{18} \times 60 \text{ km/h} = 23 \text{ km/h correct to 2 s.f.}$$

Would the speedometer read 23 km/h for the whole journey?

(*b*) The average speed for the interval 4–8 minutes is represented by the gradient of the line *PQ*. This average speed is

$$\frac{(2 \cdot 8 - 0 \cdot 7)}{(8 - 4)} \times 60 \text{ km/h} = 32 \text{ km/h, to 2 s.f.}$$

(It should be noted that $QQ_1/PQ_1 = RR_1/PR_1 = \ldots$, all represent the value of this speed.)

Would the speedometer necessarily read 32 km/h constantly during this interval?

(*c*) The answers to the last questions in (*a*) and (*b*) are almost certainly 'no'. The speedometer of a car rarely shows the same reading for any length of time.

Does the 32 km/h represent the speed shown on the speedometer at *P* (time 4 min)? Again the answer is 'no' as the 32 km/h depends on knowing the distance travelled at 8 minutes!

Make a copy of the graph and use it to find the average speeds for the intervals 4–6, 4–5, 4–4·1 minutes.

The smaller and smaller the interval from 4 minutes becomes, the nearer and nearer does the line, whose gradient represents the average speed over the interval, approach the line *PT* in the figure. The line *PT* is said to *touch* the curve at *P* and is called a *tangent* to the curve at *P* (the Latin word for 'I touch' is 'tango').

The gradient of the tangent at *P* measures the speed *at* time 4 minutes. This is the value that would be shown on the speedometer at time 4 minutes. Estimate it from your graph.

As the figure shows, this means that if after 4 minutes the speed remained constant, the car would have travelled 2·6 km at 10 min (point *V*); 5·2 km at 18 min (point *T*).

Of course every point on this curve has a tangent, the gradient of which represents the speedometer reading at that point.

Figure 15 shows some tangents drawn to the curve.

From your own graph find the values of the speed at various times. When is the car travelling fastest and what is its speed?

(*d*) On a new diagram graph 'time → speed' and so estimate the acceleration (the rate of change of speed with respect to time) at various times.

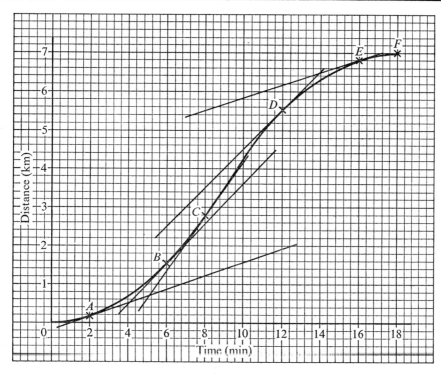

Fig. 15

Example 10

Figure 16 shows how the mass of a baby boy varied in the first 18 months of his life, assuming a smooth rate of growth.

Find the rate of change of mass at six months.

When is the rate of change of mass greatest and estimate its value?

The tangent at P is drawn and the gradient of it is found from the fraction $\dfrac{BC}{AC}$.

(These points are chosen as suitable for two reasons (i) they are lattice points on the tangent, (ii) they are far apart.)

The rate of change of mass at six months is

$$\frac{(10\cdot6-5\cdot8) \text{ kilograms}}{(13\cdot8-2\cdot1) \text{ months}} = 0\cdot41 \text{ kg/month (approximately).}$$

The rate of change will be greatest when the tangent has the largest gradient. This is obviously at time 0, at point Q on the graph. (The tangent at Q is difficult to draw and it might help to think of joining Q to another point on the curve very close to Q.)

The greatest rate of change of the mass is

$$\frac{(10\cdot6-6\cdot6) \text{ kilograms}}{(8\cdot4-3\cdot6) \text{ months}} = 0\cdot83 \text{ kg/month (approximately).}$$

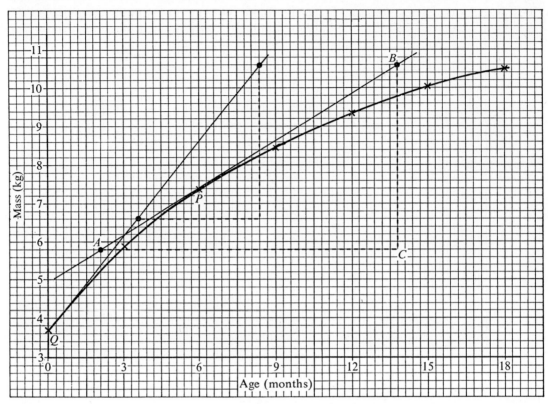

Fig. 16

Exercise F

1. The graph in Figure 17 shows the mass chart of a boxer reducing for a fight. When was he losing mass at the greatest rate? Express this rate of loss in kilograms per day.

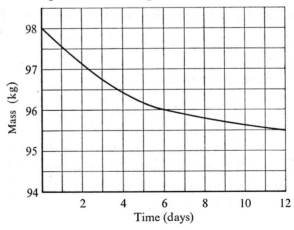

Fig. 17

228

2. Make a tracing of the graph in Figure 16. From this find the rate of increase of mass of the boy at 3 months and 9 months. Discuss whether the curve is likely to continue in the same manner in the following months.

3. A saucepan is filled with sugar and water and heated slowly. The temperature readings are as follows:

Time (min)	0	1	2	3	4	5	6
Temperature (°C)	20	21	24	29	32	34	35

At what time was the rate of temperature change greatest?
Estimate:
(a) this rate;
(b) the average rate for the first 3 min;
(c) the rate of temperature change at 5 min.

4. Figure 18 shows the graph of the height of an oat seedling. Discuss the meaning of the zig-zag part of the base line. Estimate its rate of increase of height in centimetres per day at 12 days and 15 days after planting. Use your answer to estimate the height after 30 days. What assumption do you make?

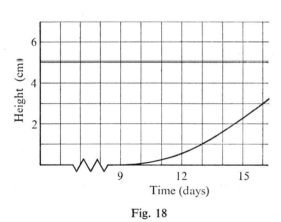

Fig. 18

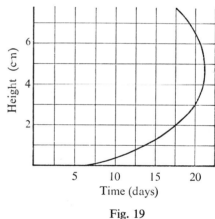

Fig. 19

5. Figure 19 shows the graph of the height of an oat seedling. Criticize it.

D6. A boy blew up a spherical balloon. The following table shows the radius at various times:

Time (s)	0	2	4	6	8	10
Radius (cm)	0	4	7	9	10	$10\frac{1}{2}$

(a) Discuss how these data may have been collected.
(b) Plot the data and then draw your estimate of the function 'time → radius' (that is, time across the page and radius up the page). When was the rate of increase of radius greatest? Discuss why this occurs. Find the rate in cm/s.
(c) Draw up a new table of values showing the times when the radius had a given measurement (see below). Take any necessary readings from the graph

Radius (cm) 0 2 4 6 8 10
Time (s)

(d) Plot the readings taken from the graph (b) and graph the function 'radius → time' (radius across the page and time up the page). At what point is the gradient of this graph smallest? Describe the geometrical connection between the two curves.

(e) Discuss the meaning of the gradient in (d). How is it related to your answers to (b)?

(f) How could these data have been collected direct? Why are so many sets of experimental data *time-based*?

7. (a) Figure 20 shows the barometric pressure taken from a recording machine. Write down the rate of change of pressure at 10.00 h and at 18.00 h. What major difference is there between your answers? Discuss how we might indicate this. What was the gradient of the graph at 15.00 h? What was the rate of change of the pressure? Why is this of interest?

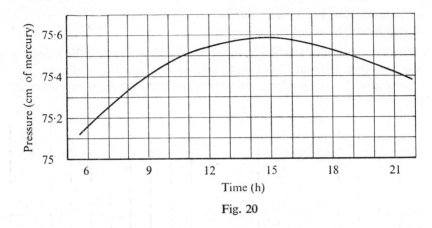

Fig. 20

(b) At 10.00 h on the morning following, the rate of change of pressure was 0·06 centimetres of mercury per hour and the pressure stood at 75·3 cm. Estimate the pressure at 11.30 h. Discuss the problem of estimating the pressure during the afternoon.

D8. Benzene diazonium chloride decomposes to form nitrogen. At 50 °C with an initial concentration of 10 grams per litre, the following results were obtained:

Time (min)	6	9	12	14	18	22	24	26	30
Nitrogen (cm³)	19·3	26	32·6	36	41·3	45	46·5	48·4	50·4

When the reaction had finished, a volume of 58·3 cm³ of nitrogen had been obtained. Graph 'time → volume'.

The rate of reaction is proportional to the gradient of the curve. Does the rate vary as the reaction proceeds?

A measure of the diazonium salt remaining is A cm³, where

A cm³ = (final volume of nitrogen) − (volume of nitrogen at a given time).

Find the value of A at times 6, 12, 18, 26 min and the gradients at these times. Graph 'gradient → A'.

Can you now say how the rate varies with the concentration of the diazonium salt?

230

Summary

Fig. 21

The gradient at a point of a curve is the gradient of a tangent drawn to touch the curve at that point (see Figure 21).

The rate of change at an instant is found by finding the rate of change represented by the tangent to the curve at that instant.

Figure 22 is the graph of the function '$q \to p$'. Figure 23 is the graph of the relation '$p \to q$'.

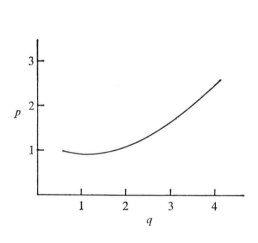

Fig. 22

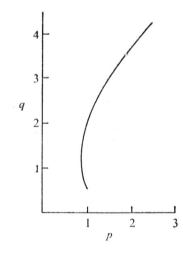

Fig. 23

11

TOPOLOGY

Stretch him out longer...
WILLIAM SHAKESPEARE, *King Lear*

1. TOPOLOGICAL MAPS

Figure 1 shows a street map of London's West End. Underground railway stations are marked like this ● and the railway itself like this ⁀⁀⁀ .

Figure 2 is a schematic map of the same area showing the underground railway only. The various lines have been drawn in different styles.

Answer the following questions where possible. You will find that some can be answered only from Figure 1, mark your answers to these *A*; some only from Figure 2 mark these *B*; and some can equally well be answered from either, mark these *E*. What else do you need to know before you can answer the remaining questions?

(*a*) If you get on at Oxford Circus and travel on the direct line to Holborn, how many stations will you pass through?

(*b*) Which is the shorter of the two routes from Piccadilly to Tottenham Court Road?

(*c*) Which stations would you pass through immediately before and after you passed through Strand?

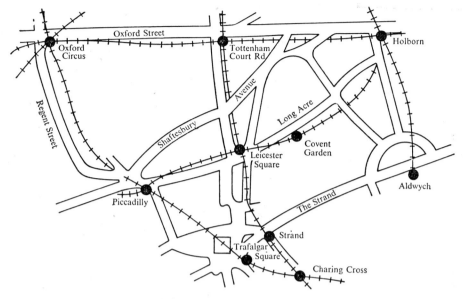

Fig. 1. London: West End.

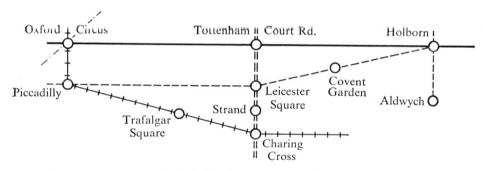

Fig. 2. London underground.

(*d*) Name any station which is a terminus.

(*e*) How long will it take from Charing Cross to Oxford Circus if trains travel at an average speed of 50 kilometres per hour and stop at each station for 1 min?

(*f*) Write down a route from Tottenham Court Road to Oxford Circus involving travelling on all four of the different lines shown.

(*g*) Do you need to change trains in going from Strand to Holborn?

(*h*) Which two stations are nearest to each other?

(*i*) On which parts of the underground can you be travelling (i) north, (ii) south-west?

Discuss the main differences between these two maps and say why the underground map is often drawn in the way shown in Figure 2.

A topological map shows the main features of a country, district, railway or building, without giving any idea of scale or showing whether routes are straight or curved.

The easiest way to picture what has happened is to think of an accurate map printed on a very thin rubber sheet. This can then be twisted about as much as you like so long as you do not actually tear it or stick two bits together. Plainly, you will only be twisting it *usefully* if you make the curved roads straight, or more simply curved. The underground map in Figure 2 shows more information than a topological map need; for example, it distinguishes between different lines.

Figure 3 shows an irregular curve drawn on a rectangular piece of rubber sheet which is pulled about so that the curve becomes circular. Note that in the process the edges of the sheet are pictured as themselves having become irregular. This is not necessarily the case.

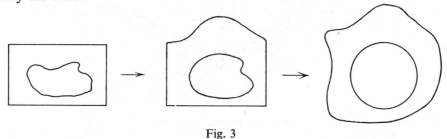

Fig. 3

Topology is sometimes called rubber sheet geometry. It is about points, lines and the figures they make. Certainly it is geometry; but distance, curvature and angle can be transformed as much as one likes. In other transformations that we have investigated in previous chapters, some or all of these things have not been allowed to vary. This kind of transformation seems to come very near to chaos!

Exercise A

1. Figure 4 shows a topological map of the international routes of a well-known airline. Which one do you think it is?

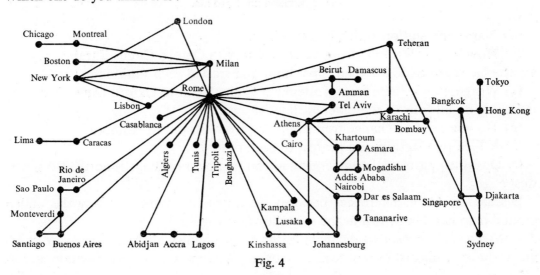

Fig. 4

How could you get from Montreal to London using this airline? Can you tell which is the greater distance, London to Rome or Abidjan to Lagos?

Make up (and, if possible, answer from the map) some more questions.

2. Figure 5 shows the principal roads in the 'heel' of Italy, the main roads being indicated by thick lines. Draw a topological map of these roads, making the roads joining Mandúria, Galátone and Lecce form a circle.

Mark these towns and also Brindisi, Mesagne, S. Vito, Nardò, Gallipoli, Máglie and Otranto. From your map describe how you would turn at junctions in making a tour from Otranto to Brindisi by way of Lecce, Galátone and Mandúria, keeping to main roads. What important point on the map does not have a name?

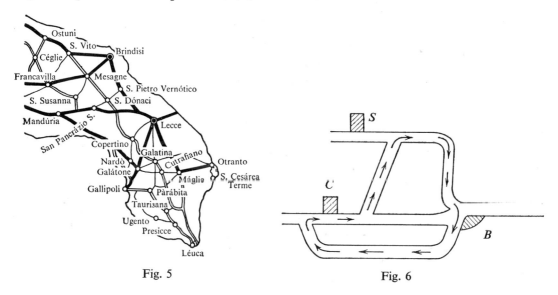

Fig. 5   Fig. 6

3. Figure 6 shows a street map of Whopham. Redraw the map so that the one-way Ring Road (marked by arrows) is a circle. Mark in the bus station (*B*), the cinema (*C*) and the school (*S*) and write down directions for getting by car from the school to (*a*) the bus station and (*b*) the cinema. Can you find the answer equally well from either map? What sort of directions should you *not* give?

4. Make a topological sketch map of your own school. Mark on it the route from the main gate to the head-teacher's study. Write out a list of directions so that a stranger visiting the school can find the headteacher.

5. Figure 7 shows a topological map of a road system. How many (*a*) cross-roads, (*b*) junctions, are there? Sketch your idea of what the actual map looks like given that the three sections marked *p* actually form a straight road running S.W.–N.E., the section *q* is winding but runs roughly N.–S. and that the sections *r* form a straight road running E.–W. Mark the *p*'s, *q*'s and *r*'s on your sketch.

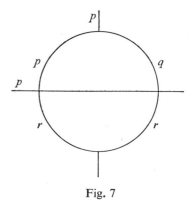

Fig. 7

235

6. A sketch of the relationships between members of a family is called a family tree. Find a family tree in an encyclopaedia, or draw your own. Answer as many of the following questions as you can:

(a) Write down two pairs (x, y) so that the relation is 'y is grandfather of x'.
(b) How old was each father when his oldest son was born?
(c) How many pairs (x, y) are there in which the relation is 'y is a cousin of x'?
(d) Does the tree help you to see at a glance who was alive in any particular year?
(e) If not, can you suggest a way of modifying it so that it does?
(f) Find a hereditary diagram in a biology book. Is this a family tree?

Is a family tree a topological diagram?

7. Let
$$A = \{(x, y) : x > 0\};$$
$$B = \{(x, y) : y > 0\};$$
$$C = \{(x, y) : x + y < 4\};$$
$$D = \{(x, y) : x + y > 4\}.$$

Draw a graph and shade $A \cap B \cap D$. Find three integers p, q and r such that (p, q) is a point of $A \cap B \cap C$ and (r, q) a point of $A \cap B \cap D$. If photographs were taken of your graph paper from various angles would (p, q) and (r, q) still appear to belong to the same sets? Could you twist your paper (but without tearing it or sticking two points together) so that (p, q) was a point of $A \cap B \cap D$?

8. In Figure 8, A, B, C, D and E represent lamps, p, q and r are switches and s is a two-way switch. Write down, if possible, an arrangement of switches under which:

(a) A and B are on, C, D and E are off;
(b) C, E and B are on, A and D are off;
(c) A, B, C and D are on, E is off.

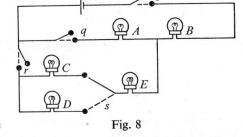

Fig. 8

Are any of the switches unnecessary? Would the answers be changed by bending the wires into new shapes? Discuss short ways of saying, in binary code, which lamps and switches are on and which are off.

9. A rectangular grid of lines 1 cm apart is printed on a thin rubber sheet and the scales are marked with 5 cm as the unit. The sheet is then twisted. Describe what happens to:

(a) your ability to pick out the point $(3, 4)$;
(b) the distance between $(2, 5)$ and $(2, {}^{-}5)$;
(c) the fact that $(1, 1)$ lies inside the circle with centre $(0, 0)$ and radius 2 units;
(d) the fact that $(1, 2)$, $(2, 5)$ and $(3, 8)$ lie on a straight line;
(e) the angle between $y = 2x$ and $y = 2$;
(f) the position of the point $(2{\cdot}1, 3{\cdot}3)$.

2. TOPOLOGICAL INVARIANTS

You will have noticed by now the sort of questions that can be answered from a topological map and the sort that cannot. Which of the following are 'topological questions' (i.e. can be answered from a topological map.)?

(a) How far?

(b) Between which points on the line?

(c) What angle?

(d) How many lines meet at this point?

(e) North or south? (Be careful!)

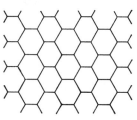

Fig. 9

Figure 9 shows a piece of wire netting. Suppose it to be twisted and bent or even stretched, but not torn nor soldered together. At the end of this it will be quite a different shape and possibly will no longer even be flat. Nevertheless, there are still some statements that you can make about the netting in its original state which are still true of it in its new state. Discuss what these can be. Think also of some properties of the original which are now quite different.

We call bending and stretching (but not tearing or sticking) *topological transformations*. The properties of the netting or underground map which remain unaltered by the transformation are, as in all transformations, the *invariants* of the transformation.

The underground map shows the result of a topological transformation on the railway lines in the street map. We can tell from it how many stations remain before we have to get off, since the order of stations on the line is invariant. We cannot tell how far we have to go to the next station—distance is not invariant.

Here is a list of the invariant facts we have already met. Let us write dashed letters for the transformed points and lines, that is, the points and lines into which we may imagine the original ones to have been twisted and stretched.

(a) If a point P lies on a line l, then P' lies on l'. (See Figure 10(a).)

(b) If $P_1, P_2, P_3, ...,$ lie in that order on l, then $P'_1, P'_2, P'_3, ...$ lie in that order on l'. (See Figure 10(b).)

(c) If n half-lines meet at P, then n half-lines meet at P'. (See Figure 10(c).)

We shall say that n half-lines meet at P if there are n different routes away from P; for example, in Figure 10(c) there are 8 half-lines meeting at P. Two special cases are worth mentioning:

(i) an end-point P becomes an end-point P'; it has just one half-line from it.

(ii) an angular point, like the corner of a square, may be smoothed out in the transformation and becomes just a point on a curve; we can still think of it as having two half-lines, meeting there.

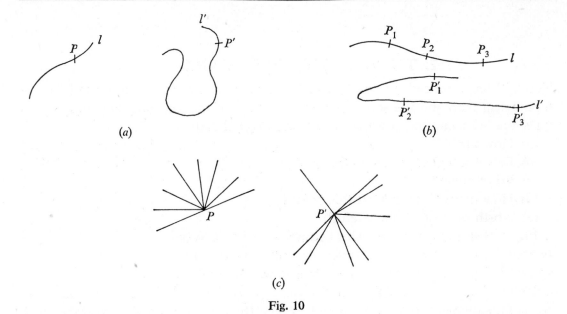

(a)

(b)

(c)

Fig. 10

3. TOPOLOGICAL TRANSFORMATIONS

Let us look a little more closely at the requirement that the 'rubber sheet' shall neither be torn nor stuck together which we have mentioned several times.

If we allowed the sheet to be torn, two points which lay on the same line before the transformation could lie on different ones afterwards. This would certainly make topological maps unsatisfactory. In more advanced work we do allow the sheet to be torn and stuck together again, but for the present we shall keep to the '*No Tearing Rule*'. Another way of putting this is to say that the transformation is *continuous*. As a train P approaches a station A in the street map, its transform P' approaches A' in the transformed map and they arrive simultaneously. (See Figures 1 and 2.)

The rubber sheet must not be folded over and stuck together so that two points of the original become one in the transformed figure. This would mean that a point not on a certain line could have its transform on the transform of that line. Oxford Circus could be transformed so that it is on the Piccadilly to Holborn line. Obviously we must exclude this.

Two figures in the same plane such that one can be transformed topologically onto the other are called *topologically equivalent*.

Exercise B

1. State whether the first of each pair in Figure 11 could be transformed topologically into the second.

2. Which of the pairs in Figure 12 are topologically equivalent? If you think a pair is equivalent, copy the second member of the pair and mark on it a possible position for A', the image of A. If more than one position is possible, mark all of them.

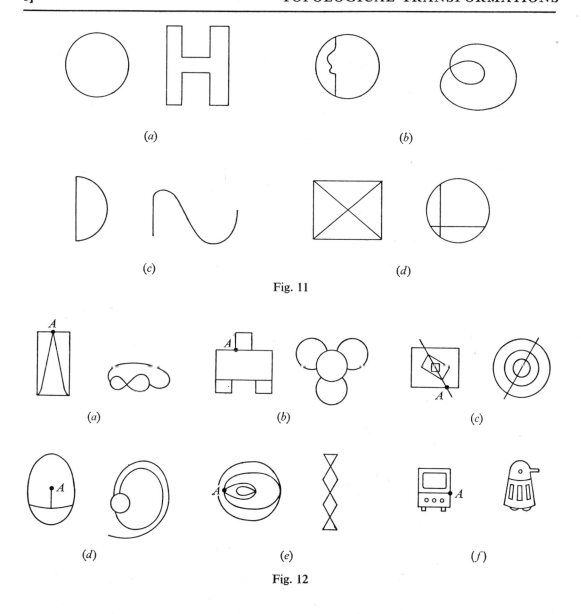

Fig. 11

Fig. 12

3. Pick out any of the line sketches in Figure 13 which are equivalent to each other.

4. *Simple closed curves*

A figure that can be transformed topologically into a circle is called a simple closed curve. Which of the following are simple closed curves:

(*a*) an equilateral triangle;
(*b*) a letter L;
(*c*) a hexagon;
(*d*) the symbol for infinity (find out what it is);
(*e*) the letter P?

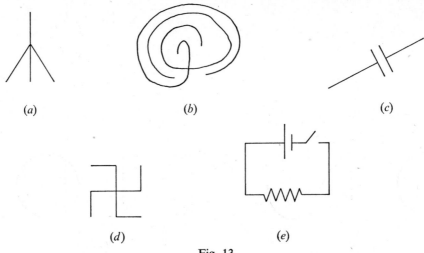

(a) (b) (c)

(d) (e)

Fig. 13

5. Inside and outside are not very easy to define in mathematical language. It is not even easy always to know which is which. Figure 14 shows a simple closed curve and three points X, Y and Z. Which two are on the inside? Discuss a definition of this and a test which can be applied.

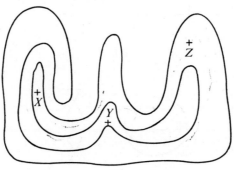

Fig. 14

6. Draw several different simple closed curves. Choose a point inside the curve and draw a line from your chosen point to the outside of the curve. Count how many times it cuts the curve. Can you discover a rule? Does the same rule apply to a curve that is not a simple closed one? Are 'inside' and 'outside' topological invariants?

7. A parallelogram $ABCD$ is transformed into a square $EFGH$. Discuss whether it is always, sometimes or never true that:

(a) $A \to E$, $B \to F$, $C \to G$, $D \to H$;
(b) $A \to E$, $B \to G$, $C \to F$, $D \to H$;
(c) $A \to G$, $B \to F$, $C \to E$, $D \to H$.

8. *Nodes*

A point where half-lines meet is called a *node*. Its *order* is the number of half-lines involved (see Figure 15).

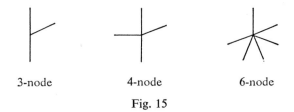

3-node   4-node   6-node

Fig. 15

Make a list of the 'small' letters in Figure 16 which have, amongst others:

(*a*) one 3-node;   (*b*) two 3-nodes;   (*c*) one 4-node.

Why do no letters have nodes of order higher than 4? Draw a 1-node and a 2-node. What is peculiar about a 2-node?

a   b   c   d   e   f   g   h   i   j   k   l   m   n   o   p   q   r   s   t   u   v   w   x   y   z

Fig. 16

9. Explain which of the rules in Section 2 (Page 237(*a*), (*b*), (*c*)) requires the number and order of the nodes to be invariant, and how this is useful in deciding whether or not two figures are equivalent. Is it possible for two figures to have exactly the same number of nodes of the same order, but not to be equivalent?

4. TOPOLOGICAL PROJECTS

Each of the five sections which follow is a project which you should attempt together with three or four other pupils (or, if you prefer, on your own).

You should try at least two of the projects, developing the theme along your own lines not necessarily sticking to those suggested, although these can be taken as guide lines.

4.1   Topological designs

Draw the capital letters of the alphabet in as simple a form as you can. For instance do not draw the fourth letter like this **D** if you think that it can be clearly recognized as a more simple shape, omitting the serifs, like this D. How many different topologically equivalent sets have you?

Draw our own alphabet as small letters and the Arabic numerals in the same way. Do you need any more sets to accommodate them?

Consider Roman numerals and any other systems and alphabets that you can find. Use any reference encyclopaedias and try to decide whether our methods in the present Western world are really as simple as they could be. How do the Chinese and Iraqis manage? How much progress has been made in the last three thousand years?

The coming of the computer and the television screen have produced new visual shapes. Find out what you can about the shapes used here. Could you improve on those normally used for instance on your parents' cheques or computerized accounts received from some large national concerns?

Produce wall charts to show your investigations and conclusions.

What you have discovered may have led you to consider what really can be done with lines on a flat surface. Draw, if possible, figures or *networks* to the following specifications.

	1-nodes	3-nodes	4-nodes	5-nodes
(a)	0	0	2	0
(b)	0	1	1	1
(c)	5	0	0	0

Continue this table making up your own specifications and trying to follow them. Are there any general rules about such networks? Can you prove that the total number of odd nodes must be even? How many of the networks that you have drawn could you produce by a single trace of the pen without going over any part twice? Are there any rules about this?

4.2   Schlegel diagrams

(a) Make from card a regular tetrahedron with edges about 5 cm long. (How many ways are there of doing this?)

(b) Cut along two edges which go to one vertex and fold back one face. Draw what you see as you look inside. Imagine your tetrahedron is made of rubber sheet. Could you flatten it out so that you see something like Figure 17? Where would you indicate the face which you folded back? All round the outside? (It was attached to all three edges, a, b, c.)

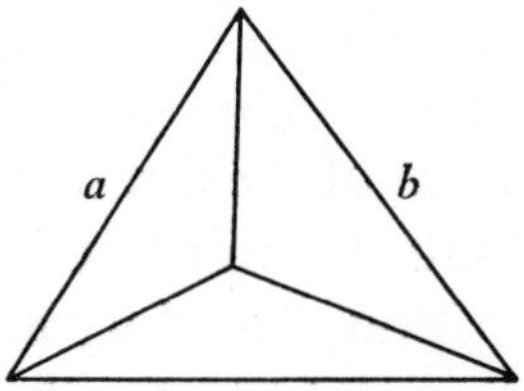

Fig. 17

(c) Make other polyhedra (start with simple ones like the cube and octahedron, but be adventurous): (see Figure 22 for the Platonic solids, each with a face removed) and do the same thing as you did in (b), drawing them on a flat surface. Find out how many completely regular polyhedra there are.

(d) Draw up a table showing how each diagram is made up.

Figure	(polyhedron)	Arcs (edges)	Nodes (vertices)	Regions (faces)	Nodes + Regions
1	Tetrahedron	6	4	4	
2	Cube	12	8	6	
3	Octahedron				
4	Dodecahedron				

What do you notice about the numbers of arcs and (nodes + regions)? Are they always related to each other in the same way?

Keep your diagrams of the polyhedra drawn flat (Schlegel diagrams) for Projects 4·3 and 4·5.

(e) Draw all sorts of doodles on your paper. Does the relationship in (d) still hold?

(f) Draw networks, in a simple way first of all, on the surfaces of a ball and on the inner tube of a motor tyre. (See Figure 18.) Does the relationship hold in these two cases?

Can you discover a rule for each?

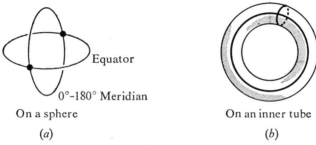

On a sphere

(a)

On an inner tube

(b)

Fig. 18

4.3   Traversability

(a) A network is said to be traversable if it can be drawn with one sweep of the pencil, without lifting it from the paper, and without tracing the same arc twice. It is easy to see that any simple closed curve, a figure eight or the word '*level*' can be so drawn. In more complicated cases you may have to try several starting places before you find the way, or convince yourself that it is impossible. Which of the networks in Figure 19 are traversable?

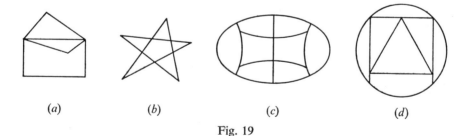

(a)   (b)   (c)   (d)

Fig. 19

(b) We have looked at railway lines, electrical circuits, road maps etc. One of the first applications of the ideas we have been using was not nearly so important. In 1737 the famous Swiss mathematician *Leonard Euler*, then working at the Court of Catherine the Great of Russia, had his attention drawn to the problem of the Koenigsberg bridges. This German university town is now found on maps under the Russian name of Kaliningrad. In the River Pregel, which runs through the town, are two islands; they are joined to the banks of the river and to each other as shown in Figure 21.

The problem was this: is it possible to take a walk so as to cross each of the bridges once and once only? Euler dealt with this problem by first reducing it to a problem about nodes and arcs. See if you can solve the problem that puzzled the best mathematicians in Europe two centuries ago, until Euler noticed the underlying principles.

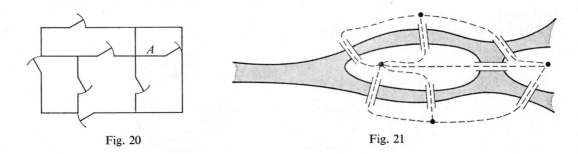

Fig. 20   Fig. 21

(c) Is the ground floor plan of your home like that in Figure 20? It is hardly likely to be! There is something particular about this plan: it can be traversed. Draw the corresponding network and discover where to start. How many such places are there? Why?

(d) 2-nodes, 4-nodes and so on are called even; 1-nodes, 3-nodes and so on are odd. Copy and complete the following table for curves (a), (b), (c) and (d) in Figure 19 and add some more of your own.

Figure	Number of even nodes	Number of odd nodes	Traversable Yes or no?
(a)			
(b)			
(c)			
(d)			

(i) Can you traverse a network which has only even nodes?

(ii) Can you traverse a network having just two odd nodes and any number of even nodes?

Go on investigating the effect of odd and even nodes on traversability.

(e) How many ends does a line segment have? When you traversed a network did you have to fix a beginning and an end, or two ends? If you started at an odd node could you finish at the same node? Why not? How many odd nodes are possible in a traversable network? Can you have only one odd node? What happens if you have none? Develop your ideas in the same way with even nodes and decide on the rules for traversability of networks.

(f) Apply your findings to the polyhedra and their edges. Are any of the five Platonic solids traversable along their edges? (See Figure 22. The Schlegel diagrams of Section 4.2 may help.)

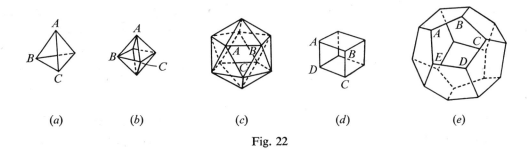

(a)　　　　(b)　　　　(c)　　　　(d)　　　　(e)

Fig. 22

4.4   Colouring regions

When colouring in this project (and the following one), use as few colours as you can. We suggest you start with one bright colour and add others only when it is absolutely necessary, that is, when two regions would otherwise be the same colour on each side of an arc. Lettering with the same letter the regions to be the same colour is as efficient as using coloured pencils, though less fun.

(*a*) Copy the patterns in Figure 23 and colour them using as few colours as possible.

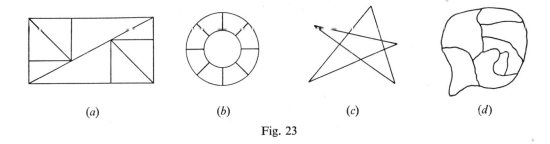

(*a*)　　　　(*b*)　　　　(*c*)　　　　(*d*)

Fig. 23

(*b*) Find line patterns which repeat such as those in Figure 24, or make up your own. How many colours do you need if we assume that the pattern goes on for ever, that is, they fill the complete plane and so contain an infinite number of tiles?

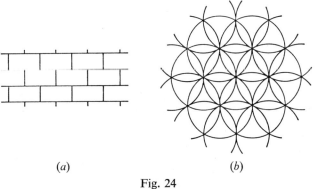

(*a*)　　　　　　　　(*b*)

Fig. 24

(c) Trace a map of northern or southern England showing the counties and colour it. Are an unnecessary number of colours used in conventional atlases? Why?

(d) Draw a *unicursal* curve, that is a continuous line which comes back to its starting point. It may cross itself as often as you like, but you must not retrace an arc already drawn (see Section 4.3). How many colours do you need to colour the regions of your design? Do not forget the outside region!

(e) Take a sheet of paper about the size of this book page and rule a line across it. There are two regions needing two colours to distinguish them but do not colour them. Rule a second line to make as many new regions as you can. How many regions now? How many colours are needed now? Continue making more regions, as many as you can each time you draw a new line across the page. Find out how the number of colours increases. Make a note of the numbers of regions as you draw more lines. Have you seen this sequence of numbers before?

(f) Have you ever had to use more than four colours for any diagrams? Can you prove that you would never need more than four?

4.5 Colouring polyhedra

(a) Make a few regular tetrahedra from white card. Colour them in such a way that no two faces with an edge in common are the same colour. Can any two separate faces be the same colour? How many colours will be needed? If four

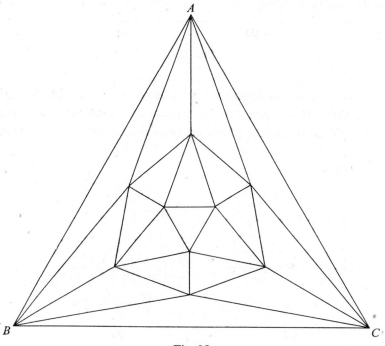

Fig. 25

246

colours were used, how many different tetrahedra of the same size could be coloured so that you could distinguish each one from the others? If the tetrahedra were not regular, would your answer be different?

(b) Try this with the other four Platonic solids (see Figure 22). You may find it helpful to draw a Schlegel diagram as described in Section 4.2(b). The Schlegel diagram for an icosahedron, (Figure 22(c)), is shown in Figure 25.

(c) Make as many other solids as you can. Colour them in different ways, always using as few colours as possible. Is it possible to distinguish otherwise identical shapes by different arrangements of the same colours?

(d) What happens if you paint a motor car inner tube with regions drawn on it? Can you find the minimum number of colours required in this case?

5. EULER'S RELATION

We can now try to show why the situation which you may have discovered in Section 4.2 always holds for any network on a flat surface. You will have discovered in that section that networks drawn on other surfaces give similar results, but that the 2 is replaced by other numbers.

For the flat surface then $R - A + N = 2,$

where R, A and N are the numbers of regions, arcs and nodes.

In the Schlegel diagram for a cube, the face $W'X'Y'Z'$ of the cube has been removed and corresponds to the region surrounding the network. (See Figure 26(a).)

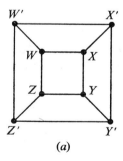

(a)   (b)

Fig. 26

The number of vertices and edges is the same for the cube and the Schlegel diagram.

Faces or regions	6
Edges or arcs	12
Vertices or nodes	8.

Thus $$F-E+V = 2,$$

or $$R-A+N = 2.$$

(a) Suppose we remove the arc $W'X'$, as in Figure 26(b). We have *one* arc less. Is the number of regions altered? Is the number of nodes altered? Is $R-A+N$ still equal to 2? Remove another arc. What happens? Figure 27(b) shows all the outer arcs removed. (W', X', Y', Z' are now 1-nodes.)

Carry on removing arcs, each time checking the value of $R-A+N$. Does removing an arc always result in decreasing the number of regions by one? What else can happen? Does $R-A+N$ always equal 2? Try removing the arcs WX and YZ from Figure 27(b). What happens to the network if you do?

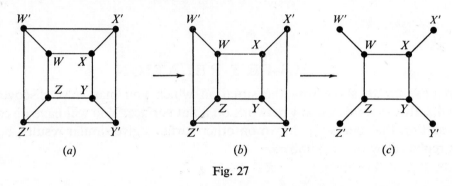

Fig. 27

Eventually we could reduce the network to the single arc WX with nodes W and X. For this simplest of networks

$$R = 1, \quad A = 1 \quad \text{and} \quad N = 2,$$

giving $$R-A+N = 2.$$

Clearly we could reduce any network in this way to a single arc.

(b) Suppose we look at the problem the other way round by building up a network, starting with a single arc, XY. Adding a second arc, as in Figure 28, A increases by one. What happens to R, N and $R-A+N$? (Note again that Z is a vertex, or 1-node.)

Fig. 28   Fig. 29

Add another arc as in Figure 29. What happens to R, N and $R-A+N$?

Could the addition of an arc have any effect other than the ones noticed in these two cases?

(c) Find $R-A+N$ for each of the networks in Figure 30. (Figure 30(b) consists of both the triangles.)

Why did you not get 2 for Figure 30(b)? What would you have to add to the figure in order to make $R-A+N$ equal to 2? Figure 30(a) is a *connected* network. Figure 30(b) is *not connected*.

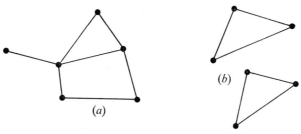

Fig. 30

Exercise C

1. Find R, A, N and $R-A+N$ for the networks in Figure 31.

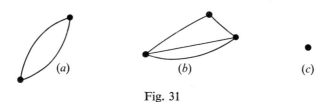

Fig. 31

2. A vertex or 2-node can be inserted at any point on a network (see Figure 32). Why does $R-A+N$ remain unaltered?

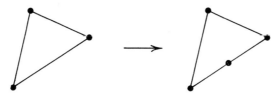

Fig. 32

3. Draw the Schlegel diagram for a tetrahedron. Can you start with a triangle and build up the network for a tetrahedron by adding nodes? If not what else needs to be done?

4. For the purposes of proving Euler's relation for a polyhedron, why have we to use a Schlegel diagram and not a net?

For example, why (a) rather than (b) in Figure 33?

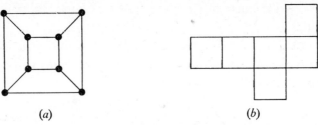

(a)   (b)

Fig. 33

5. *Project*

(a) Draw four examples of connected networks such that all the vertices are 3-nodes, that is, exactly 3 arcs meet at each vertex. Figure 34 shows an example.

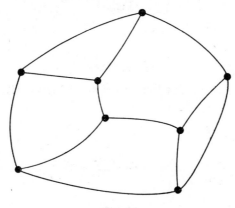

Fig. 34

Make a table showing the values of R, A and N for each of the four networks. Add an extra column showing values of $R-2$. Can you spot relations connecting (i) R and A; (ii) R and N?

(b) The 3-node network in Figure 34 has eight 3-nodes. We could say that its *node-sum* is 24. Can you explain why the node-sum is also twice the number of arcs. Would this be true for any network?

(c) The network in Figure 34 has 2 regions with 3 arcs, 2 regions with 4 arcs, and 2 regions with 5 arcs (one of these is the outside region).

If a is the average number of arcs per region then

$$a = \frac{(2 \times 3) + (2 \times 4) + (2 \times 5)}{6} = 4.$$

Find a for each of the four networks you have constructed in part (a) and check that

$$3N = 2A = aR,$$

in each case. Can you explain why this result is true for all 3-node networks?

(d) Using a as found in part (c) and R for each of the four networks you drew in part (a), copy and complete the following table:

250

R				
$\dfrac{12}{R}$				
a				

Can you spot a relation connecting a and $12/R$?

(e) Why must a 3-node network have at least one region with less than six arcs?

(f) Can you draw a network of hexagons (with curved edges) on a sphere?

6.

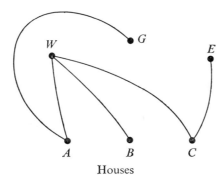

Houses

Fig. 35

There is a famous topological problem about supplying water, gas and electricity to three houses. It is required that no pipes or wires must cross each other. Figure 35 shows the situation with some of the joins made. Let us suppose we can complete the network. Why would it be a 3-node network? There would be 6 nodes and 9 arcs. Use Euler's relation to determine the number of regions. Explain why no region could be bounded by only 3 arcs. In Question 5 we found the relation

$$a = 6 - 12/R.$$

Using the value of R you have found, determine a. To what conclusion does this lead you?

Summary

A topological map shows the main features of a railway, electrical circuit, etc., without giving any idea of scale or angle. A topological transformation can be thought of as involving the twisting and stretching of a configuration without tearing or joining.

An invariant under a topological transformation is a fact about the original figure which is still true about the altered one. So far we have met the following invariant facts:

(a) if a point P lies on a line l, then P' lies on l';

(b) if P_1, P_2, P_3 lie in that order on l, then P'_1, P'_2, P'_3 lie in that order on l';

(c) if n half-lines meet at P, then n half-lines meet at P'.

A half-line starts at a finite point and continues in one direction only.

Two curves are topologically equivalent if one can be transformed into the other by a topological transformation.

A simple closed curve is topologically equivalent to a circle.

A line joining nodes is an arc.

A node is a point where half-lines meet. The order of the node is the number of half-lines. Every point of an arc is a 2-node with the exception of its ends.

A figure is traversable if it can be drawn with one sweep of the pencil, without lifting it from the paper and without tracing the same arc twice.

A figure will be traversable if it has: (*a*) two odd nodes, (*b*) no odd nodes. It may have any number of even nodes.

It is not possible to draw a figure having an odd number of odd nodes.

The area bounded by arcs is a region.

The area outside a figure is also a region.

For any connected figure, $R - A + N = 2$, where $R =$ number of regions, $A =$ number of arcs and $N =$ number of nodes.

This applies to polyhedra, where for R read Faces, for A read Edges and for N read Vertices.

For different surfaces $R - A + N = k$ where k is an invariant for one particular type of surface.

A network is *connected* if there is a continuous path (not necessarily a direct route) connecting all the vertices.

12

STATISTICS AND PROBABILITY

He uses statistics as a drunken man uses lamp-posts
—for support rather than illumination.

ANDREW LANG

1. PRESENTING INFORMATION

1.1 Bar charts

(*a*) Overleaf is a table giving the monthly rainfall recorded at a school in the south of England over the years 1956–60.

This is a *statistical* table containing data (measurements) of the rainfall. (Each observation is a *statistic*.) Can you answer the following questions?

(i) Which is usually the driest part of the year?

(ii) Was there an exceptional year?

(iii) Which was the best cricket season (May–August)?

(iv) Would it have been better if it had started and finished a month later?

(v) In what month may there have been floods?

(*b*) These questions become far easier to answer if we first illustrate the data. There are several ways of doing this, but the most striking is probably the bar

Total monthly rainfall in mm					
	1956	1957	1958	1959	1960
Jan.	125	66	89	135	110
Feb.	18	89	66	8	55
Mar.	17	59	38	121	39
Apr.	42	7	24	78	44
May	7	38	85	36	35
June	46	44	81	58	88
July	70	83	70	48	90
Aug.	120	79	65	46	90
Sept.	106	97	123	3	91
Oct.	47	65	73	57	208
Nov.	26	61	71	81	123
Dec.	116	93	91	119	120

chart. We draw a series of bars, one for each month, all the same width, but with lengths proportional to the rainfall.

Figure 1 shows a bar chart for 1956 and 1957.

Each bar has been made 0·5 cm wide and 100 mm of rainfall is represented by a height of 5 cm. Draw similar charts for the other three years. It can now be seen more readily what the pattern of rainfall was like in each year. Bar charts can be drawn with the bars going either up or across the page.

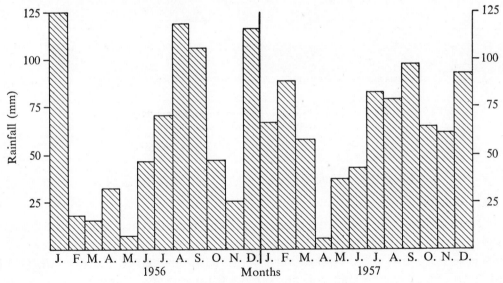

Fig. 1. Monthly rainfall in 1956 and 1957.

Exercise A

1. Count the number of cars passing a particular point per minute at different times of day. Represent your figures graphically. If you repeated this experiment on a different day, how would you expect the chart to differ? If possible, check by repeating the experiment.

2. Draw a straight line and ask 10 people to estimate its length. Show these estimates on a bar chart, and indicate on it the actual length.

254

3. Here are the figures for road deaths in Great Britain in 1960 and 1961.

Month ...	... Jan.	Feb.	Mar.	Apr.	May	June	July	Aug.	Sept.	Oct.	Nov.	Dec.
Deaths 1960	507	458	493	500	504	537	634	589	619	670	695	764
1961	552	415	581	534	580	525	607	543	612	672	613	674

Graph this information. Is there any marked difference between the months of the year? Which are the worst months.? Suggest reasons for your answer.

1.3   Other methods of presenting data

In addition to the bar charts described in Section 1, there are several other ways of representing data in an easily understood manner. For example, a survey among 200 first-formers to find the most popular subject produced the following results.

Subject	Votes		Subject	Votes
Mathematics	44		English	30
French	22		Science	36
Latin	18		History	28
Geography	22			

Total 200

Three alternative ways of presenting these data are shown here.

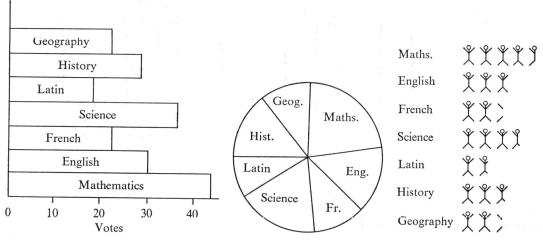

Fig. 2. A bar chart.   Fig. 3. A pie chart.   Fig. 4. A pictogram.

In Figure 3 the areas of the sectors of the circle are proportional to the numbers of votes for each subject. Thus as 30 out of the 200 voted for English the angle in degrees of the English sector is given by $\frac{30}{200} \times 360°$ which is 54°. Pie charts can be used to compare the different proportions in two different samples using separate diagrams.

If we take a little pin-drawing of a child to represent the votes of, say, 10 children, and represent a fraction of 10 votes by the same fraction of a pin-drawing, we can represent the data as in Figure 4. Such a chart is called a pictogram.

255

Although we draw the figures at equal distances apart these distances mean nothing and no scale is drawn. This is obviously not an accurate way of representing the data though it is a popular one. Pictograms can be drawn using any suitable picture.

Discuss which type of diagram you think is the most suitable for this particular type of information, giving your reasons.

Exercise B

1. Find out how many hours of a school day you spend:

(*a*) working;   (*b*) playing;   (*c*) eating;
(*d*) sleeping;   (*e*) travelling.

Make up a pie chart to show this. Make estimates and draw a similar chart for another season of the year.

2. Divide your weekly or monthly pocket money between:

(*a*) books and papers;   (*b*) sweets, ices, etc.;   (*c*) entertainment;
(*d*) savings;   (*e*) any other regular items.

Represent them by piles of coins of appropriate height.

3. The electoral swing to the Blue Party is well shown by the charts in Figure 5. Or is it?

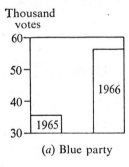

(*a*) Blue party   (*b*) Green party   (*c*) Yellow party

Fig. 5

4. The average marks scored by 1*C*, 1*M* and 1*R* in a certain school at the July examinations were 55, 45 and 42, respectively. State, with reasons, which of the two diagrams 6(*a*) and 6(*b*) presents these facts more fairly.

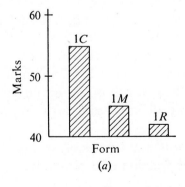
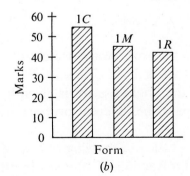

Fig. 6

2. FREQUENCY

2.1 Frequency tables and diagrams

(*a*) In a study of swans, a sample of 56 nests produced the following clutch sizes (number of eggs):

9	6	5	8	4	7	7	6
5	5	7	5	6	7	4	11
4	4	3	5	7	6	7	6
4	5	5	7	6	7	5	7
6	6	7	7	5	5	6	6
7	6	6	7	9	6	9	5
4	6	5	6	8	6	8	8

   (i)   What is the commonest clutch size?
   (ii) How many nests had 5 eggs?
   (iii) How many nests had more than 7 eggs?

(*b*) With the data as shown above, it is difficult to answer these questions. To make it easier, let us construct a table giving the number or *frequency* of each clutch size. This is called a frequency table. The simplest way of doing this, is to write down the possible clutch sizes, namely the integers 3 to 11, in a column on a piece of squared paper and then working along the data, put a stroke (or cross) against the appropriate clutch size. This is shown for the first row of the data. Copy and complete this table.

$$
\begin{array}{ll}
3 & \\
4 & /\\
5 & /\\
6 & //\\
7 & //\\
8 & /\\
9 & /\\
10 & \\
11 &
\end{array}
$$

The diagram you have obtained is not unlike a bar chart with the bars across the page whose lengths represent the frequencies of each clutch size. In this way we obtain a frequency table which represents the domain and range of the function

'clutch size → number of clutches'.

This is an example of a frequency function, where a set of possible events is mapped onto the frequency of each event.

257

Now draw the frequency diagram which is a bar chart with the bars up the page and their lengths proportional to the frequencies. As a check the sum of the frequencies should equal the sample size (56). Does it in your table? Now the questions in (a) should be easier to answer.

2.2 Grouped frequencies

If we return to the data of Section 1, we can form a frequency table representing the function 'rainfall → number of months'. However, we find scarcely any months with exactly the same rainfall. Therefore we would obviously do better to divide the possible rainfalls into a smaller number of groups (or classes). If we divide the data up into groups each covering 25 mm, the following table is obtained, with Figure 7 as the corresponding frequency diagram.

Amount falling in month	Number of months	Amount falling in month	Number of months
less than 25 mm	7	125 mm but less than 150 mm	2
25 mm but less than 50 mm	13	150 mm but less than 175 mm	0
50 mm but less than 75 mm	13	175 mm but less than 200 mm	1
75 mm but less than 100 mm	15	200 mm but less than 225 mm	1
100 mm but less than 125 mm	8		
		Total	60

Why is there no doubt about the group to which a rainfall of (say) 75 mm belongs?

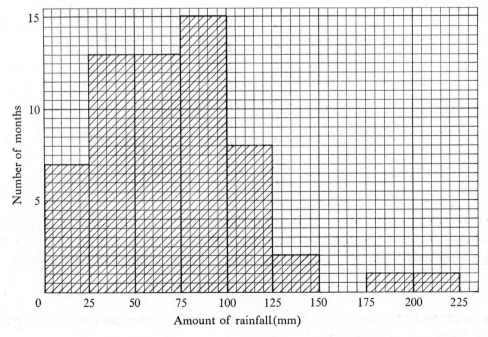

Fig. 7

Why are the boundaries between groups labelled, rather than the groups them-selves? Make a similar table, but with group intervals of 10 mm instead of 25 mm as above. Draw the frequency diagram. What differences do you notice?

2.3   Normal frequency functions

Many frequency function diagrams are similar to Figure 8(a), that is symmetrical with a central hump. They often occur when measurements, such as mass, are taken from a large number of similar items. If the sample is large and the class intervals are small, the frequency diagram approaches a bell-shape as in Figure 8(b).

Frequency functions having such a bell-shaped graph are called *Normal*.

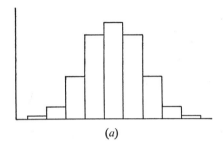
(a)

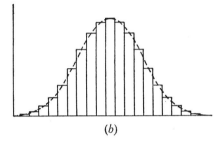
(b)

Fig. 8

Exercise C

1. Obtain the heights of all the pupils in your form. Draw the frequency diagram. Repeat for another form, from another year, and compare the two diagrams. What differences and similarities can you see? Are the frequency functions Normal?

2. Throw a pair of dice, which can be identified (for example, red and blue), and record the score as an ordered pair (red score, blue score). Repeat some 30–40 times. Compile the frequency tables for (a) scores of throws of one die (singular of dice) and (b) total score for each throw with two dice. Any comments? (Keep your results for later sections.)

3. Using a straight edge and without measuring, draw a line as nearly 7 cm long as you can. Measure and record its actual length. Show the results for the whole class in a frequency diagram. Repeat for several different lengths and compare the frequency diagrams. Does your class tend to over- or underestimate a given length?

4. Draw a line exactly 7 cm long. Now using only a straight edge try to draw a line exactly 7 cm long at right-angles to this line through its mid-point. Measure its length and display the results as for Question 3, for the whole class. How do they compare with the results for that question?

5. On 2 June 1965, 100 batsmen in first-class cricket matches made scores of less than 100 runs.

Number of runs	0–9	10–19	20–29	30–39	40–49	50–59	60–69	70–79	80–89	90–99
Number of batsmen	43	23	9	10	4	5	4	0	2	0

Graph the frequency function.

259

Compile a similar table for another day (it is best to choose a day which will be the first day of many matches; possibly a Saturday or a Wednesday; alternatively take your observations from a set of 'Sunday league' games.)

6. Discuss the likely shape of the graphs of frequency functions in the following cases.

(*a*) Number of teams scoring 0, 1, 2, 3, 4, etc., goals in football matches on a particular Saturday.

(*b*) Number of children with different shoe sizes in your class.

(*c*) Number of children in 100 families.

7. Project: draw frequency function graphs for sizes of shoes, heights, weights and ages for a group of 50–100 people in your school. Would a graph for the whole school be the same sort of shape as the graph of your sample?

8. Meteorologists measure the cloudiness of the sky in eighths. A clear sky is $\frac{0}{8}$ and an overcast sky is $\frac{8}{8}$.

Figure 9 shows the sort of graph that might be obtained if records were kept for noon on each day for a year. Explain the significance of the shape of this graph. Keep records of your own for a period, of say two months and compare the graph you get with Figure 9.

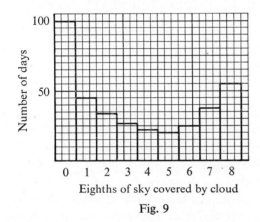

Fig. 9

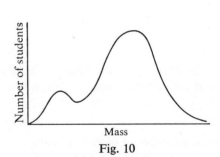

Fig. 10

9. The masses of a large group of students involved in a particular sporting activity were recorded and the graph of the function

$$\text{mass} \rightarrow \text{number of students}$$

approached the shape shown in Figure 10. What was the sport?

Could such a shape diagram be obtained by adding two normal functions?

3. AVERAGES

There are three types of averages in everyday use. The commonest idea of an average is the 'arithmetic mean', often abbreviated to the *mean*. This is obtained by adding up all the observations and dividing by the number of observations. For example the mean of the set 2, 3, 5, 7, 11, is given by

$$\frac{2+3+5+7+11}{5} = 5\cdot6.$$

In a frequency diagram, the longest bar gives the *mode* which is the most frequently occurring value or group. In the sample of swan clutches in Section 2.1, the mode is the clutch size of 6 (it occurs 16 times).

If all the observations are put in order with the smallest at one end and the largest at the other, then the *median* is the value of the middle observation. (If there are an even number of observations, there is no observation exactly in the middle, so the average of the observations either side of the middle is taken as the median.)

A darts player made the following scores in 12 throws: 11, 20, 6, 9, 17, 13, 1, 5, 14, 7, 10, 19. What was his mean score? What was his median score? Why is it not possible to have a modal score?

3.1   Calculating the mean from a frequency table

It is easy to find the mean of a small number of observations, such as the scores of a cricketer, but if the sample size is large a more systematic method is required.

Here is the frequency table for the sample of swan clutches.

Size of clutch	3	4	5	6	7	8	9	10	11	Total
Frequency	1	6	12	16	13	4	3	0	1	56

In order to calculate the mean, we need to know the total number of eggs. There is 1 clutch of 3 eggs, 6 of 4, 12 of 5 etc., which give 1×3, 6×4, 12×5 etc. eggs. This is not very neat, so we write the frequency table using columns instead of rows and add a third column formed as shown.

Clutch size	Frequency	(Clutch size) $\times$ (Frequency)
3	1	3
4	6	24
5	12	60
6	16	96
7	13	91
8	4	32
9	3	27
10	0	0
11	1	11
Totals	56	344

Thus, the mean clutch size is

$$\frac{\text{Total number of eggs}}{\text{Total number of clutches}} = \frac{344}{56} = 6 \cdot 14 \text{ (to 3 s.f.)}.$$

In this case, as the frequency diagram is almost exactly symmetrical, the mean, modal group and median (both 6) are very nearly the same.

3.2 Calculating the mean from a grouped frequency table

Refer to Exercise C, Question 5. There were 43 batsmen with scores of between 0 and 9 runs. How many runs did they score between them? It is impossible to answer this question unless we have the scores for each of the batsmen. Thus we have to make an estimate of the average score per batsman in this group. The usual estimate in this and other cases is to take the mid-interval value, which for this first group is $4\frac{1}{2}$ runs $\left(\dfrac{0+9}{2}\right)$. We can now proceed as in the previous section but with another column.

Number of runs	Frequency	Mid-interval value	Mid-interval value × frequency
0–9	43	$4\frac{1}{2}$	$193\frac{1}{2}$
10–19	23	$14\frac{1}{2}$	$333\frac{1}{2}$
20–29	9	$24\frac{1}{2}$	$220\frac{1}{2}$
30–39	10	$34\frac{1}{2}$	345
40–49	4	$44\frac{1}{2}$	198
50–59	5	$54\frac{1}{2}$	$272\frac{1}{2}$
60–69	4	$64\frac{1}{2}$	258
70–79	0	$74\frac{1}{2}$	0
80–89	2	$84\frac{1}{2}$	169
90–99	0	$94\frac{1}{2}$	0
Total	100		1990

Thus the mean score is

$$\frac{\text{Total number of runs}}{\text{Total number of batsmen}} = \frac{1990}{100} = 19.9.$$

This is only an approximate value because we had to make estimates of the average of each group; therefore we should perhaps give the mean score to the nearest whole number, that is 20.

In which group does the median occur? What is the modal group?

Exercise D

For each question indicate whether the median and mode would be more or less useful than the mean.

1. Find the mean number of children per household in Barchester at the last two censuses.

Number of children per household	0	1	2	3	4	5	6	> 6 (av. 8)	Total
Frequency (thousands of households)									
1951 census	81	25	20	10	5	3	1	0	145
1961 census	89	28	23	11	6	4	2	1	164

2. In a year's entry of 182 boys for G.C.E. O-level examinations, 6 boys had no passes; 10 boys one pass; 40 boys two or three passes; 103 boys four, five or six passes; 23 boys seven or eight passes. What is the approximate mean number of passes?

3. The masses of a group of children at 6 months were as follows:

Mass (kg)	6–7	7–8	8–9	9–10	10–11	Total
Frequency (children)	5	11	18	8	3	45

Find the approximate mean mass.

4. The ages in years and months of a form of 30 boys are as follows:

Age	13.0–13.2	13.2–13.4	13.4–13.6	13.6–13.8	13.8–13.10	13.10–14.0
Frequency (boys)	4	7	5	5	3	6

Find the approximate mean age by multiplying 13 yr. 1 month by 4 and so on.
Can you see a way of simplifying the working?

5. A timekeeper tests his reaction time with a stopwatch, and records the following results:

Reaction time ($\frac{1}{100}$ sec)	15	16	17	18	19	20	21	22	23	24	25
Frequency	2	12	25	50	92	136	98	46	23	14	2

Rewrite this table as follows:

Number of $\frac{1}{100}$ seconds by which reaction time exceeds $\frac{15}{100}$ sec	0	1	2	3	4	...
Frequency		2	12	25	...	...	...

Find an approximate value of the mean from this table.
Deduce the mean reaction time of the timekeeper.

6. Using the information you collected in Question 7 of Exercise C calculate the mean shoe size, height, mass, etc. (You may find it useful in some cases to use the technique of Questions 5 and 6.)

7. Before the final game of the season two cricketers A and B had both taken 16 wickets at 9 runs per wicket. In the final game A took 1 for 26 and B took 4 for 56. Who had the best average for the whole season?

4. PROBABILITY

4.1 Chance

'It is extremely unlikely that a Third Division team will win the F.A. Cup.'

'There's a fifty-fifty chance of a coin landing head upwards when it is tossed.'

'John has a better chance of reaching the final of the school tennis tournament than Brian.'

'The odds are on Russia beating America to Mars.'

In all these statements there is a comparison of possible future events and the chance that one is more likely than another. We make these comparisons by using words and phrases such as: 'almost certain', 'extremely unlikely', 'a good chance', 'probable', 'evens' and so on. For many purposes these phrases are sufficient, but where a comparison has to be made which involves a payment of money, whether in betting on horses or calculating the fire insurance to be paid on a house, it is necessary to be more exact.

(a) What are the chances of your school football team winning the next home match? (Very good, good, reasonable, poor.)

Before being able to answer this question what facts must you consider?

263

(b) If you look at a page of a book written in English, which of the following is the most likely:

    (i)   that there are more 'e's than 'p's,

    (ii) that there are about the same number of 'e's as 'p's,

    (iii) that there are more 'p's than 'e's?

How did you decide?

Can you give a number to say how much more likely one letter is than the other?

(c) If you kept a note of the registration numbers of passing cars, would you expect to see a 3 in the tens place more often than a 9?

(d) How would you rate your chances of playing for the school tennis team before you leave? (Good, reasonable, poor, nil.)

(e) What is the chance of throwing a '5' with a die?

(f) In which of the following would you be most likely to have an accident while travelling from Plymouth to Glasgow:

    (i) a car;   (ii) a train;   (iii) a boat;   (iv) a plane?

Before being able to answer these questions you needed to draw on some of your past experiences. To make any worthwhile judgement on the outcome of a football match it is useful to know the recent results of the two teams, whether either team has players out because of injury, what are the strong points and the weak points of each team, and so on. The more information of this kind you have, the better will be your estimate of your team's chances. However, in this situation you can never be very exact because there are so many things to consider which defy measurement.

The numbers of 'e's and 'p's in a book lend themselves far more easily to measurement and it is soon clear, from counting, that the number of 'e's far exceeds the number of 'p's. To make any more exact statement such as 'there are 20 times as many 'e's as 'p's', although tempting, would have little value.

Compare the ratio of 'e's to 'p's in four separate paragraphs.

When an experiment can be repeated many times under the same conditions, such as tossing a penny or throwing a die, it is possible to be more exact.

In an experiment a die was thrown 72 times with the results shown in Figure 11. The same die was then thrown 240 times with the results shown in Figure 12.

Score	Frequency		Score	Frequency
1	14		1	40
2	12		2	39
3	8		3	44
4	13		4	38
5	9		5	37
6	16		6	42
	Totals 72			240

Fig. 11   Fig. 12

Exercise E

1. In the experiment described in Figure 11, 6 was scored twice as often as 3. Does this mean that the chances of throwing a '6' with the die are twice as good as those of throwing a '3'?

2. In the second experiment each of the scores 1, 2, 3, 4, 5, 6, came up approximately 40 times each out of 240 throws.

(*a*) Is this what you would expect?

(*b*) Approximately how many times would you expect to score 5 out of 600 throws?

(*c*) Compare these figures with your results for Exercise C, Question 2(*a*).

3. Would you expect exactly two '3's to turn up with twelve throws of a die? If not, why not?

4. Toss two pennies and note whether 2 heads, 1 head or 0 heads occur. Repeat this experiment (i) 10 times, (ii) 50 times, (iii) 100 times. Plot the results as frequency diagrams in each case.

(*a*) What do you deduce about the relative chances of obtaining a head and a tail to 2 heads from the results of (i), (ii) and (iii)? Which result is most reliable?

(*b*) What fraction of the experiments gives 2 heads in (iii)?

(*c*) Approximately how many times would you expect 2 heads to occur if you tossed the two pennies 1000 times?

5. If 3 coins are tossed, what is the chance that they will all turn up heads? Answer this experimentally by finding what fraction of the trials you carry out produces 3 heads.

6. A match-box contains 3 red beads and 2 yellow beads which, apart from their colour, are identical. An experiment is performed by shaking the box, then, without looking, opening the box and removing a bead. The colour of the bead is recorded and the bead returned to the box. This is repeated 60 times. Approximately what fraction of the beads taken out would you expect to be yellow? Carry out this experiment using Smarties or beads in a match-box and see if the results agree with your estimate.

7. Put a selection of beads of different colours in a match-box. Use 7 beads and preferably no more than 3 colours (e.g. 2 red, 4 yellow, 1 blue). Exchange your box with a neighbour, and, by repeating the experiment described in Question 6, try to determine the contents of the box.

8. In playing games such as Monopoly it is usual to throw two dice together and total the score indicated on them. What possible scores are there?

Is the chance of scoring 2 the same as that of scoring 8?

Plot the results of Exercise C, Question 2 on a frequency diagram. Does it appear that the chances of some scores occurring are better than others? What fraction of your trials give a score of (*a*) 2, (*b*) 4, (*c*) 7?

4.2 Experimental probability

In 1654 a gambler, the Chevalier de Méré, asked a French mathematician, Blaise Pascal, to help him in deciding how to share the stake money in a game of dice. Pascal discussed the problem with Fermat, another eminent mathematician, and in solving the problem they started the theory of probability. Today the theories that were developed from a game of dice are used extensively in economics, industry, science and sociology.

To understand how these mathematicians measured the chances of a particular event happening consider Figure 12 which records the results obtained from throwing a die. The die was thrown 240 times and in these trials, the '5' turned up 37 times. This leads us to expect that if the same die were thrown 480 times the '5' would turn up 74 times. We assume, in other words, that the proportion of '5's which occur remains about the same. The validity of this assumption depends on our taking a large number of trial throws for, as you will have seen from the questions you have answered, with a small number of throws almost anything can happen. (For instance, a penny tossed 3 times might easily land head upwards each time but no one would be foolish enough to deduce from this that it would never land tail uppermost.)

The proportion of '5's which occur after a large number of trials is expressed as a fraction and called the *experimental probability* that the '5' will turn up if the die is thrown again.

In the experiment being discussed this experimental probability is $\frac{37}{240}$.

(*a*) What is the experimental probability of scoring (i) 2, (ii) 3, (iii) 1?

(*b*) What is the experimental probability that *either* a '1' *or* a '4' will turn up?

(*c*) What is the experimental probability that a '6' does *not* turn up?

(*d*) If the probability of drawing a heart from a hand of playing cards is 1, what do you deduce?

(*e*) What is the probability that a standard die will turn up a '7'?

(*f*) If a coin is tossed 1000 times and lands head uppermost 1000 times, what is the probability that if tossed again it will land head uppermost? What do you deduce about the coin?

Use the following definition of experimental probability to answer the questions in Exercise F:

$$\text{Experimental probability} = \frac{\text{the number of trials in which the event happens}}{\text{the total number of trials that have taken place}}.$$

Exercise F

1. Choose any number between 1 and 400. Look up the hymn with this number in your school hymn book and note how many verses it has. Do this for 60 trials and record your results.

(*a*) What is the experimental probability that a hymn chosen at random from the book will have: (i) 4 verses, (ii) 2 verses, (iii) more than 6 verses?

(*b*) Can you suggest any other way of working out the probability of selecting a hymn with 2 verses?

2.

Distance (km)	Less than 1000	1000 to 2000	2000 to 3000	More than 3000
Frequency	125	257	328	90

A large tyre manufacturer kept a record of the distance after which a particular kind of cycle tyre needed to be replaced. The table shows the results from 800 samples. What is the probability that if you buy a tyre of this kind:

(*a*) it will need to be replaced before it has covered 1000 kilometres;

(*b*) it will last more than 2000 kilometres;

(*c*) it will need to be replaced after it has covered somewhere between 1000 and 3000 km?

3. In a game of Whist the first player leads with a Jack of Spades while the second player plays a two of Hearts. What is the probability that the second player has no spades?

4. Deal yourself a hand of 13 cards face down. Draw a card from the hand and note its suit. Return the card to the hand. Shuffle the cards and repeat the process 40 times.

(*a*) What is the experimental probability of drawing:

(i) a heart,   (ii) a spade?

(*b*) Compare your experimental probability with the proportion of (i) hearts and (ii) spades in your hand. What do you deduce?

5. Roll a 1p coin onto a chess board and note whether it comes to rest lying completely inside a square or not. Do this a large number of times and find the experimental probability that a 1p coin rolled on the chess board comes to rest inside a square.

6. Draw a set of parallel lines across a piece of paper so that the lines are all 2 cm apart. Drop a matchstick (or pin) onto the paper and note whether it crosses a line or not. Repeat this (*a*) 10 times, (*b*) 50 times, (*c*) 150 times. What is the experimental probability in each case of a matchstick falling between the lines? If you repeated the experiment 1000 times, approximately how many times would the matchstick fall between the lines?

7. Find out the birthdays of the pupils in your year at school and from this give the experimental probability that any pupil picked out from your year has a birthday in March.

8. Throw two dice together and note the total thrown. By repeating this a large number of times find the experimental probability for each of the possible scores. In playing a game of Monopoly a boy finds that he will be 'safe' if he scores either 3, 4, 5, 6 or 10. What is the probability that he will be (*a*) 'safe', (*b*) not 'safe'?

4.3   Expected probability

In the last section you were able to determine the chance of an event happening by carrying out a series of experiments. It is likely that someone else carrying out the same experiments might obtain a different value for the probability although they are usually approximately the same. Often it is possible to deduce the chance of an event occurring by quite different means.

Consider a die which consists of a perfectly symmetrical cube. Because of its symmetry (unless it is loaded!) it is reasonable to say that the chance of a '1' turning up is the same as that of a '2' turning up and so on. There are six ways in which a die can land and, since these are equally likely, the chance that any particular one of them will turn up is 1/6. This is the *expected probability* of throwing a particular number with a die and the experimental probability usually approximates to it.

Let us look more closely at this example. The set of the possible outcomes of throwing a die is

$$S = \{1, 2, 3, 4, 5, 6\}.$$

Suppose that we are interested in the probability of throwing either a '2' or a '3'. The set of outcomes we are interested in is

$$E = \{2, 3\}.$$

Then the *expected probability* is defined as

$$\frac{\text{the number of elements in } E}{\text{the number of elements in } S}.$$

Thus the expected probability of throwing either a '2' or a '3' is $\frac{2}{6} = \frac{1}{3}$. This is usually written

$$p(E) = \frac{n(E)}{n(S)} = \frac{2}{6} = \frac{1}{3},$$

where $p(E)$ denotes the expected probability of event E, and $n(E)$ and $n(S)$ denote the *number of elements* in E and S respectively.

Example 1

Consider the probability of cutting a pack of playing cards and obtaining an ace. In this case the set of possible outcomes is

$$S = \{\text{the 52 cards in a pack of playing cards}\}$$

and the event we are interested in is

$$E = \{\text{ace of Hearts, ace of Clubs, ace of Diamonds, ace of Spades}\}.$$

The probability of obtaining an ace is

$$p(E) = \frac{n(E)}{n(S)} = \frac{4}{52} = \frac{1}{13}.$$

Earlier in the chapter when throwing two dice together you were probably surprised to find that the chance of scoring say 7 is more than scoring 12. This is where intuition leads us astray. Suppose, for argument's sake, that you have a red die and a blue die. In how many different ways can they land when thrown together?

To help in answering this, consider the numbers of ways in which the blue die can land when the red die, which has already been thrown (*a*) shows a '1', (*b*) shows a '2', (*c*) shows a '3', ..., (*f*) shows a '6'.

These possibilities can be neatly represented on graph paper as in Figure 13. Each cross represents a possible outcome. The circled cross, for example, represents a '5' on the red die and a '4' on the blue die. This can conveniently be represented as the ordered pair (5, 4).

What does (4, 5) represent?

Make a copy of Figure 13 and mark on it the ordered pairs (2, 3), (3, 1) and (6, 6). What do these represent in terms of the dice?

How many of the crosses represent a trial in which the total score is 7? (For example, (2, 5).)

What is the probability of scoring 7 when throwing 2 dice?

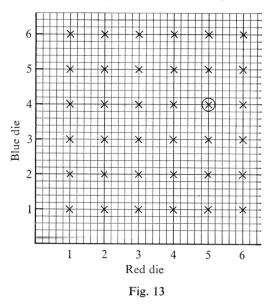

Fig. 13

Example 2

Calculate the expected probability of scoring 5 with two dice.

$$S = \{\text{the 36 different ways in which 2 dice can land}\},$$
$$E = \{(4, 1), (3, 2), (2, 3), (1, 4)\}.$$

Hence the expected probability of scoring 5 is

$$p(E) = \frac{n(E)}{n(S)} = \frac{4}{36} = \frac{1}{9}.$$

Example 3

A boy is chosen at random from Anthony, Brian, Colin, Donald, Dennis and Eric. What is the probability that the boy who is chosen has a name beginning with D?

$$S = \{\text{Anthony, Brian, Colin, Donald, Dennis, Eric}\},$$
$$E = \{\text{Donald, Dennis}\}.$$

Hence the probability that the boy chosen has a name beginning with D is

$$p(E) = \frac{n(E)}{n(S)} = \frac{2}{6} = \frac{1}{3}.$$

269

Exercise G

1. What is the expected probability of a prime number turning up when a die is thrown once?

2. A 1p and a 5p coin are tossed together. Calculate the probability that:

(a) they will both turn up heads,

(b) they will not both turn up heads.

3. Two dice are thrown together. Calculate the probabilities of each of the possible scores occurring. What is the sum of these probabilities? Compare the expected results with those in Exercise C, Question 2.

4. What is the set of possible outcomes when a penny and a die are tossed together? calculate the probability of a head and a number greater than 4 turning up.

5. What is the expected probability of selecting a picture card from a pack of playing cards by selecting a card at random?

6. Two dice are made as regular tetrahedra and have the numbers 1, 2, 3, 4 on their faces. Using ordered pairs of numbers, make a list of the possible outcomes when they are thrown together. What is the probability that the two numbers which turn up have a *product* equal to 12?

7. The names of Andrew, Betty, Christine and David are put in a hat to select two representatives for an inter-form competition. Make a list of the ways in which two names can be drawn from the hat. What are the probabilities of drawing:

(a) Andrew and David;

(c) a pair which includes Betty;

(b) a boy and a girl;

(d) a pair which does not include David?

8. A football match can end in 3 ways; the home team can win (*W*), draw (*D*), or lose (*L*). List the 9 different ways in which the results of two matches can end. Assuming that there is an equal chance of *W*, *D* or *L* for each match, calculate the probabilities that:

(a) both matches end in a draw;

(b) no match is drawn.

9. In a game of cards a tie is often decided by each player drawing a card from the pack in turn and the person who draws the highest valued card is said to have won. If the first person draws the 'nine of spades', what is the probability that the second player will draw a higher card? (That is, ten or above, with ace counting 'high'.)

10. (a) What is the set of possible outcomes when 3 pennies are tossed together?

(b) Calculate the probability that the outcome is (i) 3 heads, (ii) 2 heads and a tail, (iii) a head and 2 tails, (iv) 3 tails.

(c) What is the probability that no heads occur?

(d) What is the probability of 4 heads?

(e) In an experiment, 3 pennies were tossed 80 times. Of these trials 3 heads occurred together on 8 occasions. Do you think this is evidence that the coins are biased?

4.4   Random selection

The phrase 'at random' was used in the previous section. It is very easy to introduce bias into a random selection as you will see in some of the following examples.

270

(a) Look back at Example 3 on page 269. How would you choose one of the boys *at random*?

Can you think of methods based on:

(i) drawing from a 'hat'; (ii) throwing a die; (iii) any other ways?

(b) Ask a friend to name a number at random between 5 and 12.

In theory all numbers between 5 and 12 have an equal probability of being named, but in practice which number might be named most frequently?

(c) You want to select three pupils at random from your school for a survey about travel to school. What is wrong with going to the school entrance a few minutes before school begins and selecting a group of three who are coming in together?

(d) A firm of porridge oat manufacturers wanted to find out how popular porridge was for breakfast.

They opened a London telephone directory at random and contacted everyone on that page. They found that 80 % had porridge for breakfast.

Can you explain what had gone wrong with their random sample?

Exercise H

1. You want to predict the result of the local elections and to do this you go into the main street at 11 a.m. on a Monday morning and interview people on the pavement. Will you have a random sample?

2. 'Last night $3\frac{1}{4}$ million people watched the big fight on television.' How would this information be obtained?

3. A firm which manufactures dish-washing machines wants to know what percentage of the population uses their machines. They do this by selecting people at random from a telephone directory. Have they introduced bias into their sample?

4. 'In a random sample of 10 housewives, 9 preferred Whoosh to any other detergent.' What questions would you want to ask the advertisers about the randomness of their sample and about their method of obtaining this information?

REVISION EXERCISES

SLIDE RULE SESSION NO. 5

Give all answers to 3 s.f.

1. $2 \cdot 16 \times 1 \cdot 62$.
2. $3 \cdot 03 \times 2 \cdot 53$.
3. $2 \cdot 45^2$.
4. π^2.

5. $0 \cdot 607 \times 0 \cdot 112$.
6. $2 \cdot 16 \div 1 \cdot 62$.
7. $3 \cdot 03 \div 2 \cdot 53$.

8. $2 \cdot 45 \div 8 \cdot 1$.
9. $\pi \div 3$.
10. $0 \cdot 607 \div 0 \cdot 112$.

SLIDE RULE SESSION NO. 6

Give all answers to 3 s.f.

1. $\sqrt{10 \cdot 5}$.
2. $\sqrt{4100}$.
3. $\sqrt{0 \cdot 98}$.
4. $\sqrt{0 \cdot 098}$.

5. $8 \cdot 86^3$.
6. $0 \cdot 112^3$.
7. $\pi \times 7 \cdot 4^2$.

8. $24 \cdot 6 \times 19 \times 0 \cdot 07$.
9. $\dfrac{1 \cdot 21 \times 1 \cdot 06}{1 \cdot 12}$.
10. $(25 \cdot 2 \times 16 \cdot 1) + \left(\dfrac{1 \cdot 72 \times 0 \cdot 85}{0 \cdot 74}\right)$.

M

1. (a) One triangle has two angles which are 20° and 30°. Another triangle has two angles which are 40° and 60°. Are the two triangles similar? Why?

   (b) One triangle has two angles which are 40° and 60°, and another has two angles which are 40° and 80°. Are they similar? Why?

   (c) Two quadrilaterals each have angles of 60°, 70°, 110°, 120°, Do they *have* to be similar? Why?

2. Two triangles ABC and PQR are such that

$$\frac{AB}{PQ} = \frac{BC}{QR} = \frac{CA}{RP} = \frac{3}{2},$$

and it is given that $AB = 2$ cm, $BC = 3$ cm and $\angle ABC = 60°$.

   (a) Make an accurate drawing of triangle PQR.

   (b) Is it true that (area of triangle ABC): (area of triangle PQR) = 4:9? If not, what should the ratio be?

3. A triangle ABC has vertices $A(2, 1)$, $B(1, ^-1)$, $C(^-2, 3)$. The triangle is sheared, with invariant line $y = 0$, such that A', the image of A, is the point $(4, 1)$. Find the coordinates of B', C'.

4. A boy claimed to have found a map (see Figure 1) which needed five colours to ensure no two adjacent countries were coloured the same. Was he right? Why? How many colours would you need to paint the faces of a tetrahedron so that no two adjacent faces were the same colour?

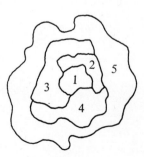

Fig. 1

5. Write down an ordered set of five integers whose mean is 4 and whose median is 3.

6. Make a frequency table for the letters occurring in the following place-name:

Llanfairpwllgwyngyllgogerychwyrndrobwll-llantysiliogogogoch.

Which are the four most frequently appearing letters? In which country do you think this place is?

7. To which of the following is the curve in Figure 2 topologically equivalent:

(*a*) a circle;   (*b*) the figure 8;

(*c*) the letter R;   (*d*) the letter X?

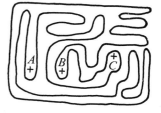

What is the name given to a figure such as this? Give a method for finding out whether each of the points *A*, *B*, *C* is inside or outside the curve.

Fig. 2

8. A man had his salary reduced by 25 %, but he protested and it was immediately increased again by 30 %. If his original salary was £1000 per annum, what did he finally gain or lose by this arrangement?

9. Express the following ratios in their simplest terms:

(*a*) 6p to £1;   (*b*) $2\frac{1}{2}$ to 100;   (*c*) a 60° turn to a complete revolution;
(*d*) a leap year to a day;   (*e*) $\frac{1}{3}$ to $\frac{1}{4}$.

10. Draw, on squared paper, the quadrilateral whose vertices are (⁻3, 4), (3, 6), (⁻3, ⁻4), (3, ⁻6). Give the equations of any lines of symmetry of the figure. Find the area of the quadrilateral.

N

1. (*a*) Is F topologically equivalent to Y?
   (*b*) The mean of 3, 7 and *x* is 9. What is *x*?

2. When a projector is 4 m away from its screen, the image of a man projected onto the screen is 1 m high. What would the distance between projector and screen have to be if the image were to be 3 m high? (Assume the projector does not have to be refocussed.)

3. Draw a figure whose sides are $\frac{1}{2}$ cm, 1 cm, $1\frac{1}{2}$ cm and 2 cm long. Draw another whose sides are 1 cm, 2 cm, 3 cm and 4 cm long, making it similar to the first figure. Draw a third figure with the same sides as the second, but *not* similar to the first one.

4. The triangle formed by joining the points *A*(2, 1), *B*(2, 4), *C*(0, ⁻2) is sheared with the line *y* = *x* invariant such that *C*′, the image of *C*, is the point (4, 2). Find the coordinates of *A*′, *B*′.

5. State the probability that a throw of a die will result in a score of 3 or more.

6. Suppose the cost of a £2·50 railway ticket is increased to £2·75 and that the cost of all other tickets is increased in the same proportion.

(*a*) What is the new cost of tickets which used to cost £0·50, £1·25 and £4·65 respectively?
(*b*) What was the old cost of a ticket which is now £3·96.

7. A musical survey in Form 2*Z* revealed the following facts about the popularity of certain recording groups:

12 liked the Earwings best of all,   4 liked the Crotchet String Quartet,
 9 liked the Caterwaulers,   3 liked the Popalongs,
 6 liked the Offbeats,   2 liked the Squareboys.

Illustrate this information by drawing: (*a*) a bar chart, (*b*) a pie chart. (*c*) Devise some other method of showing the information.

8. There are 600 boys in a certain school. 80 % of them have had measles and 65 % have had chickenpox. What is the smallest *number* of boys who could have had both?

9. If you run a kilometre in 5 minutes, what is your average speed in metres per second? What would it be if you ran the kilometre in $3\frac{3}{4}$ minutes?

10. Draw (where possible) figures with the following specifications. If impossible, say why.

1-nodes	3-nodes	4-nodes
4	0	1
1	1	0
2	1	0
0	0	1
2	0	3

Why is no mention made of the number of 2-nodes?

O

1. Figure 3 consists of six congruent rectangles. Which of the following is a centre of enlargement of *ABED* onto *DFLJ*: (*a*) *A*; (*b*) *D*; (*c*) *E*; (*d*) a point distance *DA* to the left of *A* (on the line segment *JA* produced); (*e*) the mid-point of *AB*?

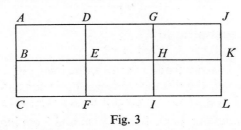

Fig. 3

2. Two mice are precisely similar in shape but different in size. Their tails are 6·5 cm and 9·75 cm long, respectively.

(*a*) How many times longer are the legs of the larger mouse than those of the other?

(*b*) If the smaller mouse has a mass of 60 g, what is the mass of the other one?

3. Give the mean and median of the numbers 4, 6, 7, 9, 10, 15, 19.

4. Find (to 2 s.f.) the numbers denoted by the letters:

$$\frac{1}{1\cdot7} = \frac{a}{2\cdot1} = \frac{b}{4} = \frac{3}{c} = \frac{d}{10} = e.$$

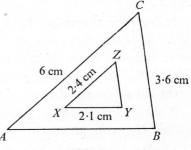

Fig. 4

5. In Figure 4, triangles *ABC* and *XYZ* are similar. $AC=6$ cm, $BC=3\cdot6$ cm, $XZ=2\cdot4$ cm and $XY=2\cdot1$ cm. Find: (*a*) the lengths of *AB* and *YZ*; (*b*) the ratio of the lengths of corresponding sides of the triangles, in the form *n*:1.

274

6. Christopher Cowshott scored 63, 7, 56, 1, 62, 0, 126, 5 in his first eight innings last season, being out each time. What was his batting average? His next innings (he was out again) brought his average up to exactly 50. What did he score? If he had been *not* out in this innings, how many runs would he then have had to make to bring his average to 50?

7. Figure 5 is not traversable. What does this statement mean, and how can you tell it is true? Make the figure traversable (*a*) by removing a line, (*b*) by adding a line.

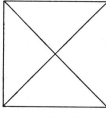

Fig. 5   Fig. 6

8. Figure 6 shows how three sticks can be joined at their ends to form three topologically different networks. Draw all the topologically different networks that you can find for four sticks of equal length.

9. The following table shows the heights, to the nearest cm, of a sample of 124 seedling fir trees:

Height	22	24	26	28	30	32	34	36	38	40	42
Number	5	9	14	18	20	17	18	13	6	3	1

Find the mean, mode, and median.

10. A bookshop buys books from a publisher at a price which is 30% less than the price marked on the books. Write down a formula for the profit the shop expects to make on a book marked at £*x*, and another formula for the price it pays for such a book.

P

1. The floor of a room is 4 m × 5 m. When the carpet is laid in the room there is an uncarpeted border of 0·5 m all the way round. Is the carpet similar in shape to the floor? Explain.

2. Copy Figure 7 (which is composed of 8 congruent triangles) and construct the centre of the enlargement which maps *PQR* onto *STU*. What is the scale factor of the enlargement?

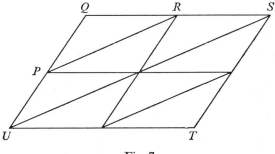

Fig. 7

275

3. Figure 8 shows a topological map of the roads connecting towns *A*, *B*, *C*, *D*, *E* and *F*. Answer as many of the following questions as possible. Where impossible, say so.

(*a*) Is it further from *A* to *B* or from *D* to *E*?

(*b*) How many roads go to *F*?

(*c*) What is the angle between the roads connecting *A* to *B* and *A* to *D*?

(*d*) Is the road from *A* to *E* straight?

(*e*) To reach *C* from *A*, what is the least number of other towns through which you must pass.

4. The earth's surface is divided up as follows: area of land, $1 \cdot 48 \times 10^8$ km²; area of sea, $3 \cdot 63 \times 10^8$ km². If a returning space capsule comes down anywhere at random on the earth's surface, what is the probability that it will arrive in the sea?

5. A survey was conducted to find out which breakfast cereals were preferred by the inhabitants of Mathsville. Measure the angle of each sector in the pie chart shown in Figure 9, and say what fraction of the people voted for each cereal.

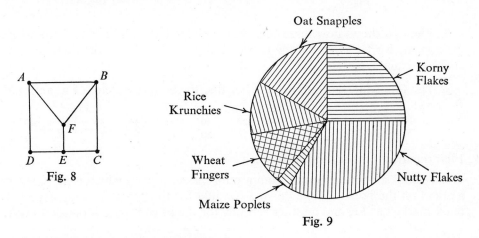

Fig. 8

Fig. 9

6. How many 1-nodes, 2-nodes, 3-nodes has:

(*a*) a line; (*b*) a half-line; (*c*) a line segment?

7. Here are the figures for the population of the village of Ambridge on 1 January of each year over the period 1957–66:

Year	1957	1958	1959	1960	1961	1962	1963	1964	1965	1966
Population (on 1 Jan.)	710	720	732	747	766	—	820	852	889	930

(*a*) Draw a line graph to illustrate this data.

(*b*) No figure was produced in 1962. Estimate it from your graph.

(*c*) Predict the population on 1 January 1967.

8. (*a*) A certain microscope is known to magnify lengths by a factor of 100. How many times does it enlarge a given area? If you wanted a microscope to magnify areas by a factor of 500, by what factor would it increase lengths?

(*b*) 'Half the size, a *quarter* the price! Buy the new small size Wosh and save money!' Comment. (See Figure 10, which is a scale drawing of the two sizes of packet.)

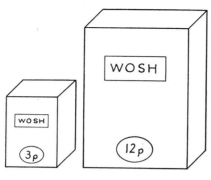

Fig. 10

9. The chart in Figure 11 illustrates information gathered about the number of sisters of the boys in a certain group. No boy had more than 5 sisters.

(a) How many boys are there in this group?
(b) What is the modal number of sisters?
(c) What is the mean number of sisters?
(d) Is it more useful to talk about the mean or the mode in this case? Why?

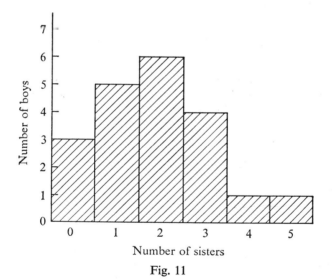

Fig. 11

10. The volume of a spherical soap bubble is increasing at a steady rate of 7 cm³/s. At a certain instant the volume is 1 cm³.

(a) If t s after that instant the volume of the bubble is v cm³, write down a formula for v in terms of t.
(b) Find the ratio of the volume when $t = 1$ to the volume when $t = 9$.
(c) Find the ratio of the radius when $t = 1$ to the radius when $t = 9$.

Q

1. Split up the following set of letters into three subsets, such that the members of each subset are topologically equivalent:

{A, E, F, I, L, R, U, T}.

2. A grocer stocks five kinds of washing powder. Figure 12 is a bar chart showing the result of a survey of the proportion of customers buying each variety. What is the probability that a customer for washing powder, chosen at random, buys Wosh?

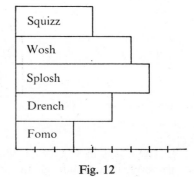

Fig. 12

3. (*a*) Calculate the mean of 1, 2, 2, 3, 5, 5, 5, 6, 7, 9.

(*b*) What is the mode of these numbers?

(*c*) What extra number would bring the mean of the numbers in (*a*) to exactly 5?

(*d*) Using the result of (*a*), write down the mean of 71, 72, 72, 73, 75, 75, 75, 76, 77, 79.

(*e*) Using the result of (*a*), write down the mean of 2, 4, 4, 6, 10, 10, 10, 12, 14, 18.

4. Triangle *DEF* in Figure 13 is an enlargement of triangle *ABC* not drawn to scale. If the scale factor is 2 which of the following are necessarily true?

(*a*) $EQ = 2BP$.

(*b*) *AB* is parallel to *DE*.

(*c*) Area of triangle $DEF = 2 \times$ area of triangle *ABC*.

(*d*) $\angle ABC = \angle DEF$.

(*e*) $OP = PQ$.

5. A machine was valued at £3000 in July 1964. Each year its value has decreased by 7 % of its value at the beginning of the year. What is its value in July 1967? (Give your answer to the nearest £10.) Is the total decrease more than, less than or equal to 21 % of its July 1964 value?

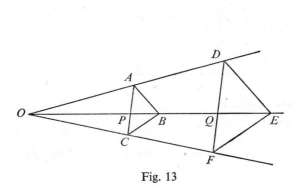

Fig. 13

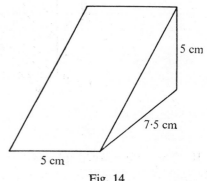

Fig. 14

6. A grocer cuts a cuboid of Double Gloucester cheese into two identical pieces and sold them at £0·15 each. One piece is shown in Figure 14. What is the cost of the cheese per cubic centimetre. How much would you expect a block 5 cm by 7·5 cm by 12·5 cm to cost? Would a rectangular piece of paper 10 cm by 12·5 cm be big enough to wrap up the piece illustrated?

7. Construct a topological map corresponding to this direct route matrix:

$$\text{from} \begin{array}{c} \\ A \\ B \\ C \\ D \\ E \end{array} \overbrace{\begin{pmatrix} A & B & C & D & E \\ 0 & 1 & 1 & 1 & 1 \\ 1 & 0 & 1 & 0 & 1 \\ 1 & 1 & 0 & 1 & 0 \\ 1 & 0 & 1 & 0 & 1 \\ 1 & 1 & 0 & 1 & 0 \end{pmatrix}}^{\text{to}}.$$

In what ways is a topological map not a true representation of, say, a network of roads?

8. Drawings for a new aeroplane are made on a scale of $\frac{1}{6}$. Copy and complete the following table:

	Drawing	Plane
Wing span	5 m	—
Length of fuselage	—	30 m
Number of seats	—	60
Surface area of wings	9 m²	—

9. Criticize the following:

(a)

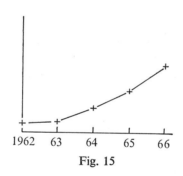

Fig. 15

'The swing is to WOSH, the wonder detergent!' (See Figure 15.)

(b) 'A recent questionnaire sent out to 50 people revealed that 68 % of the population of Great Britain live in houses graced by INFERNO Central Heating. Don't be one of the cold minority! Contact your dealer today!'

10. A car accelerates from rest, and readings are taken of its speeds at various times:

Time (s)	0	1	2	3	4	5	6	7	8	9	10
Speed (m/s)	0	3	7.5	14	18.7	21.7	24	25.5	26.7	27.5	28.2

Plot these readings on a graph (take time across the page) and so find:

(a) the speed at time 1·5 s;
(b) the time when the speed is 25 m/s;
(c) the acceleration at time 5 s.

279

R

1. Are a square and a triangle topologically equivalent?

2. Complete the following table to show the coordinates of A, B, and C after enlargement with the origin as centre and the given scale factor:

Scale factor	$A(2, 0)$	$B(0, 3)$	$C(3, ^-1)$
2			
5			
$\frac{1}{2}$			

3. If a solid sphere has a surface area of 10 cm², what is the surface area of another solid sphere with a radius three times as big? If the first has a mass of 70 g, what is the mass of the second if they are made of the same material?

4. The numbers of principal farm animals in Great Britain are: cattle, 12 million; sheep, 30 million; pigs, 6 million. Draw a pie chart to illustrate the information. State the angle for each 'slice' in degrees.

5. Two dice are made in the shape of regular tetrahedra and have the numbers 1, 2, 3, 4, inscribed on their four faces. Make a table to show all the possible combinations of the scores of the two hidden faces when the dice are thrown together. What are the probabilities of the total scores being: (a) 2, (b) 5, (c) 9?

6. (a) Write down all the subsets with three elements of the set $\{b_1, b_2, b_3, b_4\}$ and the subsets with two elements of the set $\{g_1, g_2, g_3\}$.
   (b) If three boys are chosen at random from Bryan, Basil, Benjamin and Bartholomew and two girls from Gloria, Grace and Gladys, what is the probability that neither Bryan nor Grace are chosen? What is the probability that Bryan, Basil, Gloria and Grace are chosen?

7. Calculate the gradients of the lines joining:
   (a) (1, 2) and (4, 5);   (b) (1, 2) and ($^-$2, 5);   (c) (1, 2) and ($^-$2, $^-$1);
   (d) (4, 5) and ($^-$2, $^-$1);   (e) (6, 5) and (23, 5).

8. The price of the Silver Whisper car was increased by 20% to £7200. What was the original price? A man buys one at the new price and sells it after a year for £6900. By what percentage has its value dropped?

9. Draw lines from the following information:
   (a) gradient 0 and passing through (2, 3);
   (b) gradient 2 and passing through (2, 3);
   (c) gradient $^-$2 and passing through ($^-$2, $^-$3).

10. Figure 16 illustrates the progress of a racing car starting from rest. Estimate its greatest acceleration in metres per second per second.

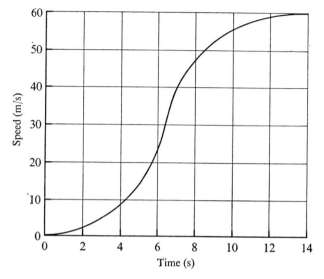

Fig. 16

13

TRANSFORMATIONS AND MATRICES

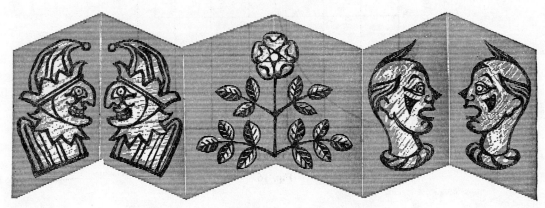

Plus ça change...

ALPHONSE KARR, *Les Guepes*

1. DESCRIBING TRANSFORMATIONS

1.1 Algebraic methods

(*a*) Suppose that under a certain transformation in the plane the triangle *ABC* in Figure 1 is transformed to a position *A'B'C'*. How much information must we have before we can describe fully the transformation involved?

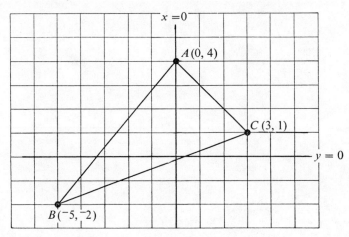

Fig. 1

(b) Copy Figure 1 (leaving room for values of x and y from ⁻6 to 6) and mark in A' at (2, 1), B' at (⁻3, ⁻5), C' at (5, ⁻2).

   (i)   What type of transformation maps $ABC \to A'B'C'$?

   (ii) How *much* movement is involved?

   (iii) Does any point of the plane remain unmoved under this transformation? If so, which point, or points?

(c) Repeat part (b) when A' is (5, ⁻1), B' is (⁻1, 4), C' is (2, ⁻4).

(d) In answering questions like (b) and (c) you will usually give such answers as

   enlargement, centre (2, ⁻3), scale factor 3;

   rotation, centre (⁻2, 0), rotation of 120°;

   translation, '5 across and 2 down'.

However, you might have described (b) as 'a transformation under which the x-coordinate of *any* point is increased by 2 units, and the y-coordinate is decreased by 3 units'. In other words, if (x, y) are the coordinates of *any* point R in the plane, then the coordinates of R', the image of R under this transformation, are $(x+2, y-3)$. Hence our description of the transformation may be abbreviated to the function form:

$$\mathbf{K}: (x, y) \to (x+2, y-3).$$

(This is read as 'the transformation **K** which maps the point (x, y) onto the point $(x+2, y-3)$'. The domain and range of the transformation (a function) are assumed to be positive and negative numbers and zero.)

Does this statement describe the transformation fully? If all the class were asked to draw the image of the triangle ABC under a transformation described in this way should they all be expected to obtain the same result?

Can you describe the transformation of (c) similarly? We shall discuss it again in Section 1.3.

1.2   Translations and vectors

(a) On the other hand, you may have described **K**, in (d) above, as 'a translation whose displacement is $\begin{pmatrix} 2 \\ -3 \end{pmatrix}$'. (See Figure 2.)

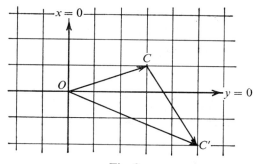

Fig. 2

Write down $\mathbf{OC}$ and $\mathbf{OC'}$ as vectors. The vector that describes the movement from the *origin* to a point is called the *position vector* of that point, and is often denoted by the corresponding lower case (or small) letter (printed, as usual for vectors, in bold type). The position vector of C is

$$\mathbf{c} = \begin{pmatrix} 3 \\ 1 \end{pmatrix}.$$

What is the relation between the position vectors of C and C' and the vector $\begin{pmatrix} 2 \\ -3 \end{pmatrix}$?

Write down the position vectors for B, B' and A, A'. Does the same relation hold?

(b) Thus, for the translation $\mathbf{K}$ described by the vector $\begin{pmatrix} 2 \\ -3 \end{pmatrix}$ the position vectors $\begin{pmatrix} x \\ y \end{pmatrix}$ and $\begin{pmatrix} x' \\ y' \end{pmatrix}$ of *any* point R and its image R' are related by

$$\begin{pmatrix} x' \\ y' \end{pmatrix} = \begin{pmatrix} x \\ y \end{pmatrix} + \begin{pmatrix} 2 \\ -3 \end{pmatrix},$$

and so the transformation may be written in function notation

$$\mathbf{K} : \begin{pmatrix} x \\ y \end{pmatrix} \rightarrow \begin{pmatrix} x \\ y \end{pmatrix} + \begin{pmatrix} 2 \\ -3 \end{pmatrix}.$$

If we write the vector $\begin{pmatrix} 2 \\ -3 \end{pmatrix}$ as $\mathbf{t}$, and $\begin{pmatrix} x \\ y \end{pmatrix}$ as $\mathbf{r}$ then the transformation can be abbreviated even further to

$$\mathbf{K} : \mathbf{r} \rightarrow \mathbf{r} + \mathbf{t}.$$

(c) $\mathbf{T}$ is a translation described by the vector $\begin{pmatrix} 3 \\ 1 \end{pmatrix}$; write it in function notation and find the image of triangle ABC in Figure 1 under this transformation.

1.3 Other transformations

(a) (i) The transformation $\mathbf{Q}$ is a positive quarter turn (anti-clockwise) about the origin. Can this transformation be expressed in the form $\mathbf{Q} : (x, y) \rightarrow ($ $)$? Consider again the points A, B and C of Figure 1 and their images under $\mathbf{Q}$. The image of $C(3, 1)$ is the point $(-1, 3)$. What are the images of A and B? Do they all satisfy $(x, y) \rightarrow (-y, x)$?

(ii) The transformation of Section 1.1 (c) was a negative quarter turn about the point $(0, -1)$. Verify, if you have not already done so in Section 1.1 (d), that this transformation may be expressed as

$$(x, y) \rightarrow (y+1, -x-1).$$

(b) (i) Find the images of A, B and C (see Figure 1) when reflected in the line $x + y = 0$, and hence describe this transformation in function form.

(ii) Verify that reflection in the line $y = x - 1$ is described by

$$(x, y) \rightarrow (y+1, x-1).$$

(c) Describe algebraically, in function notation:

(i) an enlargement, centre at origin, of scale factor 3;

(ii) an enlargement, centre at (⁻3, 0), scale factor 2. Draw a diagram.

In all these examples we have transformations of the form $(x, y) \rightarrow (x', y')$ (or $\mathbf{r} \rightarrow \mathbf{r}'$): but it is not always possible to express $\mathbf{r}'$ in terms of $\mathbf{r}$ in the same way as we did for translations. (What we have done is to express the *components*, x' and y', of $\mathbf{r}'$, in terms of the x and y components of $\mathbf{r}$.) The possibility of finding a method to express $\mathbf{r}'$ in terms of $\mathbf{r}$ for other transformations will be considered in the next section.

Exercise A

1. Copy Figure 3 onto graph paper and on it draw the images of the triangle S under the following transformations:

(a) $(x, y) \rightarrow (x, ^-y)$;   (b) $(x, y) \rightarrow (^-y, x)$;   (c) $(x, y) \rightarrow (^-x, ^-y)$;

(d) $(x, y) \rightarrow (x-5, y-4)$;   (e) $(x, y) \rightarrow (^-y+3, x)$;   (f) $(x, y) \rightarrow (2x-6, 2y)$;

(g) $(x, y) \rightarrow (x, ^-2y)$;   (h) $(x, y) \rightarrow (x, x+y-3)$.

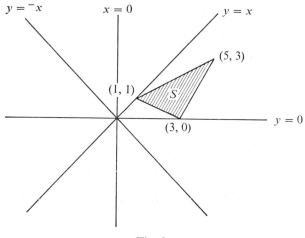

Fig. 3

(i) Describe the transformations geometrically.

(ii) Where possible, write down the transformations as functions of the position vector $\mathbf{r}$.

2. The parts of Figure 4 show an object (shaded) and its image (unshaded) under a simple transformation. In each case:

(a) describe the transformation geometrically;

(b) write down the coordinates of three object points and their corresponding images;

(c) describe the transformations in the form $(x, y) \rightarrow ($ $)$.

285

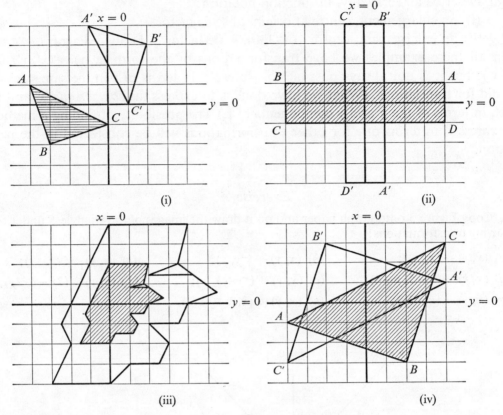

Fig. 4

2. MATRIX REPRESENTATION

(a) Work out the following:

$$\begin{pmatrix} 0 & -1 \\ 1 & 0 \end{pmatrix} \begin{pmatrix} 0 \\ 4 \end{pmatrix}; \quad \begin{pmatrix} 0 & -1 \\ 1 & 0 \end{pmatrix} \begin{pmatrix} -5 \\ -2 \end{pmatrix}; \quad \begin{pmatrix} 0 & -1 \\ 1 & 0 \end{pmatrix} \begin{pmatrix} 3 \\ 1 \end{pmatrix}.$$

What do you notice?

Work out

$$\begin{pmatrix} 0 & -1 \\ 1 & 0 \end{pmatrix} \begin{pmatrix} x \\ y \end{pmatrix}.$$

Compare these answers with the results of Section 1.3(a).

(b) From these results it appears that the matrix $\begin{pmatrix} 0 & -1 \\ 1 & 0 \end{pmatrix}$ is 'carrying out the process' of rotating through a quarter turn about the origin. (The transformation that we called **Q** in Section 1.3(a).) Compare

$$Q(B) = B'$$

286

with
$$\begin{pmatrix} 0 & -1 \\ 1 & 0 \end{pmatrix} \begin{pmatrix} -5 \\ -2 \end{pmatrix} = \begin{pmatrix} 2 \\ -5 \end{pmatrix}.$$

If we put $\begin{pmatrix} 0 & -1 \\ 1 & 0 \end{pmatrix} = \mathbf{Q}$ this latter statement may be written

$$\mathbf{Qb} = \mathbf{b'}.$$

The use of a bold capital letter for a matrix continues the practice started in Chapter 7; it is no accident that bold capital letters are used for both matrices and transformations.

(c) Work out

$$\begin{pmatrix} 0 & -1 \\ -1 & 0 \end{pmatrix} \begin{pmatrix} 0 \\ 4 \end{pmatrix}, \quad \begin{pmatrix} 0 & -1 \\ -1 & 0 \end{pmatrix} \begin{pmatrix} -5 \\ -2 \end{pmatrix}, \quad \begin{pmatrix} 0 & -1 \\ -1 & 0 \end{pmatrix} \begin{pmatrix} 3 \\ 1 \end{pmatrix}.$$

These three calculations may be combined in the one statement

$$\begin{array}{ccc} A & B & C \end{array} \qquad \begin{array}{ccc} A' & B' & C' \end{array}$$
$$\begin{pmatrix} 0 & -1 \\ -1 & 0 \end{pmatrix} \begin{pmatrix} 0 & -5 & 3 \\ 4 & -2 & 1 \end{pmatrix} = \begin{pmatrix} & & \\ & & \end{pmatrix}.$$

(Notice also that we have headed the columns with the points for which they are the position vectors.) Compare these results with those of Section 1.3(b).

If
$$\begin{pmatrix} x' \\ y' \end{pmatrix} = \begin{pmatrix} 0 & -1 \\ -1 & 0 \end{pmatrix} \begin{pmatrix} x \\ y \end{pmatrix},$$

express each of x' and y' in terms of x and y: hence complete

$$\mathbf{M}:(x, y) \to (\quad , \quad)$$

and describe the transformation geometrically.

(d) An enlargement, centre at the origin, of scale factor 3, is described by

$$\mathbf{E}: \begin{pmatrix} x \\ y \end{pmatrix} \to \begin{pmatrix} 3x \\ 3y \end{pmatrix}.$$

Express $\begin{pmatrix} 3x \\ 3y \end{pmatrix}$ as the product of a 2×2 matrix and the vector $\begin{pmatrix} x \\ y \end{pmatrix}$, and hence give the 2×2 matrix that represents the transformation $\mathbf{E}$. Verify that if this matrix multiplies the position vectors of the vertices of a triangle (for example that in Figure 1) then it does give the position vectors of the images of these vertices under the enlargement $\mathbf{E}$.

(e) We started this chapter with the transformations

$$\mathbf{K}: \begin{pmatrix} x \\ y \end{pmatrix} \to \begin{pmatrix} x+2 \\ y-3 \end{pmatrix} \quad \text{and} \quad \mathbf{L}: \begin{pmatrix} x \\ y \end{pmatrix} \to \begin{pmatrix} y+1 \\ -x-1 \end{pmatrix}.$$

Is it possible to express either

$$\begin{pmatrix} x+2 \\ y-3 \end{pmatrix} \quad \text{or} \quad \begin{pmatrix} y+1 \\ -x-1 \end{pmatrix}$$

as a product of a 2×2 matrix and the vector $\begin{pmatrix} x \\ y \end{pmatrix}$? The matrix representations of such transformations will be considered in Book 5.

It appears, therefore, that *some* transformations can be 'carried out' by 2×2 matrices; in such cases we speak of 'the transformation whose matrix is ...', or 'the transformation defined by the matrix ...', or 'the transformation which is represented by the matrix ...'. Are these matrices *information matrices*? If not, in what ways do they differ? It is sometimes helpful to think of these *transformation matrices* as 'instruction matrices'. (See Figure 15 page 296.)

Example 1

Find the image of the triangle $A(1, 2)$, $B(^-3, 1)$, $C(0, ^-1)$ under the transformation whose matrix is $\begin{pmatrix} 2 & 0 \\ 0 & -2 \end{pmatrix}$, and hence describe the transformation geometrically.

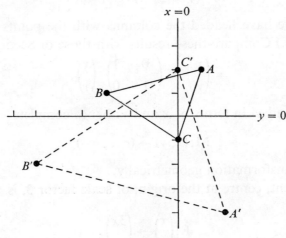

Fig. 5

(i) The position vectors of A', B', C'—the images of A, B, C—are given by:

$$\begin{array}{ccc} A & B & C \end{array} \qquad \begin{array}{ccc} A' & B' & C' \end{array}$$
$$\begin{pmatrix} 2 & 0 \\ 0 & -2 \end{pmatrix} \begin{pmatrix} 1 & -3 & 0 \\ 2 & 1 & -1 \end{pmatrix} = \begin{pmatrix} 2 & -6 & 0 \\ -4 & -2 & 2 \end{pmatrix}.$$

(ii) When these points are plotted (see Figure 5) a little consideration shows that the transformation is a combination of a reflection in the line $y = 0$ followed by an enlargement from the origin of factor 2. (Could these two transformations be applied in the opposite order to give the same image in this case?)

288

Example 2

A transformation **T** is defined by

$$\mathbf{T}: (x, y) \rightarrow (^-y, 2x+y).$$

Find the matrix for this transformation.

Writing the transformation in vector form we have

$$\begin{pmatrix} x \\ y \end{pmatrix} \rightarrow \begin{pmatrix} ^-y \\ 2x+y \end{pmatrix}, \quad \text{and arranging this as} \quad \begin{pmatrix} x \\ y \end{pmatrix} \rightarrow \begin{pmatrix} 0x+^-y \\ 2x+y \end{pmatrix},$$

we see that the image is the product of

$$\begin{pmatrix} 0 & ^-1 \\ 2 & 1 \end{pmatrix} \quad \text{and} \quad \begin{pmatrix} x \\ y \end{pmatrix};$$

hence the matrix for the transformation is $\begin{pmatrix} 0 & ^-1 \\ 2 & 1 \end{pmatrix}$.

Exercise B

In the course of this exercise the matrices for several basic types of transformation will be obtained.

1. *Isometries*

(*a*) Copy Figure 6 and draw the images of the trapezium under the transformations defined by the matrices

$$\mathbf{A} = \begin{pmatrix} 1 & 0 \\ 0 & ^-1 \end{pmatrix}; \quad \mathbf{B} = \begin{pmatrix} ^-1 & 0 \\ 0 & 1 \end{pmatrix}; \quad \mathbf{C} = \begin{pmatrix} ^-1 & 0 \\ 0 & ^-1 \end{pmatrix};$$

$$\mathbf{D} = \begin{pmatrix} 0 & 1 \\ 1 & 0 \end{pmatrix}; \quad \mathbf{E} = \begin{pmatrix} 0 & ^-1 \\ 1 & 0 \end{pmatrix}; \quad \mathbf{F} = \begin{pmatrix} 0 & 1 \\ ^-1 & 0 \end{pmatrix};$$

$$\mathbf{G} = \begin{pmatrix} 1 & 0 \\ 0 & 1 \end{pmatrix}; \quad \mathbf{H} = \begin{pmatrix} 0 & ^-1 \\ ^-1 & 0 \end{pmatrix}.$$

(*b*) What is the geometric effect of each of these transformations?

(*c*) How many lines of symmetry has the figure which is formed from all the images?

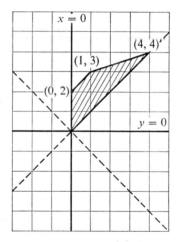

Fig. 6

2. *Enlargements*

(*a*) Find the image of the flag in Figure 7 under the transformations whose matrices are

$$\mathbf{P} = \begin{pmatrix} 1\frac{1}{2} & 0 \\ 0 & 1\frac{1}{2} \end{pmatrix}; \quad \mathbf{Q} = \begin{pmatrix} 1 & 0 \\ 0 & 1 \end{pmatrix}; \quad \mathbf{R} = \begin{pmatrix} \frac{1}{2} & 0 \\ 0 & \frac{1}{2} \end{pmatrix};$$

$$\mathbf{S} = \begin{pmatrix} 0 & 0 \\ 0 & 0 \end{pmatrix}; \quad \mathbf{T} = \begin{pmatrix} ^-\frac{1}{2} & 0 \\ 0 & ^-\frac{1}{2} \end{pmatrix}; \quad \mathbf{U} = \begin{pmatrix} ^-1 & 0 \\ 0 & ^-1 \end{pmatrix};$$

$$\mathbf{V} = \begin{pmatrix} ^-2 & 0 \\ 0 & ^-2 \end{pmatrix}.$$

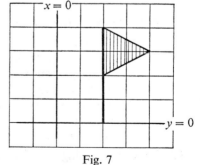

Fig. 7

(*b*) In each case state the centre and scale factor of the enlargement.

3. One-way stretches

(a) Find the image of the rectangle $A(1, 3)$, $B(1, 0)$, $C(2, 0)$, $D(2, 3)$ under the transformations given by the matrices:

$$\mathbf{U} = \begin{pmatrix} 3 & 0 \\ 0 & 1 \end{pmatrix}; \qquad \mathbf{V} = \begin{pmatrix} -1\tfrac{1}{2} & 0 \\ 0 & 1 \end{pmatrix}; \qquad \mathbf{W} = \begin{pmatrix} 1 & 0 \\ 0 & 2 \end{pmatrix};$$

$$\mathbf{X} = \begin{pmatrix} \tfrac{1}{4} & 0 \\ 0 & 1 \end{pmatrix}; \qquad \mathbf{Y} = \begin{pmatrix} 1 & 0 \\ 0 & -\tfrac{1}{3} \end{pmatrix}; \qquad \mathbf{Z} = \begin{pmatrix} 0 & 0 \\ 0 & 1 \end{pmatrix}.$$

(b) For the transformation defined by $\mathbf{U}$ are there any points (not necessarily on the rectangle) which are *invariant* (that is, not moved) under the transformation? Is there a line of invariant points?

(c) $\mathbf{U}$ can be described as a 'one-way stretch, of scale factor 3, under which the line $x = 0$ is invariant.' State the scale factors and invariant lines for each of the other transformations.

4. Two-way stretches

The transformation $\mathbf{H}$ maps the rectangle $ABCD$ onto the rectangle $A'B'C'D'$. (See Figure 8.)

(a) Express the transformation in the form $\mathbf{H}: \begin{pmatrix} x \\ y \end{pmatrix} \rightarrow \begin{pmatrix} \\ \end{pmatrix}.$

(b) Hence write down the matrix for this transformation and use it to find the image of the rectangle $ABPQ$ where P is $(-2, 0)$ and Q is $(-2, 1)$.

(c) Are any points or lines invariant under this transformation?

(d) Describe in your own words the geometric effect of this transformation.

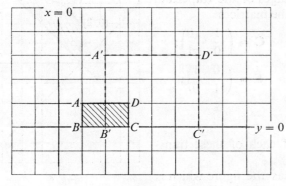
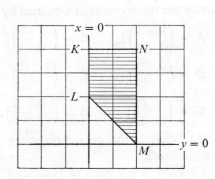

Fig. 8　　　　　　　　　　　　　　　Fig. 9

5.

(a) Write down the matrices for the following transformations:

(i) enlargement, with centre $(0, 0)$ and scale factor $2\tfrac{1}{2}$;

(ii) enlargement, with centre $(0, 0)$ and scale factor $-\tfrac{3}{4}$;

(iii) one-way stretch, $x = 0$ invariant, of scale factor $2\tfrac{1}{2}$;

(iv) one-way stretch, $y = 0$ invariant, of scale factor $-\tfrac{3}{4}$;

(v) two-way stretch, with scale factors $-\tfrac{1}{2}$ (parallel to $y = 0$) and $1\tfrac{1}{2}$ (parallel to $x = 0$);

(vi) two-way stretch, with scale factors 3 (parallel to $y = 0$) and $-1\tfrac{1}{2}$ (parallel to $x = 0$).

(b) Show on a diagram the images under each of the above transformations of the trapezium $KLMN$ in Figure 9.

6. Using values of x and y from $^-5$ to 10 draw the triangle $P(^-3, 4)$, $Q(5, 0)$, $R(7, 4)$.

(a) Find the image of PQR under the transformation whose matrix is

$$\begin{pmatrix} ^-0{\cdot}6 & 0{\cdot}8 \\ 0{\cdot}8 & 0{\cdot}6 \end{pmatrix},$$

and describe the geometric effect.

(b) Repeat for the transformation whose matrix is $\begin{pmatrix} 0{\cdot}8 & 0{\cdot}6 \\ ^-0{\cdot}6 & 0{\cdot}8 \end{pmatrix}$.

7. The transformation S is defined by $S: \begin{pmatrix} x \\ y \end{pmatrix} \to \begin{pmatrix} x+2y \\ y \end{pmatrix}$.

(a) Write $\begin{pmatrix} x+2y \\ y \end{pmatrix}$ as the product of a 2×2 matrix and the vector $\begin{pmatrix} x \\ y \end{pmatrix}$ and hence deduce the matrix for S.

(b) Show the geometric effect by copying Figure 9 and drawing in the image of $KLMN$ under the transformation S.

8. (a) If $A \begin{pmatrix} x \\ y \end{pmatrix} = \begin{pmatrix} x \\ y \end{pmatrix}$, where A is a matrix, state (i) the shape, (ii) the components of A. This matrix, which has the effect of mapping every point onto itself, is called the *identity matrix*.

(b) What is the geometric effect of the transformation defined by $E = \begin{pmatrix} 3 & 0 \\ 0 & 3 \end{pmatrix}$?

Write down a matrix F that has the opposite effect. F is said to be the *inverse* of E.

(c) Work out the products EF and FE. Comment.

9. (a) Find the image of the arrowhead (see Figure 10) under the transformation with the matrix $\begin{pmatrix} ^-2 & 0 \\ 0 & ^-2 \end{pmatrix}$.

(b) Calculate the fraction $\dfrac{\text{area of image}}{\text{area of object}}$. Comment.

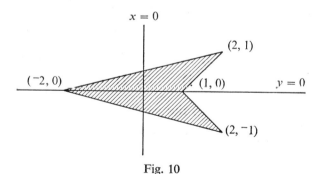

Fig. 10

10. The square $A(0, 2)$, $B(^-2, 0)$, $C(0, ^-2)$, $D(2, 0)$ is mapped by a transformation with the matrix $S = \begin{pmatrix} 1 & ^-1 \\ 1 & 1 \end{pmatrix}$.

(a) Find the image $A'B'C'D'$.

(b) Now find $A^*B^*C^*D^*$, the image of $A'B'C'D'$ under the same transformation.

(c) Describe the geometric effect of the transformation S.

(*d*) Find the image of *ABCD* under the transformation with matrix $\mathbf{T} = \begin{pmatrix} \frac{1}{2} & \frac{1}{2} \\ -\frac{1}{2} & \frac{1}{2} \end{pmatrix}$.

(*e*) How are the transformations defined by **S** and **T** related?

Without any further calculation state the image of $A'B'C'D'$ under the transformation **T**.

3. THE SHEARING MATRIX

(*a*) Figure 11 shows a tessellation of six squares sheared into six parallelograms. It is important to think in terms of every point of the diagram being mapped—we have illustrated what happens to only a few particular points.

Is it possible to represent this transformation by a 2×2 matrix?

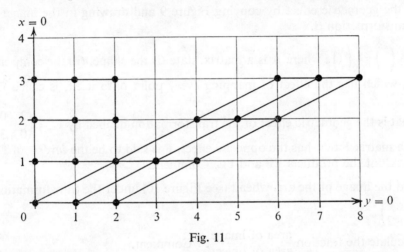

Fig. 11

Consider the vertices of the squares. They are transformed as follows:

$(0, 0) \to (0, 0)$	$(0, 1) \to (2, 1)$	$(0, 2) \to (4, 2)$	$(0, 3) \to (6, 3)$
$(1, 0) \to (1, 0)$	$(1, 1) \to (3, 1)$	$(1, 2) \to (5, 2)$	$(1, 3) \to (7, 3)$
$(2, 0) \to (2, 0)$	$(2, 1) \to (4, 1)$	$(2, 2) \to (6, 2)$	$(2, 3) \to (8, 3)$.

In each set of three, what happens to the *y*-coordinate?

Since the *y*-coordinate remains unaltered, the bottom row of the transformation matrix must be (0 1). That is
$$y = (0 \times x) + (1 \times y),$$

so we need
$$\begin{pmatrix} \cdot & \cdot \\ 0 & 1 \end{pmatrix} \begin{pmatrix} x \\ y \end{pmatrix} = \begin{pmatrix} \cdot \\ y \end{pmatrix}.$$

What happens to the *x*-coordinates?

For the first three points, for which $y = 0$, the *x*-coordinates are unaltered; for the second three, on $y = 1$, the *x*-coordinates increase by 2; for the third three, on $y = 2$, the *x*-coordinates increase by 4; and so on.

292

What would be the images of (0, 4), (1, 4) and (2, 4)?

In each case the x-coordinates *increase* by double the y-coordinates. Thus the top row of the matrix is (1   2), and the complete matrix is $\begin{pmatrix} 1 & 2 \\ 0 & 1 \end{pmatrix}$.

Apply this matrix to the position vectors of the twelve points and check that it produces the correct result.

What is the equation of the invariant line in Figure 11?

(b) Draw a diagram to show the image of the square with vertices (0, 0), (1, 0), (1, 1) and (0, 1)—we shall call this the *unit square*—under the shear with matrix $\begin{pmatrix} 1 & 0 \\ 2 & 1 \end{pmatrix}$.

What is the equation of the invariant line of this shear?

(c) Transform the rectangle with vertices $A(^-1, 3)$, $B(^-3, 1)$, $C(^-2, 0)$ and $D(0, 2)$ using the matrix $\begin{pmatrix} \frac{1}{2} & \frac{1}{2} \\ -\frac{1}{2} & \frac{3}{2} \end{pmatrix}$.

Describe your result geometrically, by drawing the rectangle and its image. Is the transformation a shear? If so, what is the equation of the invariant line? (As a help produce AD and $A'D'$ until they meet, and also BC and $B'C'$.)

Exercise C

1. Apply the shear described by the matrix $\begin{pmatrix} 1 & 1\frac{1}{2} \\ 0 & 1 \end{pmatrix}$ to the following figures. (Show each object-and-image pair on separate diagrams.)

(a) The square with vertices (0, 0), (1, 0), (1, 1) and (0, 1).
(b) The parallelogram with vertices $(^-1, 0)$, (2, 0), (0, 3) and $(^-3, 3)$.
(c) The parallelogram with vertices (4, 1), (7, 1), (6, 3) and (3, 3).
(d) The quadrilateral with vertices (0, 2), (2, 1), (4, 2) and (2, 3).

What shapes are the image figures? Find the areas of the object and image figures in each case.

2. The square with vertices (1, 1), (1, $^-1$), ($^-1$, $^-1$) and ($^-1$, 1) is to be transformed by the shear with matrix

$$\begin{pmatrix} 3 & 2 \\ -2 & -1 \end{pmatrix}.$$

By finding the coordinates of the points onto which the four vertices are mapped, find two invariant points and hence the equation of the invariant line.

3. Transform the square with vertices (0, 0), (5, 0), (5, 5) and (0, 5) using the matrix

$$\begin{pmatrix} 0\cdot6 & 0\cdot8 \\ -0\cdot2 & 1\cdot4 \end{pmatrix}.$$

Is the transformation a shear? Draw the square and its image.

If it is, find the equation of the invariant line by finding two invariant points of the transformation (that is, points that map onto themselves).

4. (*a*) Find two matrices to describe the shears that would map the unit square onto the two parallelograms in Figure 12.

(*b*) What would be the matrix of the shear that maps the left-hand parallelogram onto the right-hand parallelogram?

(*c*) What would be the matrix for the shear that has the opposite effect to that in (*b*)?

5. Find the matrix of the shear that would map the square *ABCD* onto the parallelogram *PQRS* in Figure 13.

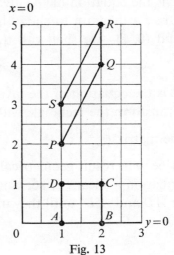

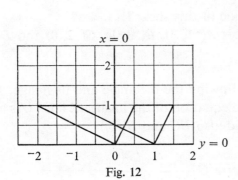

Fig. 12   Fig. 13

6. Transform the unit square by the shear with the matrix

$$\begin{pmatrix} 1 & \frac{1}{2} \\ 0 & 1 \end{pmatrix},$$

and then shear its image by the shear with the matrix

$$\begin{pmatrix} 1 & 0 \\ 1 & 1 \end{pmatrix}.$$

Is the combination of these two shears equivalent to a single shear? Does it matter in which order the shears are applied?

4. THE GENERAL 2×2 MATRIX

(*a*) In the previous sections, we have usually started with a particular transformation and obtained the corresponding matrix. Which simple transformations cannot be represented by a 2×2 matrix? Do the transformations that can be defined by a function of the form **x** → **Ax** have any properties in common? Do they all preserve shape or area? Are the directions of line segments unaltered? Are straight lines mapped onto straight lines? Is any point, or points, or line, invariant?

Conversely, does a 2×2 matrix always represent some type of transformation?

(*b*) A transformation **M** is defined by the matrix $\begin{pmatrix} 2 & -1 \\ 1 & 3 \end{pmatrix}$.

Find the image $A'B'C'D'$ of the square $A(3, 4)$, $B(3, {}^-1)$, $C({}^-2, {}^-1)$, $D({}^-2, 4)$ under this transformation.

The geometric effect of the transformation is shown in Figure 14.

    (i) Are any points invariant under **M**?

    (ii) In Figure 14 we have assumed that the line segment AB is mapped onto a *straight* line segment $A'B'$. Is this so? What are the images of the points $P(3, 3)$, $Q(3, 2)$, $R(3, 1)$ and $S(3, 0)$?

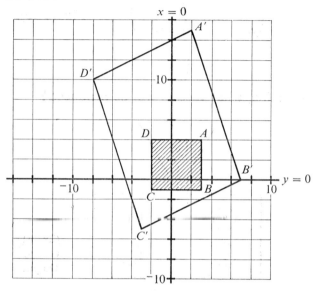

Fig. 14

    (iii) Are lengths unaltered in this transformation? What about areas?

    (iv) Are directions unaltered? Is the direction of $A'B'$ the same as that of AB? Are shapes unaltered?

    (v) What can you say about the ratios $AP:PR$ and $A'P':P'R'$; $AS:QB$ and $A'S':Q'B'$; $AB:AD$ and $A'B':A'D'$? Is the ratio of area unaltered? For example, is Area of ARD:Area of $ABCD$ = Area of $A'R'D'$:Area of $A'B'C'D'$?

(*c*) What is the image of the unit square under this transformation?

Complete the following:

$$\begin{matrix} O & I & U & J \\ \begin{pmatrix} 2 & {}^-1 \\ 1 & 3 \end{pmatrix} \begin{pmatrix} 0 & 1 & 1 & 0 \\ 0 & 0 & 1 & 1 \end{pmatrix} \end{matrix} = \begin{matrix} O' & I' & U' & J' \\ \Bigg(\qquad\qquad\qquad \Bigg). \end{matrix}$$

What happens to the origin? What is the connection between the position vectors of I' and J' and the matrix of the transformation? Sketch the image of the unit square under this transformation.

(*d*) It is sometimes helpful to picture the matrix as a machine with four dials which can be programmed to output the images of position vectors for any given transformation; a change in the settings of the dials will produce a different transformation.

For example, we can picture the matrix

$$\begin{pmatrix} 2 & -1 \\ 1 & 3 \end{pmatrix}$$

as a machine set to convert

$$\begin{pmatrix} x \\ y \end{pmatrix} \quad \text{into} \quad \begin{pmatrix} 2x - y \\ x + 3y \end{pmatrix} \quad \text{(see Figure 15).}$$

If the vectors

$$\begin{pmatrix} 2 \\ 3 \end{pmatrix} \quad \text{and} \quad \begin{pmatrix} 3 \\ 2\frac{1}{3} \end{pmatrix}$$

are fed into the machine, what vectors will be produced?

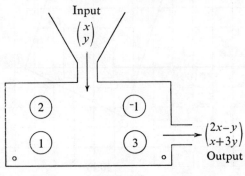

Fig. 15

Exercise D

1. Find the image of the unit square under the transformation with matrix $\begin{pmatrix} 3 & 1 \\ 1 & 2 \end{pmatrix}$. Draw a figure showing the unit square in black and its image in red.

Now find the image of the square with vertices (1, 0), (2, 0), (2, 1) and (1, 1) under the same transformation. Show your result on the same diagram, using black and red as before.

Repeat this for all the nine squares of unit area enclosed by the square with vertices (0, 0), (3, 0), (3, 3) and (0, 3).

2. Repeat Question 1 for a matrix of your own choice, and transform a tessellation of six rectangles.

3. Draw a sketch to show the result of applying the one-way stretch with matrix $\begin{pmatrix} 1 & 0 \\ 0 & 3 \end{pmatrix}$ to the circle with centre (0, 0), and radius 2. Are any of the diameters of the circle invariant? What is the name given to the image?

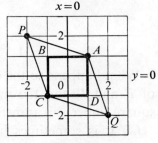

Fig. 16

4. In Figure 16, ABCD is mapped onto APCQ. What is the equation of the invariant line? Find the matrix that describes this transformation; describe also the transformation in your own words.

296

5. Find the images of the unit square under the following transformations, and hence describe these transformations in words.

(a) $\begin{pmatrix} 4 & -3 \\ 3 & 4 \end{pmatrix}$;   (b) $\begin{pmatrix} 1 & 4 \\ -2 & 2 \end{pmatrix}$;   (c) $\begin{pmatrix} 2 & -3 \\ -4 & 6 \end{pmatrix}$.

5. AREA AND MATRICES

(a) Find the image of the unit square under the transformations described by the following matrices:

(i) $\begin{pmatrix} 2 & 0 \\ 0 & 2 \end{pmatrix}$;   (ii) $\begin{pmatrix} 2 & 0 \\ 1 & 2 \end{pmatrix}$;   (iii) $\begin{pmatrix} 2 & 3 \\ 0 & 2 \end{pmatrix}$.

In each case sketch the image on graph paper and find its area using the unit square as the unit of area.

We can find the areas of these images in terms of the area of the unit square by using the formula for the area of a parallelogram. Provided one pair of the sides of the parallelogram is parallel to either $x = 0$ or $y = 0$ this is straightforward, but if this is not the case we have to fall back on other methods.

Figure 17 shows how the unit square $OABC$ is mapped onto the parallelogram $OA'B'C'$ by the transformation with matrix

$$\begin{pmatrix} 5 & 1 \\ 3 & 2 \end{pmatrix}.$$

By enclosing the parallelogram in a rectangle of area 30 units and dividing the rectangle as shown, the area of $OA'B'C'$ can be seen to be

$$30 - (2 \times 3) - (2 \times 1) - (2 \times 7\tfrac{1}{2}) = 7 \text{ units.}$$

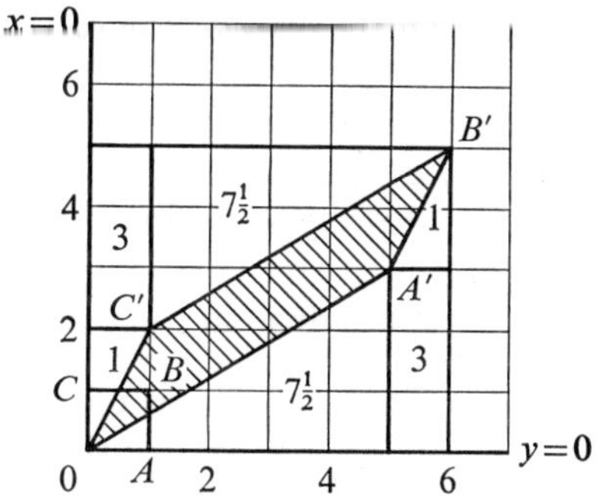

Fig. 17

(b) Repeat (a) for each of the following matrices, drawing a separate diagram for each. Find the area of the image of $OABC$ in each case.

(i) $\begin{pmatrix} 2 & 0 \\ 0 & 3 \end{pmatrix}$;   (ii) $\begin{pmatrix} 2 & 1 \\ 0 & 3 \end{pmatrix}$;   (iii) $\begin{pmatrix} 2 & 1 \\ 1 & 3 \end{pmatrix}$;   (iv) $\begin{pmatrix} 2 & 2 \\ \frac{1}{2} & 3 \end{pmatrix}$.

(c) Examine the numbers in each of these matrices and see if you can spot a connection between them and the areas of the transformed figures. For example, in (i) and (ii) you should easily see that the area is found by multiplying the two numbers in the top left- and bottom right-hand corners of the matrix; that is in (ii),

$$\begin{pmatrix} 2 & . \\ . & 3 \end{pmatrix}$$

giving $2 \times 3 = 6$. What happens in (iii) and (iv)?

Exercise E

1. Find the image of the unit square under the transformation with matrix

$$\begin{pmatrix} 4 & -1 \\ 1 & 4 \end{pmatrix}.$$

What shape is the image? What is its area? Use the area to find the length of each side and check your result by using Pythagoras's theorem.

2. A triangle has vertices (0, 0), (3, 0) and (0, 2). What is its area? Find the image of the triangle under the transformation with matrix

$$\begin{pmatrix} 2 & \frac{1}{2} \\ 1 & 1\frac{1}{2} \end{pmatrix}.$$

By referring to the numbers in this matrix find the area scale factor and hence the area of the image. Check your result by calculating the area of the image directly.

3. Find the image of the rectangle with vertices (4, 2), (10, 0), (11, 3) and (5, 5) under the transformation with matrix

$$\begin{pmatrix} 0\cdot3 & -0\cdot1 \\ 0\cdot1 & 0\cdot3 \end{pmatrix}.$$

Hence find the area of the rectangle.

4. Write down a matrix that describes:

(a) an enlargement with centre (0, 0) and scale factor h;
(b) a two-way stretch with scale factor h parallel to $y = 0$ and scale factor k parallel to $x = 0$.
Use the numbers in the matrix to find the area scale factor in each case.

5. (a) The coordinates of the vertices O, P and R of the parallelogram $OPQR$ are (0, 0), (h, k) and (u, v). What are the coordinates of Q?
What matrix would map the unit square onto $OPQR$?
What is the area of $OPQR$?
(b) Use the last result to find the areas of triangles with vertices at:

(i) (0, 0), (7, 2) and (2, 5);
(ii) (0, 0), (5, 1) and (-1, 2);
(iii) (0, 0), (2, -3) and (5, 2).

6. Draw on the same diagram the images of the unit square under the transformations with matrices

$$(a) \begin{pmatrix} 4 & 2 \\ 1 & 1 \end{pmatrix} \quad \text{and} \quad (b) \begin{pmatrix} 1 & 1 \\ 4 & 2 \end{pmatrix}.$$

What is the relationship between the two images?
Use the numbers in each matrix to find the area of each image.
What significance has the negative number that arises in (b)?

Summary

1. Many simple geometrical transformations can be described using 2×2 matrices in the form:

$$\begin{pmatrix} x \\ y \end{pmatrix} \rightarrow \begin{pmatrix} a & b \\ c & d \end{pmatrix} \begin{pmatrix} x \\ y \end{pmatrix}.$$

The matrix may be thought of as a machine that converts the position vector of an object point, $\begin{pmatrix} x \\ y \end{pmatrix}$, into the position vector of the image point, $\begin{pmatrix} ax+by \\ cx+dy \end{pmatrix}$.

2. A *two-way stretch*, with scale factor h parallel to $y = 0$ and scale factor k parallel to $x = 0$ is described by the matrix

$$\begin{pmatrix} h & 0 \\ 0 & k \end{pmatrix}.$$

If either of h or k is 1 the stretch is *one-way* only.
If $h = k$ the stretch is an *enlargement*, centre $(0, 0)$, scale factor k.

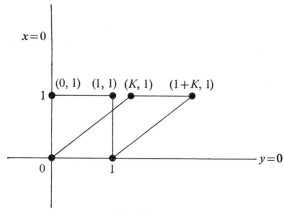

Fig. 18

3. A matrix that describes a shear keeping each point of $y = 0$ fixed is of the form

$$\begin{pmatrix} 1 & K \\ 0 & 1 \end{pmatrix}.$$

The effect of this shear is to map the *unit square* $(0, 0)$, $(1, 0)$, $(1, 1)$, $(0, 1)$ onto the parallelogram with vertices $(0, 0)$, $(1, 0)$, $(1+K, 1)$, $(K, 1)$. (See Figure 18.)

Similarly the matrix

$$\begin{pmatrix} 1 & 0 \\ K & 1 \end{pmatrix}$$

describes a shear which keeps each point of $x = 0$ fixed.

299

20-2

4. The transformation described by the general 2×2 matrix

$$\begin{pmatrix} a & b \\ c & d \end{pmatrix}$$

maps the unit square onto a parallelogram as shown in Figure 19.
    The origin is an invariant point.

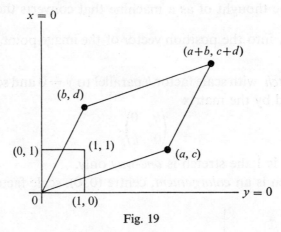

Fig. 19

5. There is an important number associated with this matrix, namely *ad–bc*. When the matrix describes a geometrical transformation, then this number gives the *area scale factor* of the transformation. In Figure 19 the area of the parallelogram is (*ad–bc*) square units.

14
FUNCTIONS AND EQUATIONS

*The principal use of the Analytic Art is to bring Mathematical Problems to Equations and
to exhibit those Equations in the most simple terms that can be.*

EDMUND HALLEY

1. FUNCTIONS

1.1 Function notation

(*a*) We met the idea of a function in Chapter 2. Part of a certain function is
shown in Figure 1.

What is 4 mapped onto by the function?

What is mapped onto $^{-}15$?

Describe the function in the form $x \to \square$

(*b*) As with transformations it is useful to have a shorthand notation to denote a
function and we often use letters such as f, g and h.

We may write the function shown above as

$$f : x \to 3x,$$

which is read: 'f is the function which maps x onto $3x$'.

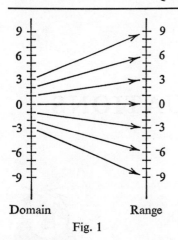

Domain   Range

Fig. 1

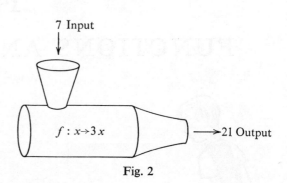

Fig. 2

You may find it helpful to picture the process as in Figure 2 which shows a machine converting members of the domain into members of the range .

An input of 7 when acted on by the function gives an output of 21 and we write $f(7) = 21$ read as 'f of 7 is 21'.

What are: (i) $f(11)$; (ii) x if $f(x) = 15$?

(c) If $g:x \to x+2$ write down the values of:

(i) $g(4)$;   (ii) $g(^-2)$;   (iii) x if $g(x) = 9$.

1.2   Composite functions

(a)

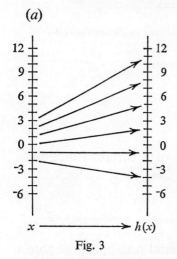

$x \longrightarrow h(x)$

Fig. 3

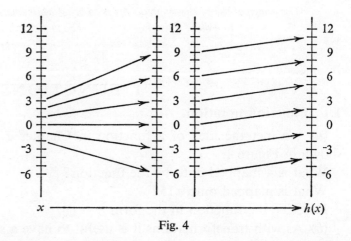

$x \longrightarrow h(x)$

Fig. 4

If h is the function partly shown by Figure 3, what are:

(i) $h(4)$;   (ii) $h(10)$;   (iii) $h(x)$?

We may see more clearly what 'h' is by splitting it into two parts (see Figure 4). The two processes are illustrated by the flow diagram (see Figure 5).

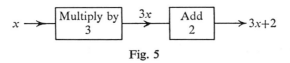

Fig. 5

So $h: x \to 3x+2$ and is the result of applying

$$f: x \to 3x \quad \text{and then} \quad g: x \to x+2.$$

You may remember that when we combined two transformations we decided to denote the result of applying   **R**   and then   **S**,

by the symbol **SR** (*not* **RS**), that is, **SR** means 'S following **R**' or 'S applied to the result of **R**'. Similarly we denote the function composed of f and then g by gf so if

$$f: x \to 3x \quad \text{and} \quad g: x \to x+2$$

then $gf: x \to 3x+2.$

(*b*) With f and g as defined above draw a flow diagram for the *composite* (i.e. combined) function fg (g and then f, or f on the result of g).

Use this to find: (i) $fg(1)$;   (ii) $fg(4)$.

Is fg the same function as gf? Justify your answer.

Write fg in the form $x \to \sqcup$.

(*c*) We have seen that two operations can lead to different results if we alter the order in which they are applied. When we write down the expression for a composite function, we must be careful to make quite clear what we have in mind. It is always possible to do this by using brackets to parcel up the parts which have to be worked out first, but we shall also make use of conventions which assign definite meanings to expressions when the brackets are omitted.

For example

$3x+2$	means	$(3x)+2$
$5+4x$	means	$5+(4x)$
$\dfrac{20}{x-7}$	means	$20 \div (x-7)$
$4x^3$	means	$4(x^3)$.

Feed $x = 12$ into each of the following and find the output:

(i) $4x-6$;   (ii) $8+3x$;   (iii) $\dfrac{x}{2}+7$;   (iv) $10-\dfrac{x}{3}$;

(v) $\dfrac{20}{x-7}$;   (vi) $\dfrac{2x+1}{5}$;   (vii) $\left(\dfrac{x}{3}\right)^2$;   (viii) $\dfrac{x^2}{3}$.

(*d*) Work out $(^-2)x$ and $^-(2x)$ when
     (i) $x = 3$;   (ii) $x = ^-5$.

What is x if $^-2x = 14$? Does it matter which of the above interpretations of ^-2x you use?

Exercise A

1. Write the following functions in the form $x \to \square$:

    (a) 'subtract 5';   (b) 'divide by 7';

    (c) 'multiply by $\frac{1}{3}$';   (d) 'square and add 2';

    (e) 'add 2 and square'.

2. Describe in words the functions:

    (a) $x \to \frac{1}{2}x$;   (b) $x \to x+2$;   (c) $x \to x^3$.

3. If $f: x \to \frac{1}{4}(x-5)$ find:

    (a) $f(9)$;   (b) $f(5)$;   (c) $f(-1)$.

4. If $f: x \to 5-2x$ find:

    (a) $f(3)$;   (b) $f(\frac{1}{2})$;   (c) $f(-2)$.

5. If $f: x \to \dfrac{x+1}{x-1}$ find:

    (a) $f(3)$;   (b) $f(-1)$;   (c) $f(0)$.

6. If $f: x \to 4^x$ find:

    (a) $f(3)$;   (b) $f(-2)$.

7. Given that f is the function which maps a number onto its highest prime factor, find:

    (a) $f(10)$;   (b) $f(20)$;   (c) $f(70)$.

8. If $f: x \to 2x+3$, find x if

    (a) $f(x) = 7$;   (b) $f(x) = 3$.

9. Is the function 'square and double' the same as the function 'double and square'?
Write both functions in the form $x \to \square$.
Distinguish between $3(x^2)$ and $(3x)^2$ by drawing flow diagrams like Figure 5.
What do you understand by $3x^2$?

10.   (a) Express the composite function shown in Figure 6 in the form $f: x \to \square$.

    (b) Find: (i) $f(1\cdot5)$;   (ii) x, if $f(x) = 6$.

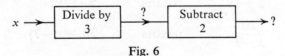

Fig. 6

11. Draw flow diagrams to show how the following functions split into simpler ones.

    (a) $x \to \frac{1}{2}(x-3)$;   (b) $x \to (3x+7)^2$.

12. Is the function 'subtract 1 and add 6' the same as the function 'add 6 and subtract 1'?
Name another pair of functions for which

$$gf = fg.$$

13. If $f: x \to 2x+1$ and $g: x \to x^2$ find:

    (a) $fg(4)$;   (b) $gf(4)$;

    (c) x, if $fg(x) = 9$;   (d) x, if $gf(x) = 9$.

14. A function f is of the form $x \to ax+b$, where a and b are numbers. Given that $f(1) = 5$ and $f(2) = 2$ find $f(3)$ by drawing a graph.

2. EQUATIONS

2.1 Inverse functions

(*a*) Think of a number. Multiply it by 4. Add 6 to the result. Exchange results with a neighbour and try to find the number he first thought of.

The flow diagram in Figure 7 shows the build-up.

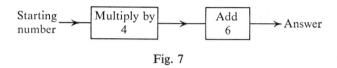

Fig. 7

Write down the flow diagram reversing the process so as to take you from the answer back to the starting number.

(*b*) Make up a similar problem involving a subtraction and a division and again give the flow diagram and the reversed flow diagram.

(*c*) Consider Figure 8.

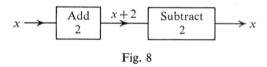

Fig. 8

The effect of adding 2 and then subtracting 2 is to bring us back to the starting number, that is $x \to x$, the *identity* function. We say that each function is the *inverse* of the other. In symbols, if $f : x \to x+2$ and $g : x \to x-2$ then g may be written f^{-1}, read as 'f inverse'. What is g^{-1}?

(*d*) If $h : x \to 3x$, what is h^{-1}?

(*e*) Figure 9 illustrates the function $f : x \to \dfrac{x+2}{3}$.

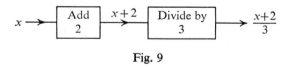

Fig. 9

The inverse function may be obtained from the reversed flow diagram in Figure 10.

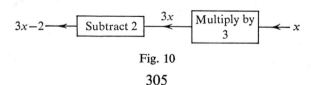

Fig. 10

305

You will notice that it is made up of the two simple functions inverse to those in Figure 9. What other important point do you notice?

We have established that if $f: x \to \dfrac{x+2}{3}$ then $f^{-1}: x \to 3x-2$.

Find: (i) $f(19)$;   (ii) $f^{-1}(7)$;   (iii) $f^{-1}(f(19))$.

2.2   Solving equations

(a) Find the value of x for which $2x+3 = 11$. Without much hesitation you will say $x = 4$. We call this 'solving the equation', or finding the 'solution set' of values of x for which the equation is a true statement. In this case, the set has only one member, 4. More briefly we say 'solve', and 'the solution is 4'. Now solve $3x+5 = {}^{-}10$.

Can you spot the answer easily?

(b) Solving by guesswork will often prove difficult and we shall look at various ways which may be used to help find the solution. One is suggested by inverse functions.

For example, to solve $3x+4 = 22,$

consider first the build up function $x \to 3x+4$. We use the flow diagram for this (Figure 11(a)) to obtain the reversed flow diagram (Figure 11(b)). Putting 22 into the latter produces the solution, $x = 6$.

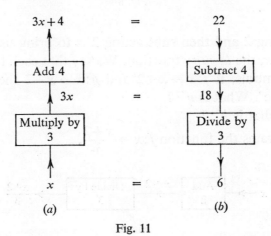

Fig. 11

The usefulness of a diagram like Figure 11 is that it shows at a glance both the simple instructions 'subtract 4' and 'divide by 3' and also the order in which we must carry them out so as to get back from $3x+4$ to the original x.

It is more usual to set the working down in the following manner. (Notice that we start at the top of Figure 11 and work downwards following the instructions

306

of Figure 11(b).)

$$3x+4 = 22,$$
(subtract 4)
$$\Rightarrow 3x = 18,$$
(divide by 3).
$$\Rightarrow x = 6.$$

Solve $2x-15 = {}^-7$ showing both the flow diagrams and the working set out separately.

(c) The following examples show other ways in which the solving of equations may be tackled.

Example 1

Solve $\qquad 3-2x = 11.$

Method (i) $\qquad 3-2x = 11,$
$$\Rightarrow 3+{}^-2x = 11, \quad \text{(because } a-b = a+{}^-b\text{)}$$
(subtract 3)
$$\Rightarrow \quad {}^-2x = 8,$$
(divide by ${}^-2$)
$$\Rightarrow \quad x = {}^-4.$$

Method (ii) $\qquad 3-2x = 11,$
$$\Rightarrow \quad 3 = 2x+11, \quad \text{(why?)}$$
(subtract 11)
$$\Rightarrow \quad {}^-8 = 2x,$$
(divide by 2)
$$\Rightarrow \quad {}^-4 = x,$$
$$\Rightarrow \quad x = {}^-4.$$

Having arrived at an answer, check to see that it fits the original equation.

Here, if $x = {}^-4$ then

the left-hand side $= 3-2x = 3-({}^-8) = 3+8 = 11 = $ the right-hand side, so

$$x = {}^-4 \Rightarrow 3-2x = 11.$$

The logical process of solving the equation can be 'reversed' at every step and so we can use the relation $\Leftrightarrow$ which means 'implies *and* is implied by' in place of $\Rightarrow$. ($\Leftarrow$ means 'is implied by'.)

Which of the relations $\Rightarrow$, $\Leftarrow$ or $\Leftrightarrow$ are appropriate links between the following pairs of statements?

(i) I have two heads. $\qquad\qquad$ I am not a man.
(ii) It is after tea. $\qquad\qquad\quad$ The time is 7 p.m.
(iii) The triangle is right-angled.   The triangle has two angles whose sum is 90°.

When you have solved an equation using $\Leftrightarrow$ between each step then it is *still* good sense to check your accuracy by substituting your solution in the original equation.

Example 2.

Solve
$$5 = \frac{20}{x+3}.$$

$$5 = \frac{20}{x+3},$$

$$\Leftrightarrow 5(x+3) = 20, \quad \text{(why?)}$$

$$\text{(divide by 5)}$$

$$\Leftrightarrow \quad x+3 = 4,$$

$$\text{(subtract 3)}$$

$$\Leftrightarrow \quad x = 1.$$

Check: if $x = 1$ then R.H.S. $= \dfrac{20}{x+3} = \dfrac{20}{4} = 5 = $ L.H.S.

Why did we start the check with the right-hand side in this case?

Example 3.

Solve
$$3x+4+4x-9 = 23.$$

$$3x+4+4x-9 = 23,$$

$$\text{(simplify)}$$

$$\Leftrightarrow 7x-5 = 23,$$

$$\text{(add 5)}$$

$$\Leftrightarrow \quad 7x = 28,$$

$$\text{(divide by 7)}$$

$$\Leftrightarrow \quad x = 4.$$

Check: if $x = 4$ then L.H.S. $= 12+4+16-9 = 32-9 = 23 = $ R.H.S.
(The step 'simplify' may need some explanation.

We regard
$$3x+4+4x-9 \quad \text{as meaning}$$

$$(3x)+4+(4x)-9;$$

also
$$3x+4x = 7x;$$

and finally $7x+4-9$ is shortened to $7x-5$.)

Exercise B

1. What are the inverse functions of:

(*a*) 'subtract 6';   (*b*) 'divide by 8';

(*c*) $x \rightarrow x+3$;   (*d*) $x \rightarrow 5x$?

2. What is f^{-1} if (*a*) $f : x \rightarrow \frac{1}{3}x$,   (*b*) $f : x \rightarrow x-4$?

3. Find, using flow diagrams, the inverses of:

(*a*) $x \rightarrow 2x-1$;   (*b*) $x \rightarrow \dfrac{x+3}{5}$;

(*c*) $x \rightarrow \frac{1}{2}(3x+7)$;   (*d*) $x \rightarrow 3\left(\dfrac{x}{2}+4\right)$.

4. Use reversed flow diagrams to solve:

(a) $3x+5 = 12$;   (b) $3(x+5) = 12$;

(c) $\frac{x}{3}+5 = 12$;   (d) $\frac{x+5}{12} = 3$.

5. Solve:

(a) $5x-8 = 12$;   (b) $3x+10 = {}^-8$;
(c) $7-2x = 1$;   (d) $9-4x = {}^-11$;
(e) $6x+19 = 17$;   (f) $8-3x = 20$.

6. Solve:

(a) $3(\tfrac{1}{2}x+7) = 27$;   (b) $\frac{30}{x+2} = 8$;

(c) $6-2x = {}^-4$;   (d) $\frac{1}{2x-3} = 1$;

(e) $\frac{24}{3x+2} = 3$;   (f) $6-(12-7x) = 1$;

(g) $14-(2x+3) = 5$.

7. Simplify and solve:

(a) $x+x+3 = 8$;   (b) $2x+3-x = 0$;
(c) $8+7x+3-5x = 14$;   (d) $4+9+10x+6x = 45$.

8. (a) The sum of two consecutive integers is 23. What are the integers?

(b) The sum of three consecutive integers is 20. By letting the integers be n, $n+1$ and $n+2$, prove that this statement is false.
State a result about the sum of any three consecutive integers.

(c) The sum of four consecutive odd integers is 192. Find the integers.

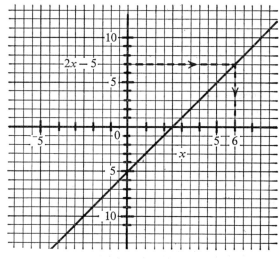

Fig. 12

9. Figure 12 shows the graph of the function

$$x \rightarrow 2x-5.$$

309

The dotted lines show how the graph can be used to solve

$$2x - 5 = 7,$$

giving $x = 6$. Use the graph to solve:

(a) $2x - 5 = 3$;   (b) $2x - 5 = 10$;   (c) $2x - 5 = 0$;
(d) $2x - 5 = {}^-7$;   (e) $2x - 5 = {}^-11$.

10. Using similar scales to those in Figure 12, graph the function

$$x \rightarrow 7 - 3x,$$

and use your graph to solve:

(a) $7 - 3x = 13$;   (b) $7 - 3x = 7$;   (c) $7 - 3x = {}^-2$;   (d) $7 - 3x = {}^-17$.

3. BRACKETS

Newspaper headline: 'Australia Routed England Win'.
Restaurant menu: 'half fresh grapefruit'.
Overheard remark: 'Have you eaten my brother William asked John'.
Comment.

3.1 Associative property

(a) Say aloud twice '36 divided by 6 divided by 2' pausing in different places so as to indicate different meanings. We can show these by using brackets.
Work out: (i) $(36 \div 6) \div 2$;   (ii) $36 \div (6 \div 2)$;
              (iii) $(12 + 7) + 3$;   (iv) $12 + (7 + 3)$;
              (v) $(12 - 7) - 3$;   (vi) $12 - (7 - 3)$;
              (vii) $(4 \times 5) \times 3$;   (viii) $4 \times (5 \times 3)$.

In which cases did the two ways of working give the same result?
(b) Let o be the operation 'take the average of'.
Work out: (i) $(20 \text{ o } 12) \text{ o } 8$;   (ii) $20 \text{ o } (12 \text{ o } 8)$.
Do you get the same result?
If an operation $*$ has the property that

$$(a * b) * c = a * (b * c),$$

for any possible elements a, b, c then it is said to be *associative*.
Use your answers above to help decide which of the operations considered (that is, division, addition, subtraction, multiplication, take the average of) are associative.
(c) When an operation is associative we may omit brackets without ambiguity.
For instance $12 + 7 + 3$ could replace (iii) and (iv) above
              and $4 \times 5 \times 3$ could replace (vii) and (viii).

Exercise C

1. Find short ways of working out:

(a) $57+9+91$; (b) $87\times4\times25$.

2. Work out:

(a) (2 to the power 3) to the power 2, that is $(2^3)^2$;

(b) 2 to the power (3 to the power 2), that is $2^{(3^2)}$.

3. $A = \begin{pmatrix} 2 & 5 \\ 1 & 3 \end{pmatrix}$; invent two other 2×2 matrices **B** and **C** using suitably small numbers for their elements.

Work out: (a) **AB**; (b) **BC**.

Use your results to work out:

(c) **(AB)C**; (d) **A(BC)**.

What do you find? Compare your results with those of a neighbour. What do these results suggest about the operation of matrix multiplication?

4. Consider the following operations:

(a) matrix addition; (b) intersection of sets;

(c) union of sets.

Can you say immediately whether or not they are associative?

If not investigate the problem choosing suitable elements as in Question 3.

5. One important operation you have met is that of *combination*, for instance combination of transformations.

In **AB**(*P*) which transformation is applied first?

Investigate whether **(AB)C** = **A(BC)**, by choosing simple transformations.

3.2 Removing brackets

(a) Figure 13 shows how we work out the value of $11-(5+2)$, the brackets being used to parcel up the part which has to be done first. Compare the result with that of $(11-5)+2$. Do you get the same answer?

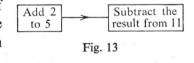

Fig. 13

If we write $11-5+2$ without brackets, we must have a convention to avoid ambiguity. The convention is that we work from left to right through the expression performing additions and subtractions as we come to them.

So $11-5+2$ means $(11-5)+2$.

(b) Work out the following pairs of expressions. Note where the same result occurs (work from left to right).

(i) $20+(12+5) = \square,$ $20+12+5 = \square$;

(ii) $20+(12-5) = \square,$ $20+12-5 = \square$;

(iii) $20-(12+5) = \square,$ $20-12+5 = \square$;

(iv) $20-(12-5) = \square,$ $20-12-5 = \square.$

In which cases do you get the same result?

In these cases it is possible to replace the expression *with brackets* by exactly the same expression *without brackets*. Where a *subtraction* sign occurs in front of the brackets, however, it is not possible to do this.

Copy and insert $+$ or $-$ so as to make correct statements:

(v) $13-(9+2) = 13$ 9   2;

(vi) $a-(b+c) = a$ b   c;

(vii) $13-(9-2) = 13$ 9   2;

(viii) $a-(b-c) = a$ b   c.

Perhaps you found the first pair easy but not the second.

Consider $100-(20-3)$ which works out as

$$100-17 = 83.$$

Copy and complete the flow diagram Figure 14.

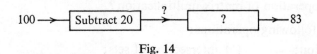

$$100 \longrightarrow \boxed{\text{Subtract } 20} \xrightarrow{\ ?\ } \boxed{\ ?\ } \longrightarrow 83$$

Fig. 14

So $100-(20-3) = 100-20+3$.

Try $30-(17-11)$ similarly.

Complete $30-(17-11) = 30$ 17   11,

$$a-(b-c) = a \qquad b \qquad c.$$

We may sum up the results so far by the four statements:

$$a+(b+c) = a+b+c, \qquad a-(b+c) = a-b-c,$$

$$a+(b-c) = a+b-c, \qquad a-(b-c) = a-b+c.$$

The point to remember is that if a $-$ sign occurs immediately before the brackets, then, when the brackets are omitted, we must change the operation, which was inside the brackets, either from $+$ to $-$ or from $-$ to $+$.

Example 4. Solve $8-(6-x) = 17.$

$$8-(6-x) = 17,$$

(remove brackets)

$$\Leftrightarrow 8-6+x = 17,$$

(simplify)

$$\Leftrightarrow \quad x+2 = 17,$$

(subtract 2)

$$\Leftrightarrow \quad x = 15.$$

Check: if $x = 15$, L.H.S. $= 8-(6-15) = 8-(^-9) = 8+9 = 17 =$ R.H.S.

Example 5. Solve $3x-(7+2x)+(x-2) = 7.$

$$3x-(7+2x)+(x-2) = 7,$$

(remove brackets)

$$\Leftrightarrow \quad 3x-7-2x+x-2 = 7,$$

(simplify)

$$\Leftrightarrow \qquad\qquad 2x-9 = 7,$$

(add 9)

$$\Leftrightarrow \qquad\qquad 2x = 16,$$

(divide by 2)

$$\Leftrightarrow \qquad\qquad x = 8.$$

Check: if $x = 8$, L.H.S. $= 24-(7+16)+(8-2) = 24-23+6 = 7 =$ R.H.S.

Exercise D

1. Copy and complete the following by inserting signs which make them true statements:

(a) $32+11+9 = 32+(11 \quad 9)$;   (b) $21-7+4 = 21-(7 \quad 4)$;
(c) $21-7-4 \ = 21-(7 \quad 4)$;   (d) $x-5+11 = x-(5 \quad 11)$;
(e) $x+5-11 \ = x+(5 \quad 11)$;   (f) $p-q+r \ = p-(q \quad r)$;
(g) $p+q-r \ = p+(q \quad r)$;   (h) $p-q-r \ = p-(q \quad r)$.

2. Copy and complete the following by inserting signs which make them true statements:

(a) $36+24-11 = 36 \ (24 \quad 11)$;   (b) $36-11+24 = 36 \ (11 \quad 24)$;
(c) $u-v+w \ = u \ (v \quad w)$;   (d) $14-(11-7) = 14 \ 11 \ 7$;
(e) $32-(5+15) = 32 \ 5 \ 15$;   (f) $l-(m+n) \ = l \ m \ n$;
(g) $a-(b-c) \ = a \ b \ c$.

3. Simplify by removing brackets:

(a) $x-(x-1)$;   (b) $6-(3+p)$;   (c) $4+(1-p)$;
(d) $(x-2)-(2-x)$;   (e) $q+4-(6-2q)$;   (f) $2a+b-(a+2b)$;
(g) $(l-m)+(m-l)$;   (h) $x^2-3x-(2-3x)$.

4. Solve the following equations for x:

(a) $x-3+2x+5 = 14$;   (b) $x+9-(17-x) = 0$;   (c) $7x-(5+3x) = 19$;
(d) $12 = 2x-(4-6x)$;   (e) $3-(2x+5)+(8-x) = 21$.

4. DISTRIBUTIVE LAW

(a) Consider the functions f and g in Figure 15. Our aim is to complete the function g so as to make it the same as f.

Use Figure 15(a) to find $f(4)$.

What number must go into the second box in Figure 15(b) so that

$$g(4) = f(4)?$$

Copy and complete Figure 15(b).

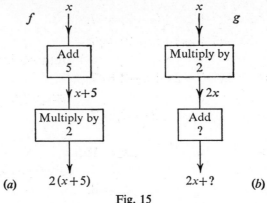

Fig. 15

Find: (i) $f(7)$; (ii) $g(7)$; (iii) $f(^-6)$; (iv) $g(^-6)$.

Does $f(x) = g(x)$ for both values of x?

(b) The function $h:x \rightarrow 4(x-3)$ can be expressed in one of the following ways:

(i) $4x-3$; (ii) $x-12$; (iii) $4x-12$; (iv) $x+1$; (v) $x-7$.

Find $h(5)$ and use this to help decide which one is correct.

(c)

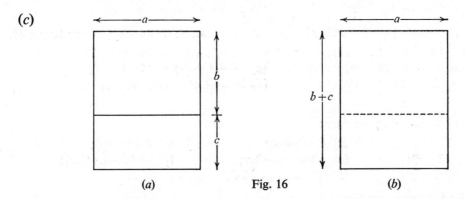

(a)   Fig. 16   (b)

Figure 16 shows the same rectangle in two ways. Figure 16(a) shows it split into two parts, one with sides a and b units and the other with sides a and c units.

What are the areas of these two parts, in square units?

Put together in Figure 16(b) the sides are now a and $b+c$ units, giving an area of

$$a(b+c).$$

So we may write

$$a(b+c) = ab+ac, \text{ which is short for } (a\times b)+(a\times c)$$

and say that we may 'remove the brackets' provided that we multiply *each* number inside them by the number outside.

Remove the brackets from

(i) $3(x+5)$;   (ii) $4(6+x)$;   (iii) $2(x+y+z)$.

314

(*d*) How does Figure 17 demonstrate that

$$a(b-c) = ab-ac?$$

This result, and the similar one above are examples of the *distributive* law (the multiplication is *distributed* over the addition or subtraction). It is of great help in simplifying expressions.

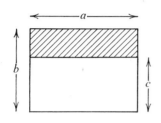
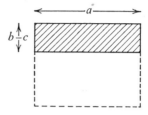

Fig. 17

Example 6.

Write without brackets:

(*a*) $7(x-3)$; (*b*) $4(3x+5)$.

(*a*) $7(x-3) = 7x-21$; (*b*) $4(3x+5) = 12x+20$.

Example 7.

Solve the equation $4(3x-5)-5x = 22$.

$$4(3x-5)-5x = 22,$$

(remove brackets)

$$\Leftrightarrow 12x-20-5x = 22,$$

(simplify)

$$\Leftrightarrow \qquad 7x-20 = 22,$$

(add 20)

$$\Leftrightarrow \qquad 7x = 42,$$

(divide by 6)

$$\Leftrightarrow \qquad x = 6.$$

Check this solution.

Example 8

Simplify $20-2(x-3)$.

This may appear ambiguous. The convention we use gives the meaning: first work out $2(x-3)$ and then subtract the result from 20.

Hence we have

$$20-(2(x-3))$$
$$= 20-(2x-6)$$
$$= 20-2x+6$$
$$= 26-2x.$$

Exercise E

1. Write without brackets and simplify if possible:

(a) $7(2x+11)$; (b) $d(d+3)$; (c) $2p(p-4)$;
(d) $3y(5y+7)$; (e) $3(a-b+c)$; (f) $2(x+6)-2(x-4)$.

2. Solve the following equations:

(a) $2(x-1)+7 = 13$; (b) $4+3(2x+5) = 43$;
(c) $4-3(5x-3) = 43$; (d) $x+2(x-3) = {}^-27$.

3. Find expressions for: (a) the perimeter, (b) the area of the shapes in Figure 18. Use brackets, if necessary, to write down your first attempt and then remove the brackets and simplify the expressions.

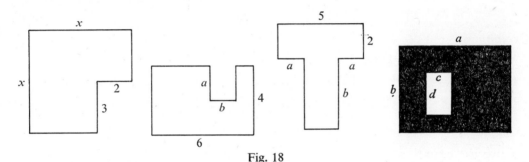

Fig. 18

4. Check that (i) and (ii) are true statements:

(i) $7\times(6+11) = (7\times6)+(7\times11)$;
(ii) $(6+11)\times7 = (6\times7)+(11\times7)$.

Which of the following are true statements?

(a) $24\div(6+2) = (24\div6)+(24\div2)$; (b) $(6+2)\div24 = (6\div24)+(2\div24)$;

(c) $\dfrac{a}{b+c} = \dfrac{a}{b}+\dfrac{a}{c}$; (d) $\dfrac{b+c}{a} = \dfrac{b}{a}+\dfrac{c}{a}$.

5. By finding the image of Figure 19 under an enlargement with centre P and scale factor 3 show that

$$3(\mathbf{a}+\mathbf{b}) = 3\mathbf{a}+3\mathbf{b}.$$

6. Work out 23×18 by long multiplication. How do you make use of the distributive property here?
(Take a, b, c as 23, 10, 8 respectively).

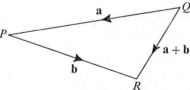

Fig. 19

7. If the draw for the F.A. Cup Semi-finals were Exeter City v Huddersfield Town and Chesterfield v Stockport County then the final would be between

(Exeter or Huddersfield) and (Chesterfield or Stockport).

Express the possible finals in more detail, using the words 'and' and 'or' correctly.

8.

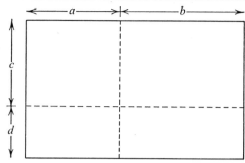

Fig. 20

(*a*) Figure 20 shows a rectangle. One side is length $a+b$ units.
What is the length of the other side?
What does $(a+b) \times (c+d)$ represent?
Show how this can be made up of four separate multiplications.

(*b*) Taking $a = 10$, $b = 2$, $c = 3$ and $d = 4$, work out
   (i) $(a+b)(c+d)$,
   (ii) $ac+ad+bc+bd$.

(*c*) Use the previous two parts to write
$$(a+b)^2 \text{ without brackets.}$$

(*d*) Write without brackets: (i) $(x+2y)^2$;   (ii) $(2m+3n)^2$.

9. We can calculate $(2 \cdot 005)^2$ in the following manner.
$$(2 \cdot 005)^2 = (2+0 \cdot 005)^2 = 2^2 + 2(2 \times 0 \cdot 005) + 0 \cdot 005^2$$
$$= 4 + 0 \cdot 02 + 0 \cdot 000025$$
$$= 4 \cdot 020025.$$

Use a similar method to calculate:

(*a*) 1002^2;   (*b*) $20 \cdot 3^2$;   (*c*) $1 \cdot 04^2$;   (*d*) 408^2.

10. (*a*) Draw a rectangle showing the area $(a+b)(c-d)$ and use it to write the expression
without brackets.

(*b*) Derive an expression for $(a+b)(a-b)$.

(*c*) Write without brackets: (i) $(d-2e)(d+2e)$;   (ii) $(2a+5b)(2a-5b)$.

11.

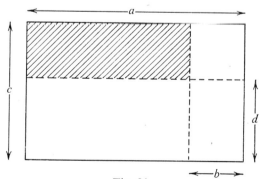

Fig. 21

(*a*) The shaded part of Figure 21 shows the area
$$(a-b)(c-d).$$

317

Show, drawing your own diagrams if necessary, how the sequence of three diagrams in Figure 22 indicates how we can express the same area in terms of the various parts, starting from the whole area ac, and find an expression for $(a-b)(c-d)$ without involving brackets.

(b) Use part (a) to find an expression without brackets for

$$(a-b)^2.$$

(c) Write without brackets: (i) $(r-2s)^2$;   (ii) $(3y-4)^2$.

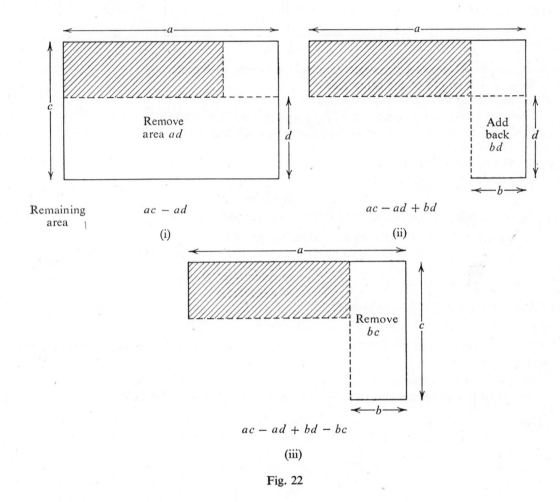

Fig. 22

12. Write without brackets, simplifying where possible:

(a) $(x+2)(y+3)$;

(b) $(x+7)(x+11)$;

(c) $(2x+5)(x+1)$;

(d) $(x+2)(x-3)$;

(e) $(x-4)(x-5)$;

(f) $(3x+7)(5x+8)$;

(g) $(x-4)(4x-1)$;

(h) $(2+a)(5-2a)$;

(i) $(x+1)(a+b+c)$;

(j) $(x+2)(2x-y+3)$.

13. (*a*) The shaded region of the Venn diagram in Figure 23 represents the set $A \cup B$, that is, the set whose members belong to either A or B, or both A and B.

Draw Venn diagrams to help you decide whether the following statements are true or false, no matter how we choose A, B and C:

(*a*) $A \cap (B \cup C) = (A \cap B) \cup (A \cap C)$;

(*b*) $A \cup (B \cap C) = (A \cup B) \cap (A \cup C)$.

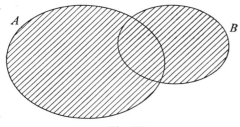

Fig. 23

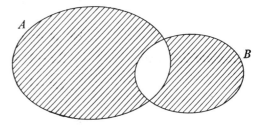

Fig. 24

(*b*) The shaded regions of the Venn diagram in Figure 24 represent the set $A \triangle B$ that is, the set whose members belong to either A or B, but *not* to both A and B.

Draw Venn diagrams to help you decide whether the following statements are true or false:

(*a*) $A \cap (B \triangle C) = (A \cap B) \triangle (A \cap C)$;

(*b*) $A \triangle (B \cap C) = (A \triangle B) \cap (A \triangle C)$.

14. (*a*) Complete the following:

(i) $(2a+b) = (a \quad b) \begin{pmatrix} \\ \end{pmatrix}$; (ii) $(3a+7b) = (\quad) \begin{pmatrix} a \\ b \end{pmatrix}$.

(*b*) Copy and complete:

$$(2a+b)(3a+7b) = (a \quad b) \begin{pmatrix} 2 \\ 1 \end{pmatrix} (3 \quad 7) \begin{pmatrix} a \\ b \end{pmatrix} = (a \quad b) \begin{pmatrix} \cdot & \cdot \\ \cdot & \cdot \end{pmatrix} \begin{pmatrix} a \\ b \end{pmatrix}.$$

By first combining $(a \quad b)$ with the 2×2 matrix and then the result with $\begin{pmatrix} a \\ b \end{pmatrix}$, find an expression for $(2a+b)(3a+7b)$ without brackets.

How does this method compare with the one we already have?

(*c*) Complete the following:

(i) $(p+3q)(5p-q) = (p \quad q) \begin{pmatrix} \\ \end{pmatrix} \begin{pmatrix} p \\ q \end{pmatrix}$;

(ii) $(a+b)^2 = (a \quad b) \begin{pmatrix} \\ \end{pmatrix} \begin{pmatrix} a \\ b \end{pmatrix}$.

5. HARDER EQUATIONS

Find, by trial, which of the following values of x fits the equation

$$5x = 3x+14.$$

(i) 2;   (ii) 5;   (iii) 7;   (iv) 10.

We have used instructions such as 'add 3', 'subtract 5', 'divide by 4' in order to solve equations. The instruction

'subtract $3x$',

319

is appropriate for the equation above. We do not need, at this stage, to know the value of x, but we must ensure that the operation is performed on the expressions both on the left-hand side and the right-hand side of the equation.

So we write
$$5x = 3x + 14$$
(subtract $3x$)
$$\Leftrightarrow 2x = 14,$$
(divide by 2)
$$\Leftrightarrow x = 7. \quad \text{Check this solution.}$$

Example 9. Solve
$$2x - 1 = 5 - x.$$
$$2x - 1 = 5 - x,$$
(add x)
$$\Leftrightarrow 3x - 1 = 5,$$
(add 1)
$$\Leftrightarrow 3x = 6,$$
(divide by 3)
$$\Leftrightarrow x = 2. \quad \text{Check this soluton.}$$

Exercise F

1. Solve:

(a) $5x + 2 = 2x + 11$;   (b) $8a = a - 14$;   (c) $2p = 3p + 6$;

(d) $8h + 9 = 15 - 4h$;   (e) $6 - 2e = 12 - 17e$;   (f) $40 - 3m = 2m - 35$;

(g) $\frac{1}{2}r + 5 = \frac{1}{4}r - 9$;   (h) $\frac{2}{3}(f - 6) = 0$;   (i) $3(b + 1) = 4(b - 1)$;

(j) $5(4 - x) = 4(x - 5)$;   (k) $25 - t = \frac{1}{2}(20t + 6)$;

(l) $2n - (3n + 4) = 4n - (3n - 2)$;   (m) $\dfrac{1}{2x - 3} = 1$;

(n) $\dfrac{8}{6 - y} = {}^-2$.

2. Find the mistakes in the following examples, Where mistakes occur, copy out the question and find the correct solution.

(a) $2x - 1 = 8 - 4x,$ (b) $14 + \dfrac{1}{x} = 7 - \dfrac{2}{x},$ (c) $3(x - 4) + 2x = 1,$

$\Leftrightarrow 6x = 9,$ $\Leftrightarrow 7 = \dfrac{3}{x},$ $\Leftrightarrow 3x - 4 + 2x = 1,$

$\Leftrightarrow x = \frac{6}{9} = \frac{2}{3}.$ $\Leftrightarrow 7x = 3,$ $\Leftrightarrow \quad 5x = 5,$

                             $\Leftrightarrow x = \frac{3}{7}.$ $\Leftrightarrow \quad x = 1.$

3. (a) Solve $\frac{1}{8}(3t + 4) = 2$.

(b) Does '$\frac{1}{8}(3t + 4)$ is a multiple of 2' $\Leftrightarrow$ 't is a multiple of 4'?

4. Solve the following equations.

(a) $2x + 3 = 2(x + 1) + x$;   (b) $2x + 3 = 2(x + 1) + 1$;   (c) $2x + 3 = 2(x + 1) + 2.$

6. ORDERINGS

An Investigation

(*a*) We saw in Chapter 8 how an ordering such as

$$x > 3,$$

can be represented by a number line diagram, as in Figure 25.

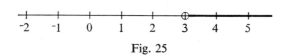

Fig. 25

Now $x+1 > 4 \Leftrightarrow x > 3$
and we could say that

$$x > 3 \text{ is the solution of the ordering}$$

$$x+1 > 4.$$

How many elements are there in the solution set?

(This will, of course, depend on what universal set we are working in; this is assumed for present purposes to be the set of all numbers.)

Solve the ordering $x-1 > 4$ representing your solution by a number line diagram.

(*b*) Using methods similar to those for the solution of equations solve,

    (i) $2x+7 > 3$,   (ii) $3x-4 > 1$,

representing the solutions on a number line.

Check that your results are correct. How can you best do this?

(*c*) Solve the following orderings:

    (i) $x-7 > {}^-1$;   (ii) $\frac{1}{2}x+1 > 9$;

    (iii) $3-4x > 11$;   (iv) $7-11x > {}^-15$;

    (v) $3-\dfrac{4}{x} > 7$;   (vi) $7+\dfrac{11}{x} > 9$.

Check each result carefully.

Will the methods used for solving equations always work in the solution of orderings? If not suggest what amendments should be made and try to explain why the need for them arises.

(*d*) Find solution sets for:

    (i) $x > 2x$;   (ii) $3x > x$;

    (iii) $x^2 > 0$;   (iv) $x^2 < x$;

    (v) $(x-1)(x-2) > 0$;   (vi) $(x-2)(x-3) < 0$.

Summary

Function notation $$f:x \rightarrow 3x+1$$

is read 'f is the function which maps x onto $3x+1$'.

$$f(x) \text{ is the result of applying } f \text{ to } x.$$

In this case, $$f(x) = 3x+1,$$

and, for example, $$f(4) = 13.$$

The composite function gf is the function 'f followed by g', or 'g following f'. If $g: x \rightarrow x^2$ then $gf: x \rightarrow (3x+1)^2$.

Brackets

The following indicate how expressions may be written when the brackets are 'removed'.

1. $a+(b+c) = a+b+c,$
   $a+(b-c) = a+b-c,$
   $a-(b+c) = a-b-c,$
   $a-(b-c) = a-b+c.$

2. $a(b+c) = ab+ac,$
   $a(b-c) = ab-ac.$

3. $(a+b)(c+d) = ac+ad+bc+bd,$
   $(a+b)(c-d) = ac-ad+bc-bd,$
   $(a-b)(c-d) = ac-ad-bc+bd.$

4. $(a+b)^2 = a^2+2ab+b^2,$
   $(a+b)(a-b) = a^2-b^2,$
   $(a-b)^2 = a^2-2ab+b^2.$

15

NETWORKS

O what a tangled web we weave!
WALTER SCOTT, *Marmion*

1. ROUTE MATRICES

In this chapter we shall see how the 'routes' on a network can be represented by a matrix, and what happens when such matrices are combined.

1.1 The matrix

$$\text{from} \begin{array}{c} \\ A \\ B \\ C \end{array} \begin{array}{ccc} & \text{to} & \\ A & B & C \\ \begin{pmatrix} 0 & 1 & 0 \\ 1 & 0 & 2 \\ 0 & 2 & 0 \end{pmatrix} \end{array}$$

represents the network shown in Figure 1. An entry of 1 (or 2) in the matrix means that there are one (or two) single stage routes *from* one point, or node, *to* another. An entry of 0 means that there is no direct route between the two points.

The routes are shown as curved lines in the figure since we are not concerned for instance with the shortest route from A to B but with the fact that there *is* a route from A to B. (There is of course no objection to drawing the routes as straight lines, but in cases where there are two separate routes, as for example from B to C, one at least will have to be curved.)

What is the sum of the entries in: (i) row B; (ii) column B?

What is the sum of the entries in: (iii) row C; (iv) column C?

What do these numbers indicate? Describe this in terms of the route map of Figure 1.

Could we see from the matrix alone that A is a 1-node?

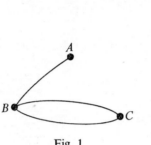

Fig. 1

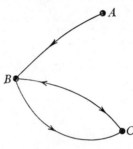

Fig. 2

1.2 Directed routes

A network can be *directed*. For example, in Figure 2, there is a route from A to B but not from B to A, and so on.

The matrix for this directed network is

$$
\begin{array}{c}
\phantom{\text{from }B} \\
A \\
\text{from } B \\
C
\end{array}
\begin{array}{c}
\text{to} \\
\begin{array}{ccc}
A & B & C
\end{array} \\
\begin{pmatrix}
0 & 1 & 0 \\
0 & 0 & 2 \\
0 & 1 & 0
\end{pmatrix}.
\end{array}
$$

(*a*) Can the order of each node still be calculated from the matrix? What is the significance of the row and column totals in the matrix for a directed network?

Why does the matrix for the map in Figure 2 have a zero total in Column A?

(*b*) Why does the *leading* diagonal, that is the one from top left to bottom right, contain only zeros in both matrices we have used so far?

What would an entry 1 mean in the top left-hand position of the matrix for Figure 2? This would mean that there would be one route *from A to A*; this would have to be indicated as a one-way looped route as in Figure 3. Can the leading diagonal contain any number? Is this so for a matrix which represents a non-directed network?

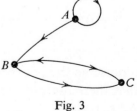

Fig. 3

(c) The matrix for the map in Figure 3 is

$$\mathbf{R} = \begin{array}{c} \\ A \\ B \\ C \end{array} \begin{array}{c} A \quad B \quad C \\ \begin{pmatrix} 1 & 1 & 0 \\ 0 & 0 & 2 \\ 0 & 1 & 0 \end{pmatrix} \end{array}.$$

Suppose that we now interchange the rows and columns in the matrix **R** giving the matrix

$$\mathbf{R'} = \begin{array}{c} \\ A \\ B \\ C \end{array} \begin{array}{c} A \quad B \quad C \\ \begin{pmatrix} 1 & 0 & 0 \\ 1 & 0 & 1 \\ 0 & 2 & 0 \end{pmatrix} \end{array},$$

where **R'** is called the *transpose* of **R**.

Draw the map corresponding to this new matrix. How does it differ from Figure 3?

What can you say about the transpose of a matrix which represents a non-directed network?

Exercise A

Where a network is not stated to be directed it is assumed to be non-directed.

1. Compile matrices for the networks in Figure 4.

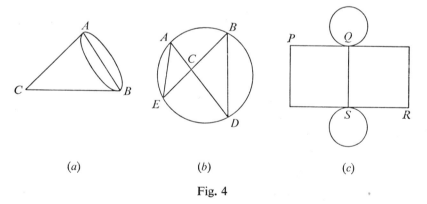

(a)   (b)   (c)

Fig. 4

2. Draw a topological map of the A-roads linking four towns in your area and compile the matrix which describes it.

3. Draw the networks described by the following matrices:

(a) $\begin{pmatrix} 0 & 0 & 1 \\ 0 & 0 & 1 \\ 1 & 1 & 0 \end{pmatrix}$,   (b) $\begin{pmatrix} 2 & 0 & 1 & 0 \\ 0 & 2 & 1 & 2 \\ 1 & 1 & 0 & 0 \\ 0 & 2 & 0 & 0 \end{pmatrix}$,   (c) $\begin{pmatrix} 0 & 3 & 1 & 0 & 1 \\ 3 & 0 & 2 & 0 & 1 \\ 1 & 2 & 0 & 1 & 0 \\ 0 & 0 & 1 & 4 & 1 \\ 1 & 1 & 0 & 1 & 0 \end{pmatrix}$.

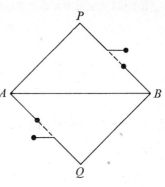

Fig. 5

4. Figure 5 shows an electrical circuit. A and B are junctions of wires and there are switches between P and B, and Q and A. Putting the rows and columns in order A, B, P, Q compile the four matrices describing the connections for all possible positions of the two switches.

5. A network consists of just two 3-nodes. By writing down all the possible matrices for such a network draw the only possible corresponding networks.

6. When is a whole row or column of a network matrix all zeros?

7. Write down the matrix which expresses the fact that every one of four points is joined to every other (except itself), and draw the network.

8. Repeat Question 7 with five points. What do you find when you draw the network? Do you think the same applies to six or more points?

9. Repeat Question 5 for a network consisting of four 3-nodes.

10. Draw the directed network for the matrix

$$\begin{array}{c} \quad A \;\; B \\ \begin{array}{c} A \\ B \end{array} \begin{pmatrix} 0 & 1 \\ 1 & 1 \end{pmatrix}. \end{array}$$

Imagine B is on a small island connected by a bridge to A, which is on the mainland. Is the ring road round the island a 'one-way street'? Which entry in the matrix would have to be changed to represent the removal of this restriction on the island, and to what would it be changed? Draw the map for the transpose of this matrix.

2. ROUTES COMBINED

2.1 Combining row and column matrices

A novel 'air-land-sea' race is arranged from town A to town D. Competitors must call at either town B or town C en route, but not at both. All three modes of travel are allowed between A and B; by air or sea between A and C, and between B and D. The only possible route between C and D across the estuary is by air. The map of the possibilities is shown in Figure 6.

Draw the topological directed route map, and consider how many different possibilities a competitor faces on the main parts of the routes, that is, ignoring the short hops from, say, airport to harbour in the same town. What are the two matrices which describe this network in two stages? First the matrix for the section from A to B or C is

$$\begin{array}{c} \quad B \;\; C \\ A \,(3 \quad 2), \end{array}$$

and then that *from B or C to D* is

$$\begin{array}{c} D \\ B \\ C \end{array} \begin{pmatrix} 2 \\ 1 \end{pmatrix}.$$

In Chapter 7 we saw how to combine two types of matrix to give a single number. Can we form the product of these two matrices? Is there only one possibility or could we form the product either way?

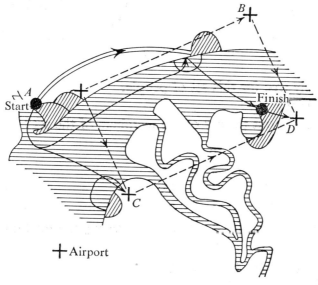

Fig. 6

If the competitor takes the choice via *B* he has three possibilities from *A* to *B*, followed by two from *B* to *D*.

For each of the three routes from *A* to *B* he has two choices for the final stage from *B* to *D*. This means that altogether he has 6 choices of route from *A* to *D* via *B*.

How many possibilities has he, if he goes via *C*?

How many choices of route are there altogether from *A* to *D*?

What we have done is to form the product of the two matrices to give the number 8.

$$(3 \quad 2) \quad \text{and} \quad \begin{pmatrix} 2 \\ 1 \end{pmatrix}$$

This combination of a row and column and the nature of the result it gives is the basis of all the remaining work on matrix combination in this chapter.

2.2 Combining square matrices

Now let us consider matrices which describe a rather different situation.

The directed networks we have been using so far are similar to the diagrams for relations used in Chapter 2.

Figure 7(a) shows the relation D, 'is the daughter of', on a set of five people and Figure 7(b) shows the relation S, 'is the sister of ', on the same set.

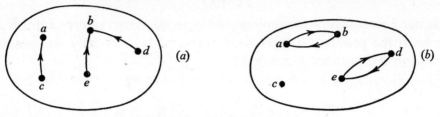

(a)

(b)

Fig. 7

(a) Draw a family tree for this set of people.

Are c and d cousins?

What can you say about the sex of the members of the set?

The matrix **D**

$$
\begin{array}{c}
 \\
 \\
\text{from }
\end{array}
\begin{array}{c}
 \\
a \\
b \\
c \\
d \\
e
\end{array}
\overset{\text{to}}{\overset{\begin{array}{ccccc} a & b & c & d & e \end{array}}{
\begin{pmatrix}
0 & 0 & 0 & 0 & 0 \\
0 & 0 & 0 & 0 & 0 \\
1 & 0 & 0 & 0 & 0 \\
0 & 1 & 0 & 0 & 0 \\
0 & 1 & 0 & 0 & 0
\end{pmatrix}}}
\quad \text{represents the relation } D,
$$

and the matrix **S**

$$
\begin{array}{c}
 \\
 \\
\text{from }
\end{array}
\begin{array}{c}
 \\
a \\
b \\
c \\
d \\
e
\end{array}
\overset{\text{to}}{\overset{\begin{array}{ccccc} a & b & c & d & e \end{array}}{
\begin{pmatrix}
0 & 1 & 0 & 0 & 0 \\
1 & 0 & 0 & 0 & 0 \\
0 & 0 & 0 & 0 & 0 \\
0 & 0 & 0 & 0 & 1 \\
0 & 0 & 0 & 1 & 0
\end{pmatrix}}}
\quad \text{represents the relation } S.
$$

Form the matrix product **DS**.

Does this product show a new relation, so that c 'is related to' b, that d 'is related to' a and e 'is related to' a? What relation is this?

(b) We have derived a relation shown by the matrix **DS**

$$
\begin{array}{c}
 \\
 \\
\text{from }
\end{array}
\begin{array}{c}
 \\
a \\
b \\
c \\
d \\
e
\end{array}
\overset{\text{to}}{\overset{\begin{array}{ccccc} a & b & c & d & e \end{array}}{
\begin{pmatrix}
0 & 0 & 0 & 0 & 0 \\
0 & 0 & 0 & 0 & 0 \\
0 & 1 & 0 & 0 & 0 \\
1 & 0 & 0 & 0 & 0 \\
1 & 0 & 0 & 0 & 0
\end{pmatrix}}}
$$

which represents the combination of two *different* relations on the *same* set. The relation represented by the matrix **DS** is not difficult to identify. What is the relation represented by the transpose of the matrices (i) **D**, (ii) **S**, (iii) **DS**?

In Section 2.1 the two matrices represented the *same* relations between elements of different sets. In Section 3 we consider the *same* relation on the *same* set.

Exercise B

1. The full arcs in Figure 8 represent the relation 'is the image after reflection in $y = 0$ of' and the dotted arcs the relation 'is the image after reflection in $x = 0$ of' on the set of four points $\{A, B, C, D\}$ of an (x, y) graph. Compile matrices **Y** and **X** to represent these relations. Work out the matrix products **XY** and **YX**, and explain why they are the same.

What relation does the resulting matrix represent?

2. Make up an example of your own in which the relation 'is brother-in-law of' appears as the combination of two other relations.

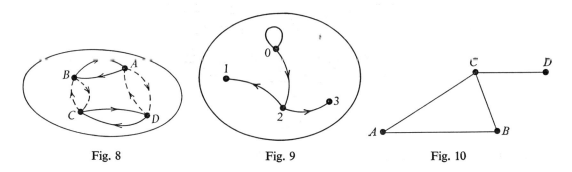

Fig. 8   Fig. 9   Fig. 10

3. Figure 9 shows the relation 'is twice' on the set $\{0, 1, 2, 3\}$ for clock arithmetic modulo 4. Compile a matrix **D** to describe this relation. Draw another figure to show the relation 'is three times' on the same set, and compile a matrix **T** to describe this relation. Work out the products of **DT** and **TD** and comment on your results.

4. With relations **D** and **S** of Section 2.2 on the same set $\{a, b, c, d, e\}$, form the matrix **SD**. What relation does this product show? Comment.

5. Inter-City rail joins A, B, C and D as shown in Figure 10. Pay-train services connect A with E, B with F and C with G. There is also a one-way Pay-train service from B to D and on to F.

Draw the topological rail map of the full services in the area and form the matrix **C** for the Inter-City services, and the matrix **P** for the Pay-train services. Express them both as matrices concerning the complete set $\{A, B, C, D, E, F, G\}$.

Form the matrices **PC** and **CP**. Use these to calculate how many journeys could be advertised consisting of 2 parts each, 1 Inter-City and 1 Pay-train service. Why is **CP** not the transpose of **PC**?

3. SQUARING THE MATRICES

On an island there are two towns, Alport and Hightown. The island bus company runs a single bus which operates on two routes:

(i) from Alport to Hightown in either direction;

(ii) a 'circular' route from Hightown through some small villages and back to Hightown.

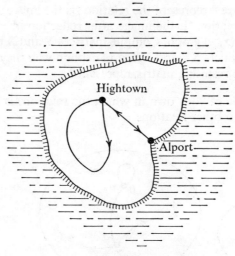

Fig. 11

(*a*) The route matrix S, to describe these routes, is

$$A \quad H$$
$$\begin{matrix} A \\ H \end{matrix} \begin{pmatrix} 0 & 1 \\ 1 & 1 \end{pmatrix},$$

and it describes the 'single stage' journeys $A \to H$, $H \to A$ and $H \to H$. What are the possible 'two-stage' journeys?

Suppose the bus starts at Alport. After going to H it has two possibilities for its second stage, either to go back to A or round the island back to H. What are the possibilities if the bus starts at H?

Compile a matrix showing 'two-stage' journeys. There is, for example, one route from A to H and back to A, that is *one* two-stage journey from A to A. Similarly there is one from A to H, that is $A \to H \to H$. You should find that the complete matrix is

$$\text{to}$$
$$A \quad H$$
$$\text{from} \quad \begin{matrix} A \\ H \end{matrix} \begin{pmatrix} 1 & 1 \\ 1 & 2 \end{pmatrix}.$$

330

(*b*) Take the first matrix, **S**, for one-stage journeys and multiply it by itself: that is

$$\begin{pmatrix} 0 & 1 \\ 1 & 1 \end{pmatrix} \begin{pmatrix} 0 & 1 \\ 1 & 1 \end{pmatrix}.$$

What do you notice about the resulting matrix, S^2?

Can you see why the two-stage matrix at the end of (*a*) above is equal to S^2?

Consider two-stage journeys starting and finishing at H. These are

$$H \to A \to H$$

and

$$H \to H \to H.$$

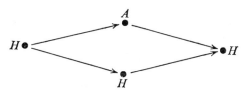

Fig. 12. Two-stage journeys from H to H in Figure 11.

The 2 in S^2 indicates these two possibilities and is found by multiplying the numbers in the second row by those in the second column as we did in Section 2.1, the situation being illustrated by the diagram in Figure 12.

This clearly shows

$$\begin{pmatrix} \text{number of routes} \\ H \to A \end{pmatrix} \times \begin{pmatrix} \text{number of routes} \\ A \to H \end{pmatrix} + \begin{pmatrix} \text{number of routes} \\ H \to H \end{pmatrix} \times \begin{pmatrix} \text{number of routes} \\ H \to H \end{pmatrix}$$
$$(1 \times 1) \qquad\qquad + \qquad\qquad (1 \times 1)$$
$$= 2.$$

(*c*) Calculate S^3 and find out what information it gives.

(*d*) The relation 'is perpendicular to' on the set of lines $\{l, m, n\}$ is illustrated in Figure 13.

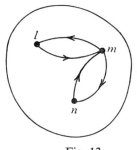

Fig. 13

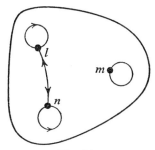

Fig. 14

Compile the matrix **R** to describe this relation.

Why is this matrix symmetrical about the leading diagonal.

R' is the transpose of **R**. Is **R'** = **R**? Find R^2. What relation does it represent?

Figure 14 illustrates the relation described by R^2. Is your matrix R^2 correct? Would the interpretation 'is parallel to' for R^2 be correct if $\{l, m, n\}$ is a set of lines in 3D space?

What restrictions would you have to place on the lines, and also on the meaning of the relation 'is parallel to'?

22-2

Summary

A 'route' matrix which represents a *non-directed* network is symmetrical about the *leading diagonal* and must contain zero or even numbers in each position on the leading diagonal.

A *directed* network is represented by a matrix which is usually not symmetrical about the leading diagonal.

The *transpose* of a matrix is obtained by interchanging rows and columns.

The transpose of a symmetrical matrix is equal to itself.

If a matrix represents a particular relation, then its transpose represents the *inverse* relation.

If a directed route is represented by a route matrix S which shows all possible single-stage journeys, then S^2 will show all possible two-stage journeys, and so on.

Exercise C

1. Write down the route matrix S for the network in Figure 15. Find S^2 and explain why the first columns of S and S^2 contain only 0's.

2. Compile a matrix S to describe the route system in Figure 16. Find S^2 and S^4. How many four-stage routes are there:
   (*a*) starting and finishing at the same point;
   (*b*) starting and finishing at different points?

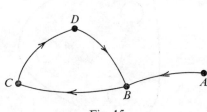

Fig. 15

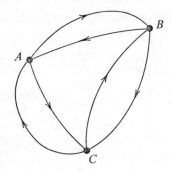

Fig. 16

3. Compute the matrices S^3 and S^4 for the matrix

$$S = \begin{pmatrix} 0 & 1 \\ 1 & 1 \end{pmatrix}.$$

Do you see a well-known sequence emerging in the numbers that occur in these matrices? Can you guess what S^5, S^6 and S^7 are?

4. If a matrix S describes a network in which there is no more than one route between any two points and no route from a point to itself (for example, the network in Figure 15 but not that in Figure 16) explain why the leading diagonal of S^2 contains only 0's. Under what circumstances would this also be true for S^3?

5. If a matrix **R** represents the following relations on a set of numbers, what relation is represented by (i) **R²**, (ii) **R'** in each case:

(*a*) 'is twice'; (*b*) 'is a factor of'; (*c*) 'is the square of'?

6. Figure 17 shows the relation 'is on the left of' for three people sitting round a table. Compile a matrix **R** to represent this relation. Find **R²**. What relation does it represent?

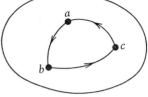

Fig. 17

4. DOMINANCE MATRICES

We can apply the ideas in Section 3 in an interesting, though not too serious manner, to the results of games.

Suppose four boys Alan, Brian, Charles and David play each other at table tennis. If Alan beats Brian then we represent the game as shown in Figure 18.

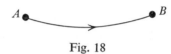

Fig. 18

We can say that this shows a relation between *A* and *B*:

$$A \text{ 'beats' } B,$$

or *A* 'is a victor over' *B*.

Figure 19 shows a network indicating the results of all the matches.

$$\text{The matrix, } \mathbf{T} = \begin{matrix} & \begin{matrix} A & B & C & D \end{matrix} \\ \begin{matrix} A \\ B \\ C \\ D \end{matrix} & \begin{pmatrix} 0 & 1 & 0 & 1 \\ 0 & 0 & 1 & 1 \\ 1 & 0 & 0 & 0 \\ 0 & 0 & 1 & 0 \end{pmatrix} \end{matrix}$$

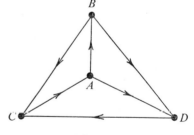

Fig. 19

describes this network, or the relation 'beats' as defined on the set {*A*, *B*, *C*, *D*}.

If we add the numbers in any particular row, what does it tell us?

If we add the numbers in any particular column, what does it tell us?

Both Alan and Brian won 2 games and lost 1.

Can you say whether Alan or Brian is the better player?

Can you say whether Charles or David is the better player?

Now Alan beat Brian, and Brian beat Charles. We say that Alan has 'two-stage dominance' over Charles. Has Alan two-stage dominance over Charles in any other way? Finding the possible two-stage dominances is similar to finding two-stage journeys when dealing with route matrices, or combining relations.

Verify that the matrix for two-stage dominance is:

$$
\begin{array}{c}
\begin{array}{cccc} A & B & C & D \end{array} \\
\begin{array}{c} A \\ B \\ C \\ D \end{array}
\begin{pmatrix}
0 & 0 & 2 & 1 \\
1 & 0 & 1 & 0 \\
0 & 1 & 0 & 1 \\
1 & 0 & 0 & 0
\end{pmatrix}.
\end{array}
$$

By totalling the 1st row and then the 2nd row we see that Alan has 3 'two-stage dominances' over other players, whereas Brian has only 2.

Alan has, then, a total of 5 'one- and two-stage dominances' and Brian has 4.

On this basis Alan could claim to be the better player. Who could claim to be the better of Charles and David?

Exercise D

1. Figure 20 shows the results of ten matches played between five boys. Find the totals of one- and two-stage dominances. Hence put the boys in an order of merit.

2. A tennis tournament between six girls, in which each girl was to play every other girl, was rained off after each had played four matches.

> Anne beat Catherine;
> Betty beat Anne, Freda and Catherine;
> Catherine beat Freda and Daphne;
> Daphne beat Anne and Edna;
> Edna beat Anne, Betty and Freda;
> Freda beat Daphne.

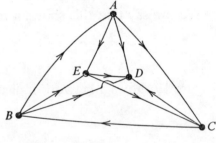

Fig. 20

Compile matrices showing one- and two-stage dominances and hence put the girls in order of merit.

3. If a dominance matrix T has a row or column of 0's explain the meaning of this.

Show that T^2 will also have a corresponding row or column of 0s and explain this in terms of matches played.

4. If a person has three-stage dominance over himself, then in the network there will be a triangular arrangement of arcs such as BE, EC and CB in Figure 20. In that example, B, E and C each have three-stage dominances over themselves.

Are there any other such triangular arrangements of paths in Figure 20?

Find by examining different cases, the maximum number of triangular paths of this kind in tournaments involving (a) four players, (b) five players, in which each person plays every other person once, and no draws occur.

5. INCIDENCE MATRICES

We shall now investigate another way of compiling matrices to describe networks, a way which has important applications in electrical circuit theory.

The matrix **R**

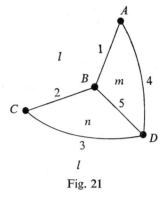

Fig. 21

$$\text{nodes} \begin{array}{c} A \\ B \\ C \\ D \end{array} \overset{\displaystyle \overset{\text{arcs}}{\begin{array}{ccccc} 1 & 2 & 3 & 4 & 5 \end{array}}}{\begin{pmatrix} 1 & 0 & 0 & 1 & 0 \\ 1 & 1 & 0 & 0 & 1 \\ 0 & 1 & 1 & 0 & 0 \\ 0 & 0 & 1 & 1 & 1 \end{pmatrix}}$$

describes the network in Figure 21 in the following way: the 1 in the first row and first column indicates that node A is 'incident' on arc 1; A is also on arc 4 but not on arcs 2, 3, or 5.

(*a*) We can also have incidence matrices which show the relationship between arcs and regions and between nodes and regions.

Copy and complete these two incidence matrices, for the network in Figure 21.

$$\text{arcs} \begin{array}{c} 1 \\ 2 \\ 3 \\ 4 \\ 5 \end{array} \overset{\displaystyle \overset{\text{regions}}{\begin{array}{ccc} l & m & n \end{array}}}{\begin{pmatrix} & & 0 \\ & & 1 \\ & & 1 \\ & & 0 \\ & & 1 \end{pmatrix}} = \mathbf{S}$$

$$\text{nodes} \begin{array}{c} A \\ B \\ C \\ D \end{array} \overset{\displaystyle \overset{\text{regions}}{\begin{array}{ccc} l & m & n \end{array}}}{\begin{pmatrix} & & \\ & & \\ 1 & 0 & 1 \\ & & \end{pmatrix}} = \mathbf{T}$$

Some entries are shown to help you. For example.

in **S**, region n has arcs 2, 3 and 5 on its boundary;

in **T**, node C is on the boundary of regions l and n.

(*b*) Now work out the matrix product **RS**. The result of combining, for example, the second row of **R** with the third column of **S** is shown below.

$$\begin{array}{ccc} \mathbf{R} & \mathbf{S} & \mathbf{RS} \end{array}$$

$$2\text{nd row} \begin{pmatrix} \cdot & \cdot & \cdot & \cdot & \cdot \\ 1 & 1 & 0 & 0 & 1 \\ \cdot & \cdot & \cdot & \cdot & \cdot \\ \cdot & \cdot & \cdot & \cdot & \cdot \end{pmatrix} \begin{pmatrix} \cdot & \cdot & 0 \\ \cdot & \cdot & 1 \\ \cdot & \cdot & 1 \\ \cdot & \cdot & 0 \\ \cdot & \cdot & 1 \end{pmatrix} = 2\text{nd row} \begin{pmatrix} \cdot & \cdot & \cdot \\ \cdot & \cdot & 2 \\ \cdot & \cdot & \cdot \\ \cdot & \cdot & \cdot \end{pmatrix}$$

3rd column   3rd column

$$(1 \times 0) + (1 \times 1) + (0 \times 1) + (0 \times 0) + (1 \times 1) = 2.$$

Compare carefully the matrix you obtain with the matrix **T** in (*a*). What do you notice?

Try to explain why the multiplication produces the result you have observed. Would it be correct to say $RS = 2T$?

(c) When dealing with a route matrix we multiplied it by itself. However, in this case the matrix R in (a) above has four rows and five columns, that is, it is a 4×5 matrix. Why is it then not possible to find R^2? We can get over this difficulty by forming the transpose R' of R by interchanging rows and columns. That is R' is

$$
\begin{array}{c}
 \\
1 \\
2 \\
3 \\
4 \\
5
\end{array}
\begin{array}{cccc}
A & B & C & D \\
\end{array}
\left(
\begin{array}{cccc}
1 & 1 & 0 & 0 \\
0 & 1 & 1 & 0 \\
0 & 0 & 1 & 1 \\
1 & 0 & 0 & 1 \\
0 & 1 & 0 & 1
\end{array}
\right)
$$

R' is a 5×4 matrix and is compatible for multiplication with R.

Find the matrix RR'.

(d) Now compile a route matrix M for the network in Figure 21.

$$
\begin{array}{c}
 \\
A \\
B \\
C \\
D
\end{array}
\begin{array}{cccc}
A & B & C & D \\
\end{array}
\left(
\begin{array}{cccc}
 & & & \\
 & & & \\
 & & & \\
 & & &
\end{array}
\right).
$$

Compare this matrix with RR'. What features do they have in common? Why does RR' produce these features? Explain the significance of the numbers on the leading diagonal of M.

Summary

For a given network, three incidence matrices can be compiled. If R, S and T stand for

$$
\text{nodes} \begin{array}{c} R \quad\quad \text{arcs} \\ \left(\quad\quad\quad \right) \end{array}
\quad
\text{arcs} \begin{array}{c} S \quad\quad \text{regions} \\ \left(\quad\quad\quad \right) \end{array}
\quad
\text{nodes} \begin{array}{c} T \quad\quad \text{regions} \\ \left(\quad\quad\quad \right) \end{array}
$$

and if the route matrix M is

$$
\text{nodes} \begin{array}{c} \text{nodes} \\ \left(\quad\quad\quad \right) \end{array},
$$

then (a) $RS = 2T$,

(b) RR' differs from M only in the leading diagonal.

336

Exercise E

1. Draw networks described by the following incidence matrices:

(*a*)   arcs

    nodes $\begin{pmatrix} 1 & 1 \\ 1 & 1 \end{pmatrix}$;

(*b*)   arcs

    nodes $\begin{pmatrix} 1 & 0 & 0 & 1 & 1 \\ 1 & 1 & 1 & 1 & 0 \\ 0 & 1 & 1 & 0 & 1 \end{pmatrix}$;

(*c*)   regions

    arcs $\begin{pmatrix} 1 & 1 \\ 1 & 1 \\ 1 & 1 \end{pmatrix}$;

(*d*)   regions

    nodes $\begin{pmatrix} 1 & 1 & 1 & 0 \\ 1 & 1 & 1 & 1 \\ 0 & 1 & 0 & 1 \end{pmatrix}$.

2. Find **R**, **S** and **T** for the network in Figure 22 and verify that **RS** = 2**T**.

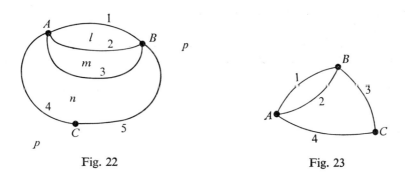

Fig. 22   Fig. 23

3. (*a*) Find the product **R′R** for the network in Figure 22. Is it the same as **RR′**?

   (*b*) Compile a matrix to describe the network as follows:

$$
\begin{array}{c}
\\
\\
\text{arcs} \;
\begin{array}{c} 1 \\ 2 \\ 3 \\ 4 \\ 5 \end{array}
\end{array}
\begin{array}{c}
\text{arcs} \\
1 \quad 2 \quad 3 \quad 4 \quad 5 \\
\left(
\begin{array}{ccccc}
\; & \; & \; & \; & \; \\
\; & \; & \; & \; & \; \\
\; & 2 & \; & \; & \; \\
\; & \; & \; & \; & \; \\
\; & \; & \; & \; & \;
\end{array}
\right).
\end{array}
$$

The '1' shown indicates that arcs 3 and 2 have *two* nodes in common.
Compare this matrix with **R′R**.

4. Repeat Question 3 for the network in Figure 23. What is the meaning of the numbers on the leading diagonal of **R′R**?

5. (*a*) Find **TT**′ for the network in Figure 21.

   (*b*) Compile a matrix as follows:

$$
\text{nodes}\quad
\begin{array}{c}
\\ A \\ B \\ C \\ D
\end{array}
\begin{array}{c}
\text{nodes} \\
A\ \ B\ \ C\ \ D \\
\left(\begin{array}{cccc}
 & 2 & & \\
 & & & \\
 & & & \\
 & & & \\
\end{array}\right).
\end{array}
$$

An entry in this matrix means that two nodes are on the boundary of the same region, for example, A and B are both on the boundary of regions l and m. What features does this matrix have in common with (i) **TT**′, (ii) **RR**′?

6. Find **S**′**S** and **T**′**T** for the network in Figure 21.

(*a*) Explain the meaning of the common leading diagonal.

(*b*) Explain the meaning of the other numbers in each matrix.

7. (*a*) Find **R**, **RR**′ and the route matrix **M** for the network in Figure 24. Check that if the numbers in the leading diagonal of **RR**′ are replaced by 0's then we get **M**.

(*b*) Find out whether **RS** = 2**T** for this network. Discuss how you might overcome any difficulties that arise.

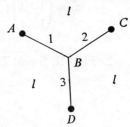

Fig. 24

16

COMPUTERS AND PROGRAMMING

Dear reader, this notice will serve to inform you that I submit to the public a small machine of my own invention by means of which you alone may, without any effort, perform all the operations of arithmetic, and may be relieved of the work which has often times fatigued your spirit, when you have worked with the counters or with the pen.

PASCAL

1. THE LANGUAGE OF COUNTING

In the past, when scientists, engineers or astronomers were faced with problems, failure to arrive at a solution was frequently due to the impossibility of carrying out the necessary hundreds or thousands of computations in the time available. With the means available, even a life-time devoted to computing was often too short.

Charles Babbage (1792–1871) saw clearly that a device was wanted to carry out relatively simple computations at high speed. His attempts to build such machines were unsuccessful largely because engineers of his day were unable to make parts with the precision that he required. Now, with great advances in the understanding of electrical circuitry and developments in the reliability of valves, relays and transistors, it is possible to build electronic computers whose speeds of computation far outstrip the wildest dreams of men like Babbage.

Although problems have become more and more complicated as civilization has developed, it is their length—the enormous number of computations involved—rather than their complexity that is fully within the scope and speed of operation of the modern machine. The computer *enables* men to compute; it can do nothing unless men tell it both what to use as data and how to use that data. The thinking and the giving of instructions still has to be done by man. All the skilled work is done on a problem before the computer receives it. This 'thinking' work of man is broadly speaking:

(*a*) to put the problem into a mathematical form and to decide what sums need to be done to reach a solution; and in what order they shall be done;

(*b*) to put the instructions for the calculation into a code or 'language' that the computer can 'understand'.

To appreciate a little of what is involved in these two 'thinking' processes we must first become familiar with the computer's language.

Let us look at an ancient—but very efficient—computer.

1.1 The spike abacus

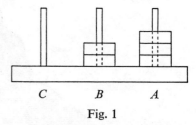

Fig. 1

Figure 1 shows a spike abacus. Discs with holes through them are placed on the spikes. It is a practical way of representing numbers and doing sums. Other versions of the abacus are still in use in parts of the East.

Only a certain number of discs will go on each spike. In the abacus shown in Figure 1, the number is five. We will agree that we will count 1, 2, 3, 4, 5 by placing discs on spike *A*. We will represent 6's by discs on spike *B*. For example, one disc on *B* and two discs on *A* will represent one six and two, that is, the number 8 and so on.

(*a*) What number is represented in Figure 1?

(*b*) What is the largest number that can be represented, using spikes *A* and *B*?

(*c*) What do you think a single disc on *C* would mean?

(*d*) What is the largest number that could be represented by discs on all three spikes?

(*e*) What would we have to do, if even larger numbers were required?

(*f*) Sketch a spike abacus with three discs on *C*, four on *B* and none on *A*. What number is it representing?

(*g*) How many discs will there be on spike *A* of an abacus showing

(i) 18; (ii) 34; (iii) 43?

1.2 Place value

Look again at Figure 1. It represents the number 15, but what it shows is

$$(2 \times 6) + 3.$$

We are using the *principle of place value*. Suppose we write this as 23. We shall mean 'two sixes and three' and not 'two tens and three' as in ordinary arithmetic. We can indicate this by writing the *number base* after the number and a little lower, like this:

$$23_6 \text{ means } (2 \times 6) + 3,$$

$$23_{10} \text{ means } (2 \times 10) + 3.$$

Note. For clarity we shall write the number base in base 10.

A number written after another number and a little lower down is called a *suffix*. We use the suffix 6 to show that the number is to be read in base 6. Our ordinary numbers are in base 10, which we call the *decimal* system.

Example 1

Convert 45_6 into decimal form.

45_6 means four 6's and 5,

that is, $45_6 = (4 \times 6) + 5,$

$$= 24 + 5,$$

$$= 29_{10}.$$

Example 2

Convert 19_{10} into base 6.

6)19

3 remainder 1,

$$19_{10} = (3 \times 6) + 1$$

$$= 31_6.$$

Exercise A

1. Convert into ordinary decimal numbers: 15_6, 30_6, 23_6, 2_6.

2. Convert into base 6: 9_{10}, 19_{10}, 32_{10}, 5_{10}.

3. What does the 2 mean in: 12_6, 24_6?

4. Write down in words the meaning of 40_6. Why is the 0 there?

5. Convert into decimal: 114_6, 231_6, 104_6, 540_6.

6. Explain the difference between 14_6 and 41_6.

7. Write in words the meaning of 15_7, 123_4, 20_8.

8. Convert into decimal: 33_5, 14_5, 10_5.

9. Convert into base 5: 18_{10}, 11_{10}, 20_{10}, 26_{10}.

10. Convert into decimal: 13_4, 25_7, 54_9, 108_9.

1.3   The language of 0 and 1

The work of a computer is to add and subtract. Multiplication is just repeated addition. Division is done by continued subtraction. Essentially the business of preparing a problem for a computer to solve consists of splitting it into lots of small steps, consisting mainly of addition or subtraction.

Electronic computers work, of course, by electricity. A lamp is either on or off, according to whether the current is, or is not, flowing, The flow is controlled by a switch, which can break the circuit.

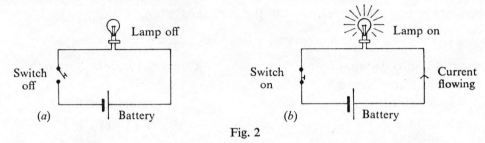

Fig. 2

Figure 2(a) shows the circuit with no current flowing. In a computer this can represent the number 0. Figure 2(b) shows current flowing; this can represent the number 1. No other numbers need be used. The arithmetic uses only the numbers 0 and 1. This is arithmetic in base 2 called *binary*. (Remember that just as in base 10 there is no single symbol for ten, so in base 2 there is no single symbol for two.)

In binary
$$10_2 = \text{(one two and zero)} = (1 \times 2) + 0 = 2_{10},$$
$$11_2 = \text{(one two and one)} = (1 \times 2) + 1 = 3_{10},$$
$$100_2 = \text{one four} = (1 \times 4) + (0 \times 2) + 0 = 4.$$

The columns will be headed as follows:

...	Eights	Fours	Twos	Units
	I	I	O	I

The number shown is $(1 \times 8) + (1 \times 4) + (0 \times 2) + 1 = 13_{10}$.

In lamps, this would look like Figure 3 (reading from left to right).

Fig. 3

In simple home-made computers, numbers are represented by lamps in this way. In large computers, the work is done with valves or transistors instead of switches. The numbers are not now represented by lamps, but the ideas are much the same.

1.4 Binary arithmetic

Example 3. Calculate $1011 + 110$.

Remember the usual carrying rule, and that $1 + 1 = 10$.

$$
\begin{array}{r}
1011\ + \\
110 \\
\hline
10001
\end{array}
$$

Example 4. Calculate $1011 - 110$.

Remember, when we come to $0 - 1$, we must 'borrow'.

$$
\begin{array}{r}
1011\ - \\
110 \\
\hline
101
\end{array}
$$

Example 5. Calculate 1011×110.

Use the usual layout for long multiplication and remember the importance of starting each line under the number we are multiplying by.

$$
\begin{array}{r}
1011\ \times \\
110 \\
\hline
000 \\
10110 \\
101100 \\
\hline
1000010
\end{array}
$$

Binary multiplication is extremely easy since the only numbers to multiply by are 1 and 0. In fact, there are virtually no multiplication tables at all: it is scarcely necessary to learn the one times table!

Let us check our answer by converting the whole problem into decimal in the usual way.

$$1011 = (1 \times 8) + (0 \times 4) + (1 \times 2) + 1 = 11_{10},$$
$$110 = (1 \times 4) + (1 \times 2) + 0 = 6_{10},$$
$$1011 \times 110 = 11_{10} \times 6_{10} = 66_{10}.$$

Check for yourself that $1000010_2 = 66_{10}$.

The figures in a binary number are called *binary digits*, or *bits* in computer language. Although the arithmetic of binary is so simple, you will have noticed a snag. It took 7 bits to write the number 66. It will take more to represent 100_{10} and 1000_{10}. Find out how many bits are needed. This fact is of small importance in a machine, but it makes paper arithmetic rather tedious.

Exercise B

All numbers are in binary, unless otherwise stated.

1. Calculate:

(*a*) $1111 + 1001$;

(*c*) $1110 + 101 + 10010$;

(*b*) $1001 + 10101$;

(*d*) $1 + 11 + 111 + 1111$.

10. Calculate:

(*a*) $111 - 101$;

(*c*) $10110 - 101$;

(*b*) $1000 - 110$;

(*d*) $101011 - 11011$.

11. Calculate:

(*a*) 110×11;

(*c*) 10110×1000;

(*b*) 10110×101;

(*d*) 111×111.

100. Calculate the following in decimal; then convert into binary and repeat the calculation, checking your answer by converting back into decimal.

(*a*) $3_{10} + 7_{10}$;

(*c*) $17_{10} - 9_{10}$;

(*b*) $11_{10} + 6_{10} + 13_{10}$;

(*d*) $15_{10} \times 9_{10}$.

101. Convert 11011 into decimal. Hence find its factors and express them in binary. Check your answer by multiplication in binary.

110. Illustrate 1011001 in lamps.

111. Arrange the following numbers in order of increasing size:

$$1010, \ 1001, \ 110, \ 1011, \ 1111.$$

1.5 Binary division

Division, as we have already noted, is really repeated subtraction. This is obvious when the division is done in binary.

Example 6. *Calculate* $11011001101 \div 101$.

We adopt the usual long division arrangement, but the work is simplified by the fact that 101 either divides once, or not at all. Remember positional notation and that we have to insert a 0 if it is necessary to bring down two figures or more. (Why?)

```
           101011100  remainder 1
     101 )11011001101
           101
           ---
           111
           101
           ---
          1000
           101
           ---
           111
           101
           ---
           101
           101
           ---
           001
```

344

Exercise C

All numbers are in binary

1. Calculate $101101 \div 11$.   10. Calculate $11011 \div 101$.

11. Find a binary number a such that $a \times 1101 = 11010$.

100. Find a binary number b such that $110 \times b = 11001$.

101. Divide 1011010 by (*a*) 10, (*b*) 100, (*c*) 110.

110. Write down a division to which the answer is 11.

111. Find three final digits of the number $110010...$ if it is to be exactly divisible by

(*a*) 101,   (*b*) 111.

Exercise D (*Miscellaneous*)

1. Convert into decimal: (*a*) 26_{12}, (*b*) 109_{12}.

2. If these numbers are octal, that is in base 8, calculate:

(*a*) $27 + 4 + 116$;   (*b*) 373×2;   (*c*) $1516 - 247$.

3. Convert: (*a*) 10110_2 into octal, (*b*) 253_8 into binary.

4. Calculate: (*a*) $110110_2 - 110_2$, (*b*) $110111_2 + 1011_2$.

5. Calculate $101011_2 \times 1101_2$.

6. If a mysterious race in the Amazonian jungle calculates that $21 + 14 = 40$, do you think it most likely that they have only one hand, or three? Give your reason.

7. Is there any sort of animal which, if it were intelligent enough, might reckon that $31 - 25 = 5$?

8. Is it true or false that $8_{10} = 10_8$, and that $12_{13} = 13_{12}$? Find some other similar examples.

9. Can your find any number that cannot be expressed

(*a*) in base 10,   (*b*) in base 2,   (*c*) in base 7?

10. Is it always true that, in binary, any number ending in 0 can be divided exactly by a smaller number ending in 0? Give examples.

11. Say what you mean by an even number. How can you tell an even number from an odd number (*a*) in base 6, (*b*) in base 5?

12. Explain why multiplying by 10 simply 'adds a 0' in any base.

13. Calculate 11×11 where the base is (*a*) 2, (*b*) 5, (*c*) 8. Is there an answer applicable to all bases?

14. Convert 189_{10} into (*a*) binary; (*b*) octal. Set out your answers in a table like this:

	128's	64's	32's	16's	8's	4's	2's	1's
189 in base 2								
		64's			8's			1's
189 in base 8								

What do you notice? How does this help you to convert other decimal numbers into binary?

15. Describe the relation between: (*a*) base 2 and base 4, (*b*) base 2 and base 16. Convert 1011011 into: (i) base 4, (ii) base 16.

2. HIGH-SPEED DIGITAL COMPUTERS

Before we look at the main parts of a modern computer, consider the way you might set about using a desk calculator to compute the total mass of the boys in your class.

To start with, you would need to find the mass of each boy in the form and store this information by writing it down on a piece of paper. These numbers are the *data* required for the computation.

Next, you would read a number from the paper, build it up on the setting register and wind it into the product register by turning the handle once (see Figure 4).

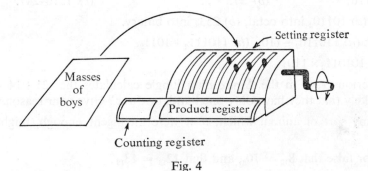

Fig. 4

This would be repeated until all the numbers had been added. The product register would then be storing a number corresponding to the total mass of the class and this could be noted. (What number would be stored in the counting register?) We can divide the solution of this problem into various steps as shown in Figure 5.

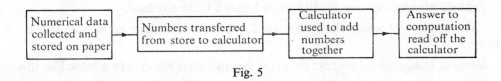

Fig. 5

Every step is controlled by the person operating the calculator.

The designers of modern computers have retained this basic pattern and Figure 6 shows the distinct parts into which a computer can be divided.

346

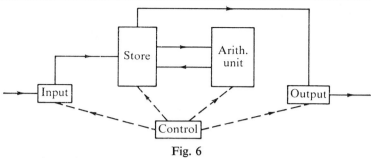

Fig. 6

2.1 The arithmetic unit

This part of a computer corresponds to the desk calculator. It carries out all the arithmetical work at a very high speed (for example, it could carry out 10000 decimal additions a second) using transistorized circuits (see Figure 7) in which binary digits

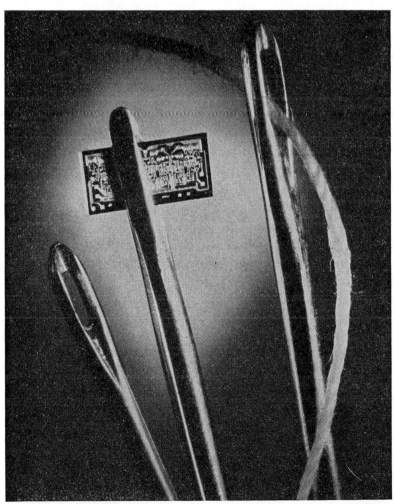

Fig. 7. A micro-integrated electronic circuit pictured passing through the eye of a no. 5 sewing needle. These miniature circuits not only save space but increase the operating speed of a computer.

are represented by electrical pulses. Numbers in binary form are fed into the arithmetic unit from the *store* as sequences of electrical pulses and the result of the computation is then returned to the store.

2.2 The store

The store corresponds to the piece of paper on which numbers are written, or to a filing cabinet containing information. The numbers now, however, have to be stored in a way which can quickly be converted to electrical pulses and this is usually done magnetically, in much the same way as magnetic tape is used in tape recorders. Numbers are transferred from the store to the *arithmetic unit* and back as required. This is a fully automatic process which depends on the instructions given to the computer. These instructions, called the *program**, must tell the computer in detail which numbers to operate on, what operations to perform, and where to store the result. The program is stored by the computer in the same way as the numbers. Once the data and the program have been fed into the store, the computer can be left to carry out all the computations required without any outside assistance.

2.3 The control unit

This unit takes the instructions, one by one, from the store and issues the appropriate orders to the different parts of the computer. It acts rather like a foreman who conveys instructions that have been given to him by his manager, to the workmen. The 'manager' in this case is the computer programmer, the person who writes out the list of instructions to be followed.

2.4 The input and output

Inside a computer, numbers and instructions are represented by magnetic intensities or electrical pulses. It is the job of the input and output units to convert the kind of instructions which we can write on paper into these forms. This is done in various ways but one of the easiest to understand is the use of patterns of holes punched in paper tape. Each letter of the alphabet and each digit is represented by a pattern of holes (see the 5-hole tape in Figure 8). The pattern for 'letter shift' indicates a change

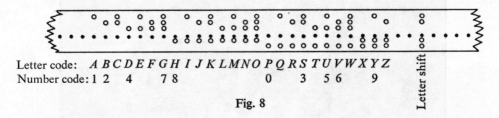

Letter code: *A B C D E F G H I J K L M N O P Q R S T U V W X Y Z*
Number code: 1 2 4   7 8   0   3 5 6   9

Fig. 8

---

* It is now common practice to use this spelling to denote a set of instructions to a computer, rather than 'programme' which, of course, is the spelling we use for the information sheet we buy at a theatre or a football match.

from the letter code to the number code or vice versa. This enables us to use the same pattern of holes for numbers or letters.

The program is put onto tape using a tape punch which resembles a typewriter. The tape is then fed into the tape reader which uses photo-electric cells to convert the patterns of holes into electrical pulses. Before the program is put onto tape, however, the instructions must all be written in a way that can be understood by the computer. In the next section, you will see how this is done.

The output is just the reverse of the input. Electrical pulses corresponding to the results of the computation are used to drive a tape punch and the paper tape is then fed into a teleprinter which types out the results.

Class projects

1. Collect newspaper cuttings and magazine articles about computers and their uses and display them in the classroom.

2. Make a list of the different types of industries that are advertising in this week's papers for computer operators or programmers.

3. Draw up a time chart showing the historical development of computing aids.

4. Find out what you can about the following:

(a) punched cards;	(b) relays;
(c) magnetic core stores;	(d) analogue computers;
(e) data processing;	(f) Algol;
(g) line printer;	(h) teleprinter.

5. Write a message to a friend and code it using the 5-hole tape shown in Figure 8.

6. Can you discover the code that is used in the tape design on the cover of this book?

3. FLOW DIAGRAMS

At the beginning of this chapter it was mentioned that one of the tasks of the programmer is to decide the order in which computations should be carried out. To help him organize the order in which things have to be done, the programmer often begins his work by drawing a *flow diagram*.

Flow diagrams can be useful in tackling many problems which are not mathematical, as shown in the following examples.

Example 7

Figure 9 is a flow diagram showing how to telephone a friend on a private S.T.D.-type phone.

You will notice in this flow diagram that there are two kinds of boxes:

1. Rectangular boxes which contain operations to be carried out. These boxes have only one route leading from them.

2. Diamond shaped boxes containing questions that can be answered *yes* or *no*. These decision boxes have two routes out, depending on the answer to the question in the box.

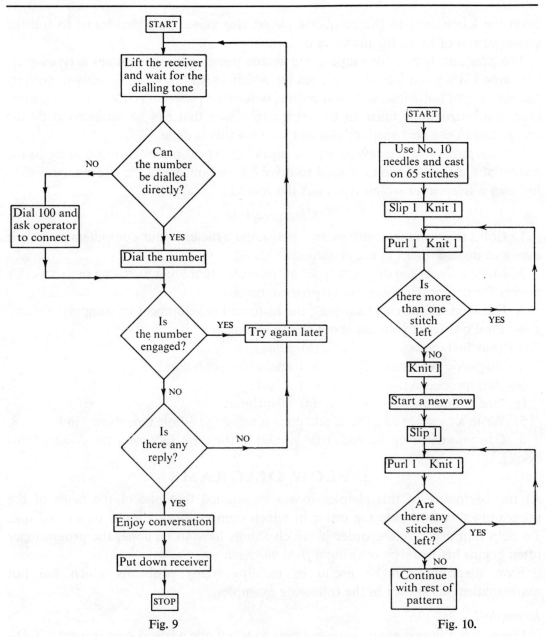

Fig. 9

Fig. 10.

Example 8

In a knitting pattern for a pullover the instructions for knitting part of the rib start as follows:

'Using the No. 10 needles cast on 65 sts.,

1st row: Sl. 1, K. 1, *P. 1, K. 1 repeat from * to the last stitch, K. 1;

2nd row: Sl. 1, *P. 1, K. 1 repeat from * to end of row.'

A flow diagram incorporating these instructions is shown in Figure 10.

350

In both of the above flow diagrams there are lines which leave decision boxes and then re-enter the diagram at an earlier point. What is the effect of this? Lines like these are called *loops* and they play an important part in programming.

Exercise E

1. Draw flow diagrams to show someone how to:

(*a*) cross a road;
(*b*) buy a new dress or suit;
(*c*) use a record player;
(*d*) bake a cake;
(*e*) take your turn in a game of Whist;
(*f*) look for a word in a dictionary;
(*g*) sing a hymn with a refrain.

2. Construct a flow diagram to show someone how either (*a*) to start a car, or (*b*) to make a dress.

3. Work through the flow diagram in Figure 11 taking $a = 0$ and $b = 1$. What name is given to the sequence of numbers you have written down?

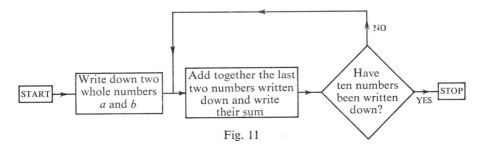

Fig. 11

4. (*a*) Work through the flow diagram in Figure 12. Repeat the work twice more placing V in a different position each time. What do you notice about the length CD in each case?

(*b*) Try the same flow diagram with different distances between A, B and C, and see what you can discover.

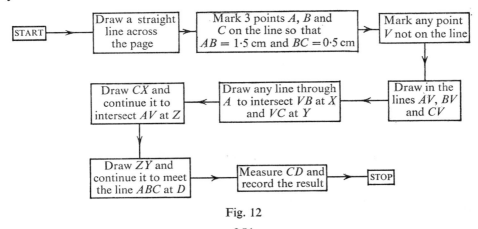

Fig. 12

351

5. Given that $\triangle A'B'C'$ is the image of $\triangle ABC$ under a rotation, construct a flow diagram to show how the centre of rotation can be found using a ruler and compass.

6. A group of explorers arrive at the left bank of a river that has to be crossed. The only means of transport is a boat being rowed by two boys. If the boat can carry at most two boys or one explorer, how can the explorers cross the river? Draw a flow diagram to illustrate your solution.

7. Construct a flow diagram to show how to find whether or not a number is prime.

8. Construct a flow diagram to show the second player in a game of noughts and crosses how to avoid defeat.

4. PROGRAMMING

Programming Simon

Simon (see Figure 13) is a simple digital computer capable of carrying out the four operations: addition, subtraction, multiplication and division. Like most computers, he has only a very restricted vocabulary and only responds to a limited number of types of instruction. He can only obey binary operations—he cannot operate on more than two numbers at a time. (Can you?)

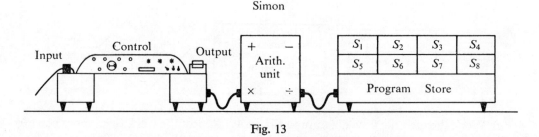

Fig. 13

To communicate with Simon, programs have to be written in a special language, Simpol, which can be punched onto paper tape and then fed into the computer. The store in Simon can hold as many instructions as are ever likely to be wanted, but there are only eight compartments for storing numbers. Each of these eight compartments has an address, one of S_1, S_2, ..., S_8, and stores only one number at a time.

There are three kinds of basic instructions that can be written in Simpol and these are illustrated in the following examples.

(1) *Input* 3·5 to S_3.

The term 'input' is restricted to the feeding of data into Simon. This instructs Simon to replace whatever number is in the compartment whose address is S_3, by the number 3·5.

(2) *Replace S_5 by $S_2 + S_4$.*

352

The term 'replace' is used when what is in one store, often the result of some calculation done by Simon, is moved into another store.

This instructs Simon to look at the numbers stored in S_2 and S_4, to add them together and then put the result in S_5. It has the effect of displacing whatever number was in S_5 before the instruction was carried out. S_2 and S_4, however, still contain the same numbers.

For example, suppose that, before the instruction, the stores held the following numbers: S_2, 20; S_4, 11; S_5, 99. Then after the instruction 'replace S_5 by S_2+S_4' the stores would hold the numbers: S_2, 20; S_4, 11; S_5, 31.

This instruction can also be used with the same store repeated, for example,

$$\text{'replace } S_5 \text{ by } S_2+S_2\text{',}$$

$$\text{'replace } S_5 \text{ by } S_5+S_5\text{',}$$

or with the operations $-$, $\times$ and $\div$,
for example,
$$\text{'replace } S_6 \text{ by } S_2 \div S_1\text{'.}$$

(3) *Output* the number in S_7.

This instructs Simon to punch out in tape the pattern of holes representing the number stored in S_7, and then to pass this through the teleprinter equipment to be typed out.

We can now write down a complete program for Simon.

Example 9

Write a program for Simon to find the average of the three numbers 7·6, 3·2 and 8·1.

To find this average Simon must be given instructions to compute $\dfrac{7\cdot6+3\cdot2+8\cdot1}{3}$.

Figure 14 shows a suitable program:

(1) Input 7·6 to S_1 ⎫
(2) Input 3·2 to S_2 ⎪ These instructions read the data for the computation into
(3) Input 8·1 to S_3 ⎬ Simon's store.
(4) Input 3 to S_4. ⎭

(5) Replace S_5 by S_1+S_2 ⎫
(6) Replace S_6 by S_5+S_3 ⎬ These instructions tell Simon what calculations to make.
(7) Replace S_7 by $S_6 \div S_4$ ⎭
(8) Output the number in S_7.

Fig. 14

It is useful to check a program by using a table with one column corresponding to each store compartment used in the program and, acting as the computer, carrying out the instructions one at a time without actually making the calculations. This is known as 'dry checking' and it is a very necessary part of the art of programming (see Figure 15).

Number of instruction	Numbers stored in Simon's store						
	S_1	S_2	S_3	S_4	S_5	S_6	S_7
I	7·6	—	—	—	—	—	—
2	7·6	3·2	—	—	—	—	—
3	7·6	3·2	8·1	—	—	—	—
4	7·6	3·2	8·1	3	—	—	—
5	7·6	3·2	8·1	3	7·6+3·2	—	—
6	7·6	3·2	8·1	3	7·6+3·2	7·6+3·2+8·1	—
7	7·6	3·2	8·1	3	7·6+3·2	7·6+3·2+8·1	$\dfrac{7·6+3·2+8·1}{3}$

The contents of S_7 are punched out

Fig. 15

Notice that instruction (5) uses a new store S_5. The instruction could have been 'Replace S_1 by S_1+S_2'. This would have the effect of saving S_5 for some other occasion. Even though the data in S_1 is lost we can still complete the program, altering instructions (6) and (7), using far fewer stores.

Remember that Simon can only add, subtract, multiply or divide two numbers at a time so that instructions (5), (6) and (7) cannot be replaced by a single instruction, although this may seem more natural.

The time taken by a computer to read in data is comparatively long compared with the time taken to perform an arithmetic calculation, so it is usual to arrange, whenever possible, that all the input instructions come together at the beginning of a program.

Exercise F

1. (a) How much of the program in Example 9 will need to be altered if we wish to find the average of 2·65, 4·72, and 3·68?

(b) Write a program for Simon to compute

$$\frac{326·7}{68·73}+\frac{29·35}{0·0876}.$$

2. Write down the computations Simon will carry out if given the following programs:

(a) Input 28·3 to S_1,
Input 19·7 to S_2,
Input 485 to S_3,
Replace S_4 by S_1-S_2,
Replace S_5 by $S_3 \times S_4$,
Output the number in S_5.

(b) Input 36·2 to S_1,
Input 18·9 to S_2,
Input 21·4 to S_3,
Input 14·7 to S_4,
Replace S_5 by S_1+S_2,
Replace S_6 by S_3-S_4,
Replace S_7 by $S_5 \div S_6$,
Output the number in S_7.

3. Write programs in Simpol to compute the following (flow diagrams will be helpful):

(a) $(3·6 \times 6·2)+(2·8 \div 0·94)$;

(b) $12·6-\dfrac{4·7 \times 6·2}{13·9}$;

(c) $6·7^5$;

(d) $65·1-[23·7 \div (218 \div 189)]$.

In each case dry check your program.

354

4. (*a*) What is the minimum number of store compartments that are needed to find the average of three numbers?

(*b*) Write out a program to find the average of the five numbers 2·3, 3·6, 4·2, 5·1 and 2·9 using as few store compartments as possible.

(*c*) Dry check your program.

5. Write a program to compute $4·6 + 4·6^2 + 4·6^3$ using (*a*) four stores, (*b*) three stores, (*c*) two stores. Check each carefully. Are there *any* advantages of (*b*) or (*a*) over (*c*)?

5. PROGRAMS FOR FORMULAE

Consider the program in Figure 16 for computing $3·4^5$ and the corresponding dry check:

	S_1	S_2
(1) Input 3·4 to S_1	3·4	—
(2) Replace S_2 by $S_1 \times S_1$	3·4	$3·4^2$
(3) Replace S_1 by $S_1 \times S_2$	$3·4^3$	$3·4^2$
(4) Replace S_1 by $S_1 \times S_2$	$3·4^5$	$3·4^2$
(5) Output the number in S_1		

Fig. 16

This program shows clearly how the use of stores can be kept to a minimum by re-using store compartments containing numbers that are not required again. With many calculations, particularly in statistics, there is much data and it is vital that stores be used over and over again.

(*a*) How will the program have to be altered if we wish to find 62^5, $0·0058^5$ or $32\,671\,853^5$?

This program is really a set of instructions for computing x^5 from x. Whatever number is fed into S_1 in the first instruction, it will be raised to the fifth power and printed out.

Once a program has been written to make a computation, then it can be used to make many others simply by adjusting the input data.

Example 10

Write a program to compute the volume, V, of a cylinder when its radius, r, and height, h, are known (see Figure 17).

The volume of the cylinder is given by the formula

$$V = \pi r^2 h,$$

and a possible program to compute it is as follows:

Fig. 17

	Dry check		
Program $V = \pi r^2 h$	S_1	S_2	S_3
(1) Input r to S_1	r	—	—
(2) Input h to S_2	r	h	—
(3) Input π to S_3	r	h	π
(4) Replace S_2 by $S_2 \times S_3$	r	πh	π
(5) Replace S_1 by $S_1 \times S_1$	r^2	πh	π
(6) Replace S_1 by $S_1 \times S_2$	$\pi r^2 h$	πh	π
(7) Output the number in S_1			

355

Note: (i) All input instructions come together at the start of the program.

(ii) As few stores as possible have been used.

(iii) Store S_3 still contains π, i.e. the computer is ready to compute any number of volumes of cylinders simply by changing data input to S_1 and S_2.

(iv) Although we can prepare a program in this way, Simon will only calculate with numbers—letters mean nothing to him; that is, we can, say, store 2, 3, or 6 in S_1, but not a symbol, such as v, that might take any of these values. In other words, Simon cannot perform algebraic operations but only arithmetical ones.

(*b*) Compute the volume of a cylinder of radius 3·7 m and height 5·9 m by following through the program in Example 10 and using your slide rule for instructions (4), (5) and (6).

Exercise G

(Units have been omitted from most of these questions. The units set against the output, where this refers to a measurement, will therefore depend upon the units of the data. You should sketch a flow-diagram before preparing a program.)

1. Prepare programs, using as few stores and instructions as possible, for the following:

	Compute	Data	Formula
(*a*)	S	π, r, h	$S = 2\pi r(r+h)$
(*b*)	K	a, p	$K = \dfrac{(a^2+p)^3}{ap}$
(*c*)	T	C, R	$T = \dfrac{C(R^4-1)}{R-1}.$

2. Using as few stores as possible write a program to compute:

(*a*) x^7; (*b*) x^8.

(*c*) What powers of x could be computed if only one store were available?

3.

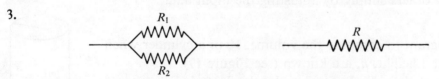

Fig. 18

Two resistances R_1 and R_2 in an electrical circuit (see Figure 18) can be replaced by a single resistance R where

$$R = \frac{R_1 R_2}{R_1 + R_2}.$$

(*a*) Write a program to compute R when R_1 and R_2 are known.

(*b*) Work out R, using your program, when $R_1 = 2$, and $R_2 = 6$. Check that this gives the same result as working R out from the formula directly.

4. $\{(x, y): 2x+y = 7\}$ and $\{(x, y): 5x-3y = 1\}$ can both be represented on graph paper by straight lines which intersect at the point (2, 3). Sets of this kind are always occurring

wherever mathematics is used but the numbers involved are not usually as simple as those in this example. The point of intersection can usually be computed, however, by using the fact that the intersection of the sets

$$\{(x, y): ax+by = c\} \quad \text{and} \quad \{(x, y): px+qy = r\}$$

is

$$\left\{\left(\frac{cq-rb}{aq-pb}, \frac{ar-pc}{aq-pb}\right)\right\}.$$

(a) Show that this formula gives the correct point of intersection in the example given.

(b) Write a program to compute the point of intersection when a, b, c, p, q and r are given.

(c) Check that your program gives $(1, 2)$ when $a = 1, b = 1, c = 3, p = 3, q = {}^-1$ and $r = 1$.

(d) The statement of this question says 'can usually be computed'. Under what circumstances does this formula break down?

5.1   Rearrangement of formulae

So far we have restricted our attention to using formulae in the form in which they are given. It is, however, often easier to make a calculation if the letters in a formula are rearranged.

In Question 1(a) in Exercise G you wrote a program to compute S (the total surface area of a closed, cylindrical can), given the values of r and h (the can's radius and height).

To find the height h of a can when the values of S and r are known, we need to rearrange this formula in the form

$$h = ...$$

This is known as 'making h the subject of the formula'.

We can view the formula as a function mapping $h \to S$ as follows:

Reversing this flow diagram we have the function mapping $S \to h$.

Alternatively the flow diagram can be dispensed with and we write:

$$S = 2\pi r(r+h) \quad \Leftrightarrow \quad \frac{S}{2\pi r} = r+h$$

$$\Leftrightarrow \quad \frac{S}{2\pi r} -r = h.$$

That is,

$$h = \frac{S}{2\pi r} -r.$$

Example 11

Make P the subject of the formula $R = 3Q + 2P$. (Draw a flow diagram for the given formula, mapping $P \to R$, reverse it and then compare your result with the following working)

$$R = 3Q + 2P,$$
$$\Leftrightarrow \quad R - 3Q = 2P,$$
$$\Leftrightarrow \tfrac{1}{2}(R - 3Q) = P,$$
$$P = \tfrac{1}{2}(R - 3Q).$$

Exercise H

Flow diagrams should be used if necessary.

1. Make c the subject of the formula $f = \tfrac{9}{5}c + 32$. To what do you think this formula relates?

2. Make c the subject of Einstein's energy formula $E = mc^2$. Find c, when $m = 0\cdot03$ and $E = 2\cdot7 \times 10^{19}$.

3. Make l the subject of the formula $T = 2\pi\sqrt{\dfrac{l}{g}}$ and write a Simpol program to compute l given π, g and T.

4. Figure 19 shows a *closed* box.

(*a*) Calculate P, the total length of its edges. Show that
$$b = \frac{P - 8a}{4},$$
and arrange the formula in the form $a = \dots$.

(*b*) Calculate S, the total area of the faces of the box. Make b the subject of this formula.

(*c*) Calculate V, the volume of the box. Make a the subject of this formula.

(*d*) Use the formulae to find:

    (i)   the value of b when $P = 32$ and $a = 1\tfrac{1}{2}$;

    (ii) the value of b when $S = 137$ and $a = 5$;

    (iii) the value of a when $V = 80$ and $b = 5$.

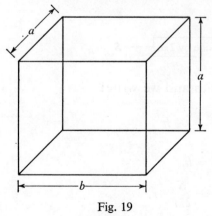

Fig. 19

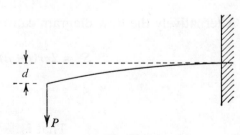

Fig. 20

5. A mass P is hung from the end of a beam of length l embedded in a wall (see Figure 20). The deflection d of the end of the beam is given by the formula

$$d = \frac{Pl^3}{3EI},$$

where I depends on the shape of the cross-section of the beam and E on the material of which it is made.

Make E the subject of this formula.

Find E when $P = 150$, $l = 40$, $I = 1$ and $d = 1{\cdot}5$.

6. When a current of I amps flows through a resistance of R ohms, then the difference in potential V volts, across the resistance is given by $V = IR$.

(a) If a current of 5 amps flows through a resistance of 480 ohms, what is the voltage difference across the resistance?

(b) Make I the subject of the formula.

(c) The power consumed, W watts, is given by $W = IV$. Combine the two formulae to find W, first in terms of I and R, then in terms of V and R.

(d) An electric fire consumes 2000 watts and the voltage of the power supply is 250 volts. What are: (i) the resistance of the fire, and (ii) the current consumed? If standard fuse wires melt when currents of 5, 10 or 15 amps flow through them, what fuse wire would you use with this fire to guard against overloading?

5.2   Repetition

Many practical problems cannot be solved exactly and one must be satisfied with approximate answers. These latter are often found by methods that can best be described as 'organized trial and error'.

Example 12

A bridge is to be constructed across a river using reinforced concrete construction. The cost £C, of building such a bridge is given approximately by the formula

$$C = 20000 \left(\frac{12}{n} + n \right),$$

where n is the number of spans in the bridge. What number of spans will give the lowest cost? (Can you explain how this formula arises?)

This problem can be solved by systematically trying different values for n and finding which value of n gives the lowest cost. The costs corresponding to some values of n are shown in the table below.

n	$\dfrac{12}{n} + n$	C	
1	13	260000	
2	8	160000	
3	7	140000	} Lowest cost
4	7	140000	
5	7·4	148000	
6	8	160000	

It can be seen from the table of values that the lowest cost can be achieved by building a bridge of either 3 or 4 spans. The formula is such that as n increases above 4 so does the cost C.

In this example, we see which value for n is required by calculating a sequence of values for C. The next example also requires a sequence of calculations, but in this case each calculation yields a better approximation than the previous one.

Example 13

Show how $\sqrt{24}$ can be found to any required degree of accuracy.

Stage 1. First take any number (for example, 4) and divide 24 by it. 24 divided by 4 is 6 and since

$$4^2 < 4 \times 6 < 6^2,$$

taking the square root gives $\qquad 4 < \sqrt{24} < 6.$

We have found two limits for $\sqrt{24}$.

Stage 2. Next take 5, the mean of 4 and 6, and divide 24 by it. 24 divided by 5 is 4·8, hence, by the same reasoning

$$4 \cdot 8 < \sqrt{24} < 5.$$

Stage 3. Take 4·9, the mean of 4·8 and 5, and divide 24 by it. 24 divided by 4·9 is approximately 4·898, hence,

$$4 \cdot 898 < \sqrt{24} < 4 \cdot 9.$$

Each stage of this process produces two numbers between which $\sqrt{24}$ must lie. The difference between these numbers decreases at each stage and gives a measure of the accuracy obtained. For example, at the end of Stage 2, $\sqrt{24}$ is squeezed between 4·8 and 5 so we could say that

$$\sqrt{24} = 4 \cdot 9 \pm 0 \cdot 1.$$

At the end of Stage 3 we could similarly say

$$\sqrt{24} = 4 \cdot 899 \pm 0 \cdot 001.$$

We could find $\sqrt{24}$ by starting with 12 instead of 4 at the first stage. No matter which number we divide 24 by, each step of the process will yield a better approximation to the square root.

This method, which is now widely used for computers, was first developed by Hero, a Greek mathematician who lived in the early part of the third century A.D.

Hero's method for finding an approximation to $\sqrt{N}$ can be expressed neatly as a flow diagram. Try to draw the diagram.

Exercise I

Answer the following questions by 'trial and error' methods.

1. A shell is fired from an anti-aircraft gun. t seconds after the gun is fired, the height h metres is given by the formula
$$h = 1000t(6-t).$$

By computing the value of h when $t = 1, 2, 3, 4$, and 5 estimate the maximum height reached by the shell.

2. If x is an approximate cube root of 30, then
$$y = \frac{1}{3}\left(2x + \frac{30}{x^2}\right)$$
is a better approximation.

Show that 3 is an approximate cube root of 30 and use the formula once to obtain a better approximation. Cube the approximation you obtain to see by how much it differs from 30.

5.3   Square roots and loops

You will probably have realized that so far throughout this chapter Simon's very basic computing language of 'input', 'replace' and 'output' has limited the number of problems that we can solve with his aid.

In particular his basic language does not:

(i) enable us to call for a square root;

(ii) provide us with a way of producing a loop, such as we used in Section 3.

To overcome these difficulties Simon has now been developed so that he can understand the following instructions:

(i) 'replace S_4 by $\sqrt{S_5}$' (whatever is in S_4 is to be replaced by the square root of the number in S_5). Note that the number in S_5 remains unchanged in that store after this instruction has been obeyed: only S_4 changes.

(ii) 'if $S_4 < S_6$ go to (3)',

This instruction in a program plays the same role in Simon's working out of a problem as the decision box in a flow diagram. Simon understands that if S_4 is less than the number in S_6 he must return to the third instruction of his program—(3).

Note that, if S_4 is equal to or greater than S_6, Simon will move on to obey the next instruction in the program. This new instruction places a condition upon the computer's progress.

Simpol allows the use of the following *conditional* instructions:

$$<, >, \leqslant, \geqslant =.$$

Remember that the computer will not follow the route of the loop unless the condition is satisfied.

The following examples illustrate the use of these two new instructions.

Example 14

The periodic time T for a simple pendulum of length l is given by the formula

$$T = 2\pi \sqrt{\frac{l}{g}},$$

where g is the acceleration due to gravity.

Write a Simpol program to compute T given l and g.

(Sketch a flow diagram and compare it with the following program)

	Notes
(1) Input π to S_1	—
(2) Input g to S_2	—
(3) Input l to S_3	—
(4) Replace S_1 by $S_1 + S_1$	$2\pi \to S_1$
(5) Replace S_3 by $S_3 \div S_2$	$l/g \to S_3$
(6) Replace S_3 by $\sqrt{S_3}$	$\sqrt{(l/g)} \to S_3$
(7) Replace S_1 by $S_1 \times S_3$	$T = 2\pi\sqrt{(l/g)} \to S_1$
(8) Output the number in S_1	

Example 15

Construct a program to compute the remainder when a number b is repeatedly subtracted from a number c.

To see what this entails consider the case where $b = 23$ and $c = 95$.

We start with 95 and successively obtain:

$$95 - 23 = 72,$$
$$72 - 23 = 49,$$
$$49 - 23 = 26,$$
$$26 - 23 = 3.$$

What has not been written down is the fact that before we take away 23 we assess whether the difference left from the previous subtraction is bigger than 23. If the difference is bigger than (or equal to) 23 we subtract again, if not we have the remainder.

The Simpol program for this is very short:

(1) Input b to S_1
(2) Input c to S_2
LOOP ⎡ (3) Replace S_2 by $S_2 - S_1$ (i.e. $c - b \to S_2$)
⎣ (4) If $S_2 \geqslant S_1$ go to (3)
(5) Output the number in S_2.

Dry check this program with $b = 17$ and $c = 63$.

How many times did you repeat instructions (3) and (4)? What does this tell you about 63 and 17?

Exercise J

1. Referring to Example 10 and Figure 17, make r the subject of the formula and write a Simpol program to compute r, given V and h.

2. Write a Simpol program for solving the problem of Example 12. Dry check your program. (Hint: does your program give a complete solution? You may find a flow diagram helpful.)

3. When measured at a depth of h metres below the surface of the water the minimum thickness of a dam is given by the formula

$$T = \sqrt{\left(\frac{5h^3}{900+3h}\right)},$$

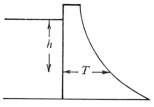

where T metres is the thickness.

   (a) Write a program to compute T given h.

   (b) What is the minimum thickness at the base of a dam when the total depth of water is 110 m?

   (c) Can there be a value of h equal to that of T?

Fig. 21

4. Work through the following program letting $N = 100$ and $a = 3$.

   (1) Input N to S_1
   (2) Input a to S_2
   (3) Input a to S_3
   (4) Input 1 to S_4
   (5) Input 1 to S_5
   (6) Replace S_2 by $S_2 \times S_3$
   (7) Replace S_4 by $S_4 + S_5$
   (8) If $S_2 < S_1$ go to (6)
   (9) Output the number in S_4.

What meaning can you attach to the number computed?

5. Write a program to compute x^n.
Check your program when $x = 3$ and $n = 5$.

6. If £A is invested at $r\,\%$ compound interest, the amount accumulated after n years is given by the formula

$$A = P\left(1+\frac{r}{100}\right)^n.$$

   Write a program to compute A.

REVISION EXERCISES

The following two sets are to be attempted *without* a slide rule, to help you know your book of tables.

TABLES SESSION NO. 1

1. $42 \cdot 3^2$.

2. $0 \cdot 615^2$.

3. $\sqrt{(1470)}$.

4. $\sqrt{(0 \cdot 0426)}$.

5. $\dfrac{1}{4 \cdot 25}$.

6. $\dfrac{1}{0 \cdot 963}$.

7. $\dfrac{8}{52 \cdot 6}$ $\left(\text{i.e. } 8 \times \dfrac{1}{52 \cdot 6}\right)$.

8. $\sin 46 \cdot 8°$.

9. $\cos 24 \cdot 3°$.

10. $\dfrac{1}{\sin 10 \cdot 3°}$.

TABLES SESSION NO. 2

In Questions 3, 5, 10 *find the value of x.*

1. 173^2.

2. $0 \cdot 0124^2$.

3. $\cos x° = 0 \cdot 910$.

4. $\sqrt{(13600)}$.

5. $\sin x° = 1 \cdot 23$.

6. $\sqrt{(0 \cdot 000174)}$.

7. $\dfrac{5}{\cos 24°}$.

8. $\dfrac{7}{\sin 65°}$.

9. $\left(\dfrac{6}{\sin 10°}\right)^2$.

10. $x^2 = 4 \cdot 61^2 + 3 \cdot 82^2$.

S

1. Transform the square with vertices (0, 0), (4, 0), (4, 4), (0, 4) using the matrix

$$\begin{pmatrix} \tfrac{1}{2} & \tfrac{1}{2} \\ -\tfrac{1}{2} & 1\tfrac{1}{2} \end{pmatrix}.$$

Identify the transformation, as fully as you can.

2. Transform the quadrilateral $ABCD$, where A is (8, 5), B is (28, 17), C is (34, 21), and D is (14, 9), using the matrix

$$\begin{pmatrix} 2 & -3 \\ -3 & 5 \end{pmatrix}.$$

What is the area of the image figure $A'B'C'D'$?
Hence find the area of the figure $ABCD$.

3. A transformation is defined by the matrix $\begin{pmatrix} 2 & -1 \\ 1 & 2 \end{pmatrix}$. Find the image, $P'Q'R'S'$, under this transformation of the square $PQRS$ whose vertices are (2, $^-$1), (2, 3), ($^-$2, 3), ($^-$2, $^-$1). Draw a diagram showing the square and its image. How would you describe the transformation?

4. Use reversed flow diagrams to solve:

(a) $3x - 4 = 7$;

(b) $2 - 4x = 8$;

(c) $\dfrac{12}{3 - 2x} = 6$.

364

5. If $g:x \rightarrow 1+\frac{x}{2}$, express the function g^{-1} in the same form and find the value of $g^{-1}g^{-1}(2)$. Check your answer by finding the function gg and using the fact that $(gg)^{-1} = g^{-1}g^{-1}$. For what value of x does $gg(x) = 101$?

In Questions 6, 7, 8, state the letter corresponding to the correct answer, or answers.

6. The statement $(x+3)^2 = x^2+6x+9$ is true for:

(a) all values of x;
(b) only two values of x;
(c) only one value of x;
(d) no value of x.

7. It is given that $T = (3W-a)/2a$. Then:

(a) $T = (3W/2a)-\frac{1}{2}$;
(b) $T = 0$ when $a = 3W$;
(c) $W = \frac{2}{3}aT+a$;
(d) if $a = W$, then $T = 1$.

8. If $p = 12_5$, $q = 24_5$ and $r = 24_{10}$, then:

(a) $q < r$; (b) $q = 2p$; (c) $p+q = 42_5$; (d) $5q = 240_5$.

9. Figure 1 shows three networks. List the numbers of nodes (N), arcs (A) and regions (R), and so verify Euler's relation $R+N = A+2$ for each network. Does this relation hold for Figure 1 considered as a whole?

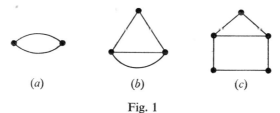

(a)   (b)   (c)

Fig. 1

10. (a) Construct the network corresponding to this direct route matrix:

$$\begin{array}{c} \text{to} \\ \begin{array}{cccccc} & A & B & C & D & E \\ \text{from} \begin{array}{c} A \\ B \\ C \\ D \\ E \end{array} & \begin{pmatrix} 0 & 1 & 0 & 1 & 1 \\ 1 & 0 & 1 & 0 & 1 \\ 0 & 1 & 0 & 1 & 1 \\ 1 & 0 & 1 & 0 & 1 \\ 1 & 1 & 1 & 1 & 0 \end{pmatrix} \end{array} \end{array}.$$

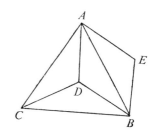

(b) Construct the route matrix corresponding to the network shown in Figure 2.

Fig. 2

T

1. By studying their effect on a suitable figure, describe fully the transformations represented by the matrices:

(a) $\begin{pmatrix} -2 & 0 \\ 0 & -2 \end{pmatrix}$;

(b) $\begin{pmatrix} 1 & 0 \\ 0 & 2 \end{pmatrix}$;

(c) $\begin{pmatrix} 0 & -1 \\ -1 & 0 \end{pmatrix}$.

2. The quadrilateral with vertices $A(8, 6)$, $B(20, 15)$, $C(38, 30)$, and $D(20, 16)$, is transformed using the matrix

$$\begin{pmatrix} 2\frac{1}{2} & ^-3 \\ ^-1\frac{1}{2} & 2 \end{pmatrix},$$

forming the images A', B', C', D'.
What is the area of $A'B'C'D'$?
Hence find the area of $ABCD$.

3. Construct the image of the letter V whose vertices are $(2, 2)$, $(4, ^-2)$, $(6, 2)$ under the transformation whose matrix is

$$\begin{pmatrix} 0 & ^-\frac{1}{2} \\ \frac{1}{2} & 0 \end{pmatrix}.$$

4. Solve the equations:
(a) $4 - (2x - 3) = 5$;
(c) $3 - 4(5 - x) = 11$;
(b) $7 + 2(4 - x) = 5$;
(d) $(4 - x)(3 - x) = (x + 1)(x + 2)$.

5. Represent on number lines, the intersection of the solution sets for each of the following pairs:
(a) $3 - x < 5$ and $5 < 4 - x$;
(b) $\dfrac{x+1}{2} \leqslant 1$ and $1 \leqslant x + 3$.

6. The statement $(x + 3)^2 = x^2 + 4x + 6$ is true for which of the following alternatives:
(a) all values of x;
(c) only one value of x;
(b) only two values of x;
(d) no value of x.

7. (a) Add 35_6 and 14_6, giving the answer in the same base.
(b) Express 230_n in terms of n.
(c) State two factors of 230_n.

8. 10022 is a number in base three and the same number is represented as 155 in another base. What is the base?

9. Philip, Quentin, Rupert and Simon played one another at chess. P beat Q and S; Q beat S; R beat P and Q; S beat R. Complete matrices showing one- and two-stage dominances and hence put the players in an order of merit.

10. (a) Compile a direct-route matrix S for the network shown in Figure 3. Why is it not symmetrical about the leading diagonal?
(b) Find the matrix S^2. What does it represent?

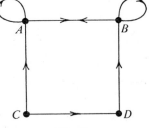

Fig. 3

U

1. Transform the square with vertices $(1, 1)$, $(^-1, 1)$, $(^-1, ^-1)$ and $(1, ^-1)$ using the matrix

$$\begin{pmatrix} 0 & ^-2 \\ 2 & 0 \end{pmatrix}.$$

Describe the transformation, as fully as you can.

366

2. Transform the triangle with vertices (1, 0), (0, 1) and (⁻1, ⁻1) using the matrix

$$\begin{pmatrix} 3 & 0 \\ 0 & 1 \end{pmatrix}.$$

Describe the transformation as fully as you can.
The image is further transformed using the matrix

$$\begin{pmatrix} 1 & 0 \\ 0 & 2 \end{pmatrix}.$$

Describe this transformation.
Can you suggest a name for the resultant effect of the two transformations?

3. If the matrices $X = \begin{pmatrix} 1 & 1 \\ 0 & 1 \end{pmatrix}$ and $Y = \begin{pmatrix} 1 & 0 \\ 1 & 1 \end{pmatrix}$ represent transformations, find the matrices X^2, Y^2, XY and YX. Use a graph to show the effect of these combined transformations on the unit square.

4. Find the solution set of each of the following:
(a) $2(3x-1)+8 = 6(x+1)$; (b) $1-(4-x) > x-2$; (c) $(x-3)^2 = x(x-6)$.

5. Give the number-base of the correct addition $27+6 = 34$.

6. Remove the brackets and simplify the following expressions
(a) $(2x+y) +(x+y)$;   (b) $(2x-y)-(x-y)$;
(c) $(m+n)+(2m+n)-(3m+n)$;   (d) $(r-s)+(r-2s)+(r-3s)$;
(e) $2(x+y)+(3x+y)$.

7. Draw a diagram to illustrate the relation 'is a prime factor of' over the set {2, 3, 4, 6, 8}. Construct the matrix R for the relation and write down R^2 and the transpose R'. What relations do these matrices represent?

8. A regular polyhedron has 12 edges and 6 vertices. Using Euler's relation $F+V = E+2$, find its number of faces. What is this polyhedron called? Draw a net which could be used for constructing it.

9. (a) Write a program to compute S where $S = r(\pi r+2h)$.
(b) Check your program, given $\pi = \frac{22}{7}$, $r = 4$, $h = 3$.

10. Draw a flow diagram to find the 15th term of the sequence 1, 1, 2, 3, 5, 8, ... where each term (except the first two) is the sum of the two preceding terms. (This is known as a Fibonacci sequence.)

V

1. Transform the unit square using the matrix

$$\begin{pmatrix} 0 & 1 \\ 1 & 0 \end{pmatrix},$$

and hence describe the transformation.
Transform the image obtained above, using the matrix

$$\begin{pmatrix} 0 & 1 \\ -1 & 0 \end{pmatrix},$$

and so describe the second transformation.
What single transformation is equivalent to these two successive transformations?

2. Is the set of enlargements, centre the origin, closed under the operation 'following'?

3. Is {translations, rotations, reflections} closed under the operation 'following'?

4. The symbol $[x]$ stands for 'the largest integer which does not exceed x'.
For example $[2\cdot3] = 2$, $[\pi] = 3$, $[^-2\cdot1] = ^-3$, etc.
Sketch the graph of the function $x \rightarrow [x]$, for $^-3 \leqslant x \leqslant 3$.
Is the inverse of this function itself a function? Give reasons for your answer.

5. Say whether the following functions have inverses which are functions:

(a) person → name;   (b) person → home address;
(c) person → fingerprints;
(d) car engine number → car registration number;
(e) name of newspaper → price of newspaper.

6. This is an extract from a boy's exercise book.

$$\tfrac{1}{3}(x+2) - \tfrac{1}{4}(x-3) = 1,$$
$$\Rightarrow 4(x+2) - 3(x-3) = 12,$$
$$\Rightarrow \quad 4x+8-3x-9 = 12,$$
$$\Rightarrow \quad\quad 4x-3x = 13$$
$$\Rightarrow \quad\quad\quad x = 13.$$

Where had the boy gone wrong? What answer should he have obtained?

7. (a) Convert the numbers 15_{10} and 29_{10} into the binary scale.
(b) Convert the numbers 1101_2 and 11010_2 into the decimal scale.
(c) Work out $1101_2 \times 101_2$. (Answer in binary.)

8. Solve the following equations. All the numbers are in base 5. Give the answers in that base.

(a) $x+4 = 23$;   (b) $2x+4 = 40$;   (c) $\tfrac{1}{2}(x-1) = 4$;   (d) $x^2+1 = 32$.

9. Figure 4 shows the net for the construction of a certain polyhedron. Describe the polyhedron. State its number of faces (F), vertices (V), and edges (E). Show that Euler's relation $F+V = E+2$ is satisfied.

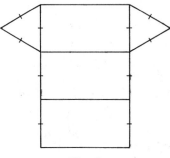

Fig. 4

10.

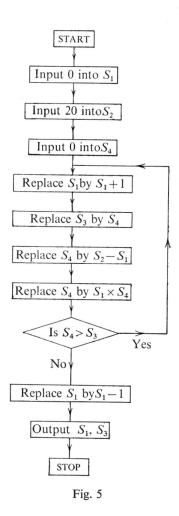

START

Input 0 into S_1

Input 20 into S_2

Input 0 into S_4

Replace S_1 by $S_1 + 1$

Replace S_3 by S_4

Replace S_4 by $S_2 - S_1$

Replace S_4 by $S_1 \times S_4$

Is $S_4 > S_3$

Yes

No

Replace S_1 by $S_1 - 1$

Output S_1, S_3

STOP

Fig. 5

Figure 5 shows a program to find the maximum value of A, where $A = x(20 - x)$. A will be the output from S_3, x from S_1.

See if you can find the value of x which makes A a maximum.

W

Miscellaneous

1. If $2 \leqslant x \leqslant 5$, $\quad 15 \leqslant y \leqslant 22$ and $t = y/x$, state:

(*a*) the maximum value of t;

(*b*) the minimum value of t.

2. $a \theta b$ denotes the remainder when $a \times b$ is divided by 11.

(*a*) Evaluate $4 \theta 5$.

(*b*) Find an integer y such that $8 \theta y = 1$.

3. A model of a boat is made on a scale of 1 to 10.

(*a*) If the boat is 50 m long, how long is the model?

(*b*) If there are k m² of deck area in the model, what is the deck area on the boat?

(*c*) The boat's fuel tank holds 2000 litres of fuel. How much could the model's tank hold?

(*d*) If the boat has a speed of 30 knots, can you say anything about the speed of the model?

4. Express 265_{10} (*a*) in the base of two; (*b*) in the base of eight.

5. What is the gradient of the straight line whose equation is $5x+2y = 7$?

6. Write down the coordinates of two points on the straight line $x-2y = 4$, and find their images under the transformation defined by the matrix

$$\begin{pmatrix} 1 & ^-1 \\ 1 & 1 \end{pmatrix}.$$

Find also the image of the line, stating any general assumptions you make about the image of a straight line under such a transformation. What is the angle between the line and its image?

7. Which of the following are correct?

(*a*) $7-2 = 2-7$;   (*b*) $2+3+4 = 4+3+2$;

(*c*) $12 \times (4 \div 2) = (12 \times 4) \div 2$;   (*d*) $5-(4-3) = (5-4)-3$;

(*e*) $2+(3 \times 4) = (2+3) \times 4$.

8. In each of the following cases say which member of the set is an ' odd man out ' and why:

(*a*) $\{1, 4, 9, 25, 36, 47, 64\}$;   (*b*) $\left\{\frac{1}{2}, \frac{3}{6}, \frac{2}{5}, \frac{9}{18}, \frac{1\frac{1}{2}}{3}\right\}$;

(*c*) $\{(0, 0), (4, 2), (3, 1), (16, 8), (9, 4\cdot5)\}$;

(*d*) $\{9_{10}, 1001_2, 100_3, 13_6, 23_4\}$;

(*e*) {triangle, quadrilateral, tetrahedron, pentagon, hexagon}.

9. An icosahedron has 20 faces. A regular icosahedron, equally likely to come down on any of its faces when thrown, is tossed twice in succession. If we call the faces upon which it lands F_1 and F_2, what is the probability that F_1 and F_2 are:

(*a*) the same face;

(*b*) parallel faces exactly opposite each other;

(*c*) adjacent faces;

(*d*) neither the same nor adjacent?

10. A forest fire spreads so that the area covered is doubled every hour. At mid-day the area is 100 m². Draw a graph to illustrate the extent of the fire between mid-day and 5 p.m. Measure the rate at which it is spreading at 2 p.m., 3.30 p.m., 4.40 p.m.

X

Miscellaneous

1. Examine the transformation

$$\begin{pmatrix} x \\ y \end{pmatrix} \rightarrow \begin{pmatrix} 3 & 6 \\ 2 & 4 \end{pmatrix} \begin{pmatrix} x \\ y \end{pmatrix}$$

by considering its effect on a suitable figure.

What is the determinant of the matrix?

Will the inverse transformation be a function? Explain your answer.

2. If $a\mu b$ means $a+2b$, find the values of:

(a) $(2\mu 3)\mu 4$; (b) $2\mu(3\mu 4)$.

Is the operation associative?

Is it commutative?

3. If $f(x) = 3x+1$, find $f^{-1}(x)$.

Find $ff(x)$ and $f^{-1}f^{-1}(x)$.

4. The table shows the number of goals scored in 43 Football League matches on the same day.

Total number of goals in a match	0	1	2	3	4	5	6	7
Number of matches with this total score	4	10	5	9	7	5	1	2

(a) Find the mode of this distribution.

(b) Find the median.

(c) Calculate the mean number of goals scored per match. (Answer to the nearest 0·1.)

5. (a) Find a and b if $\begin{pmatrix} a \\ 13 \end{pmatrix} = \begin{pmatrix} 2 & 0 \\ 1 & b \end{pmatrix} \begin{pmatrix} 4 \\ 3 \end{pmatrix}$.

(b) Use your slide rule to calculate $\sqrt{(2 \cdot 5 \div 0 \cdot 76)}$.

6. (a) A number written in the usual way to base 10 appears as $abcd$. Express the number algebraically as a function of the four digits.

(b) Subtract this number from the number formed by reversing the digits and show that the difference is divisible by 9.

(c) Show that, if the two middle digits of the given number are equal, the difference is also divisible by 37.

(d) What result similar to (b) would you expect for numbers expressed to base n?

7. A trapezium $PQRS$ has SR parallel to PQ. $PQ = 12 \cdot 2$ cm, $QR = 8$ cm, $RS = 3 \cdot 8$ cm, $SP = 5 \cdot 2$ cm.

The translation which maps S onto P maps R onto T.

(a) State, with reasons, the lengths of RT and TQ.

(b) Construct accurately triangle QRT and hence $PQRS$.

(c) Measure the distance between the parallel sides and find the area of the trapezium.

8. If n is an integer such that $n < 7\sqrt{(0 \cdot 8)} < n+1$, which of the following is the correct value of n?

(a) 2; (b) 5; (c) 6; (d) 7.

9. You are given that $Ps = \frac{1}{2}mv^2 - \frac{1}{2}mu^2$. Which of the following are correct:

(a) $\dfrac{2Ps}{m} = u^2 - v^2$; (b) $s = \dfrac{m(v^2 - u^2)}{2P}$;

(c) $P = 0$ and $m \neq 0 \Rightarrow v$ is numerically equal to u;

(d) $^{-}v = u \Rightarrow P = 0$ or $s = 0$.

10. If x belongs to the set of real numbers, state which of the following are correct:

(a) $(x-2)(x+3) > 0 \Rightarrow x > 2$; (b) $x > 2 \Rightarrow (x-2)(x+3) > 0$;

(c) $x > 2 \Rightarrow \dfrac{1}{x} > \frac{1}{2}$; (d) $x^2 > x \Rightarrow x > 1$.

371

INDEX